Contributions to Economics

For further volumes:
http://www.springer.com/series/1262

Pablo Coto-Millán • Miguel Ángel Pesquera
Juan Castanedo
Editors

Essays on Port Economics

Physica-Verlag

Editors
Prof. Dr. Pablo Coto-Millán
Universidad de Cantabria
Avenida de los Castros s/n
39005 Santander
Spain
cotop@unican.es

Prof. Dr. Miguel Ángel Pesquera
Universidad de Cantabria
Avenida de los Castros s/n
39005 Santander
Spain
mpesquera@unican.es

Prof. Dr. Juan Castanedo
Universidad de Cantabria
Avenida de los Castros s/n
39005 Santander
Spain
castanjm@unican.es

ISSN 1431-1933
ISBN 978-3-7908-2424-7 e-ISBN 978-3-7908-2425-4
DOI 10.1007/978-3-7908-2425-4
Springer Heidelberg Dordrecht London New York

Library of Congress Control Number: 2010930042

Cover design: WMXDesign GmbH, Heidelberg, Germany

Printed on acid-free paper

Physica-Verlag is a brand of Springer-Verlag Berlin Heidelberg
Springer-Verlag is a part of Springer Science+Business Media (www.springer.com)

Contents

List of Figures

List of Tables

Introduction to Essays on Port Economics

Pablo Coto-Millán, Miguel Ángel Pesquera, and Juan Castanedo

The aim of *Essays on Port Economics* is to offer further reading for specialist and postgraduate courses, Master's degrees and doctorates in Economics, Business Administration, Engineering, the merchant navy and other port professionals. The editors all teach at the University of Cantabria: Pablo Coto-Millán has a Ph.D. in Economics and is a Professor of Transport Economics and Microeconomics; Miguel Ángel Pesquera has a Ph.D. in Engineering and is a Professor of Port Management and Operations, and Juan Castanedo has a Ph.D. in Marine Sciences and Merchant Shipping and is an Associate Professor of Port Management and Operations.

The text is divided into the traditional five parts of an economics handbook: demand, supply, economic impact on the port, regulation and efficiency and lastly, cost-benefit analysis and ecological footprint of the port.

The first part defines the demand for port services using an empirical approach with modern econometric techniques allowing predictions to be made. Price and product elasticity values are useful for finding out the response sensitivity of demand to price variations of transport products and services.

The second part analyses the supply of services using the production and cost functions of the various shipping companies.

The third part combines the two previous parts on supply and demand to offer a global configuration of the market. One way of looking at the market as a whole is to use input/output methodology, which, when applied to ports, also allows us to estimate the direct, indirect and induced economic impact on jobs and the Gross Domestic Product of a port in its city, region or country. Moreover, by combining several input/output tables from the different regions over which a port can exert its influence, we can estimate the direct, indirect and induced economic impact on the port hinterland.

P. Coto-Millán (✉), M. Ángel Pesquera, and J. Castanedo
University of Cantabria, Santander, Spain
e-mail: cotop@unican.es

P. Coto-Millán et al. (eds.), *Essays on Port Economics*, Contributions to Economics,
DOI 10.1007/978-3-7908-2425-4_1,

The fourth part analyses the regulation from the various types of technical, allocative and economic efficiency. This analysis is carried out using traditional econometric methods as well as other modern econometric techniques such as DEA and distance functions applied to ports.

Finally, the fifth part uses more modern methodology to assess the feasibility of building a new port or expanding an existing one. This is called cost-benefit analysis methodology and requires a knowledge of techniques for estimating demand and building up a good supply that is efficient in technical, allocative and economic terms. In cost-benefit analysis, benefits and imputed costs must also be clearly distinguished from the economic impacts generated by the construction, operation and expansion of a port. Thus, a good knowledge of the limitations of input/output methodology is required when using it to estimate economic impact. The various regulations have an impact on efficiency so the analysis of efficiency and regulation can form part of the cost-benefit analysis. Lastly, modern cost-benefit analysis methodology increasingly incorporates more monetary evaluations of intangibles, such as the value of time and the value of pollution. These and similar issues are incorporated into the cost-benefit analysis and we conclude with a chapter introducing the concept of the Port Ecological Footprint, applying it to a specific case.

This fifth part takes information from the previous four parts. Each of the previous sections is self-contained and uses an introductory and modern approach to discuss issues of supply, demand, economic impact and regulation and efficiency with sufficient clarity for these to be reproduced for other cases in different countries and regions. The last section, however, requires a good knowledge of the previous four to be fully understood and applied. It also requires additional knowledge, which is explained in a straightforward way so that it can be reproduced with open topics allowing various interpretations, giving individual researchers the opportunity to discover original, fertile and unexplored lines of study for Master's research projects and doctoral theses.

Part I
Demand

Port Marketing Strategies and the Challenges of Maritime Globalization

Fernando González Laxe

Abstract Ports use port marketing as a guideline to face the demands in view of more open and competitive markets; hence, port marketing is considered as a high-value tool in order to search and capture new markets and products.

This chapter collects those elements that shape the market plans and, notably, the base concepts related to the strategic mission of a Port Authority.

In order to do that, the existing indicators of the variables of the strategy are disaggregated: product, price, allocation criteria and promotion. At the same time, the different strategic plans are assessed. On the one hand, we classify the *offensive* positions, including those actions aimed at increasing the services in the markets, investing to improve the competitive positions, and investing to gain access to new markets. On the other hand, we include the *defensive* positions in the strategic plans, that is, those actions aimed at protecting the achieved position and discerning other strategies of disinvestment.

Finally, the axes enabling to measure the potential of each Port Authority are included.

1 The Objectives and Challenges of Marketing

The greatest challenge in marketing is to attract the highest number of clients possible to the market. Thus, in order to capture a substantial number of users to the market, marketing strategies focus on two aspects: firstly, the need to increase market share in an open and competitive market and secondly, to attract new clients, thus obtaining a better (or good) growth of potential benefits.

The marketing strategy must be in keeping with the company's profitability or activity. Consequently, as we show in the following diagram, in order to increase

F. González Laxe
University of La Coruña, A Coruña, Spain
e-mail: laxe@udc.es

P. Coto-Millán et al. (eds.), *Essays on Port Economics*, Contributions to Economics,
DOI 10.1007/978-3-7908-2425-4_2,

Marketing Contribution = [*global demand from the market* · *company market share* · (*revenue per client* – *variable cost per client*)] – *marketing expenditure*

Fig. 1 Growth strategies orientated to the market

profitability, the focus is on increasing market demand, companies' market share, client revenue and, finally, to reduce variable costs for the client or marketing expenses (Fig. 1).

In other words, marketing's contribution to the functioning of an activity or business is linked to strategies to increase market demand, with the option to enter or abandon the market and it is also related to increasing the market share. This is conditioned by strategies to generate increased buying (on the clients' part) depending on changes in variable costs. Finally, it has an inverse relationship with strategies to improve the efficiency in the use of marketing resources.

Strategies intended to increase the market share and client revenue need a mature market so that companies benefit from a significant and relevant market share.

Taking into account these assumptions, clients are the best strategic assets and a needs analysis could reveal the operating capacity of improvements both as regards products and services offered. (This eventually has an impact on the value of client revenue).

The other way to increase the benefits of the activity (or company) involves reducing variable unit costs. Generally, this strategy anticipates increasing margins and maintaining customer satisfaction levels. If customer satisfaction levels are not maintained then this may contribute to an overall reduction in total benefit.

Thus, one could choose to reduce marketing expenses (although only if the company is totally focused on its target market and if it has a very dominant position in the market).

1.1 Market Development Index

In order to increase market share and be able to attract new clients, companies must reduce prices, offer wider product ranges and focus on expanding their distribution. In this way, they increase the availability of their offer and positioning in an open and global market.

The market development index is defined as the quotient between the market's natural demand and its maximum potential. To reach such an index, one must overcome barriers and current limits. To do this, one must have development strategies which consider processes of offer differentiation. This can make it very difficult and costly to imitate the competition.

The calculation of market growth is defined by market potential (maximum number of clients who can enter the market), the penetration of the potential market (number of clients) and the rhythm of market development (i.e. the dynamics and speed in which new clients access the market).

Marketing strategy → Production strategy → Positioning strategy → Location strategy → Services strategy → Market share index

Or, more simply: *visibility* · *interaction* · *availability* · *services*

Fig. 2 Phases of the marketing strategy

Being very dispersive and heterogeneous, the port market shows various differentiating aspects. Firstly, there are many innovative ports and leader ports faced with a market which needs integrated solutions. Secondly, the growth rate of ports is variable as it depends on the characteristics of clients, positioning, on products and services offered, its involvement in the surrounding areas of routes and other influences. For this reason, the port specification emphasizes the relative advantages of entering the market, the relative prices, the availability of services offered, integrated services, the reduction of risks and safety.

The market share will be defined by market share development indexes which are, in these cases, the result of the intersection of various marketing actions (Fig. 2).

1.2 Market Segmentation

Creating a new positioning of attractive products and reaching the desired level of market share requires a decisive, firm and clear effort with regard to defining the product strategies. On one hand, one must define and clarify a strategy of business positioning (depending on the need to cover a market objective) and on the other hand, one must move towards differentiation of services offered.

In this way, the stages of the process of segmentation of markets include the following states:

(a) Segmentation based on needs; this groups the clients depending on their needs.
(b) Identification of segment integration: this determines the behaviour of goods, services and products in each segment.
(c) Attractiveness of the segment : this calculates the global attractiveness of each segment.
(d) Profitability of the segment: this estimates the profitability of each segment.
(e) Positioning of the segment; this defines the value proposition and product/price positioning.
(f) Marketing strategy: this develops positioning strategies.

One must analyse the positioning depending on its market share and prices in order to place each service or product in an open and competitive market.

In principle, the market share is the result of the product's position and marketing effort. Firstly, the following variables are inferred; price, product differentiation, the size and range of product lines, the quality of goods or services, brand image and the existence of competitors' products and substitutes. Secondly, the marketing effort is conditioned by the following variables: sales, physical and logistical distribution, marketing and publicity.

In this way, markets are very attentive to the benefits from the producer's position and respond to differentiation of products, services and brand. Efforts to reduce costs and obtain benefits come from prices and transaction costs.

1.3 The Strategic Map

This is planned depending on vertically defined aims (according to strategic lines) and is segmented horizontally according to market perspectives and dimensions. As the profile of a modern port focuses on a triple dimension (economic, social and environmental), we see that the port turns into a node of logistic channels since it integrates itself, is orientated to the client, generates value for the territory and carries out its activities whilst respecting its surroundings.

One way of specifying the strategic map of the port is to collate a host of aims which simultaneously respond to an economic perspective, to clients, processes and resources. Strategic lines combine the value, efficiency, integration and profitability.

In this way, a port combines two functions. Firstly, to define instruments to facilitate the business and secondly, to facilitate the insertion of other modal systems and sectorial integration.

As regards the effects of businesses, ports aim to reach three basic aims: (a) attain better competitiveness, (b) show high levels of quality, and (c) try to generate the best added value possible of services rendered.

In the first case, improved competitiveness is achieved through actions related to infrastructure and specialised/basic equipment, through fair competition and transparency in the market of port services (avoiding any distortions to the aforementioned competition). It is also reached through tariffs and costs of port services provided, and through the efficiency of human resources.

In the section on defining characteristics of quality, the key elements are technification, workers' professional capacity, quality standards, safety and reliability of logistical port systems, the protection of the environment and maintaining peace in the workplace.

Finally, one must develop value added activities in port services, create an appropriate environment for investment, promote port facilities of high added value, and automise processes.

As regards transport and sectorial integration, we refer to modal and logistical integration. This involves the concentration of services, the development of cabotage, modal systems and suitable port/city relations. Sectorial and national integration refer to the integration of ports in national and international logistical chains, logistical development in ports, the development of logistical areas of activity and the promotion of intra-port communication.

Thus the actions of the strategic map are limited to the following dimensions:

Economical: Be profitable and generate added value to the client and territory;

Clients: Grow as regard to traffic, offer of regular sea/land lines, be competitive as regards price and tariffs, have operational capacity, enable logistical/industrial activities, promote ports in the network and show respect to those around.

Processes: Productivity of facilities and operations, orientate quality processes, manage costs and investments, achieve an integrated port community, links with other ports and transport means, and promote the port.

Resources: Physical infrastructures, port innovation, teams.

Consequently, strategic planning of a port takes heed of the necessary methodology to study economical perspectives of investment and the necessary resources to develop (in competitive conditions) one or several port businesses, taking into consideration long-term strategic scenarios, whether they be offensive or defensive (see Graph 1).

OFFENSIVE STRATEGIC PLANS

BASIC STRATEGY I	BASIC STRATEGY II	BASIC STRATEGY III
Invest in order to increase sales and services to existing markets	Invest in order to improve the competitiveness	Invest in order to enter new markets
↓	↓	↓
Grow in existing markets	Improve margins	Diversify growth
↓	↓	↓
• Growth of market share • Growth of client revenue • Enter new market segments • Increase market demand	• Improve customer loyalty and purchasing levels • Improve the advantages of differentiation • Reduce marketing costs	• Enter new related and unrelated markets • Enter new emerging markets • Develop new markets

DEFENSIVE STRATEGIC PLANS

BASIC STRATEGY I	BASIC STRATEGY II	BASIC STRATEGY III
Protect the position	Optimize the position	Disinvest
↓	↓	↓
Maintain benefits	Maximize benefits	Obtain cash-flow
↓	↓	↓
• Protect market share • Develop customer loyalty	• Maximize net contribution of marketing • Focus on the approach	• Manage in order to obtain liquid assets • Disinvest and obtain liquid assets

Graph 1 Strategic plans

The basis of this analysis is reflected in the processes of integration and globalization of the international economy, in the classification of the multiple opportunities which open up through increased international businesses and investment and via the establishment of global networks (both sectorial and the insertion of intermodal networks of transport).

In this sense, the port's potential will be set through six axes: (a) transport infrastructures, (b) logistical infrastructures, (c) logistical basis, (d) conditions related to legal and institutional aspects, (e) the functions of consumption and costs, and (f) the info-structure.

2 The Strategies of Port Marketing

The mission of a marketing strategy stems from the port's organization of services which maintain a high level of efficiency and which ensure the fulfilment of defined economical, social and environmental objectives considered in the strategic plan.

The points of analysis refer to previous studies of the SWOT analysis. As regards the analysis of external contexts of activities, we will refer to the Opportunities and Threats. With respect to the examination of internal contexts, our reference analyses are Strengths and Weaknesses.

Under these criteria, the strategic analysis defines certain factors of competitiveness, the factors of differentiation and factors referred to value chains. The combination of this opens the doors to various options of strategies as valid alternatives although acting on its quantification, evaluation and possible economic, social and environmental effects can bring about different typologies and consequently final results can be very different from set objectives.

For this reason, one can distinguish various marketing strategies: those which analyze options from market perspectives and others which visualize objectives from the competition's environment.

From the market's point of view, we consider the following as basic strategies: (a) penetration options, (b) the intensity and speed of expansion, (c) levels of specialization, (d) diversification, and (e) reconversion. From the competition's perspective, options are summarized as (a) offensive, (b) defensive, (c) differentiated, and (d) agreed.

Classically, marketing strategies have four different axes which determine options carried out. These axes are product, price, place and promotion.

(a) The port product. This emphasizes the characteristics of buildings, equipment and organizational means. Thus its functions are related to the services of sea/land interface; they ensure the conditions of transporting goods, have space for storage, take on the concepts of commercial and industrial areas and set out the performance criteria of commercial services and various activities. The port product thus identifies the vocation of all businesses and outlets around the port and defines the attractiveness of maritime businesses, contributing to its growth and expansion. In the same way, it forms new relations with the market, with transport systems and international businesses. For this reason, the elements which refer to geographic localization, geo-strategic position, access/insertion of transport networks, technical characteristics, logistical development and the development of the concept of the port brand are basic elements (marketing options can be inferred from the product's policy to reach the product's concept, product lines, the brand, the product's life cycle and the development of new products). There is also the concept of the *augmented product* which considers not only the product itself but also a series of attributes which increase both the specific variety and additional guarantee; i.e. the augmented product is equal to the base product plus associated products or services.

(b) The port price. Amongst the factors which are involved in setting the price, are those which regard costs, commercial viability, economic sensitivity, qualities and competition of services and the availability and occupation of infrastructures and equipment. Moreover, one must take into account offer/demand, client focuses and competitors' strategies, in short the weight which acquires port tariffs per type of merchandise and product. Various problems encountered by port facilities include those linked to delocalization of ships or merchandise or to differing tariffs for the same function or service. For this reason, when it comes to shaping prices, the aims refer to simplicity, transparency, publicity and facility. The aims are obvious – to attract clients (providing there is excessive offer and available capacity) using the best possible quality of service provision and the best internal organization possible, making it competitive. Campaigns to attract clients are set up using activities which integrate and improve the port value chain, whether they concern vessels, safety of goods, services in the quay, efficiency of storage or the speed in processing information and the abundance of room for storage and processing.

(c) The port place/distribution. This refers to physical access (goods and ships) or economic access (links and network connectivity). It also regards the integration to various means of transport. In this way, the economical activity of the port is in direct relation to the port links with means of transport and accessibility (extending distribution beyond its own hinterland). This presumes that the concept of distribution is associated to economic flows of goods and services in the direction of markets where various operations and functions occur. One can distinguish two operations, firstly distribution channels, i.e. means (of transport, storage, retention) and secondly, the means by which goods are transported to the client i.e. inbound and outbound, in line with localisation in the logistical system of a certain product/respective client. In these conditions, marketing aims to focus on the negotiation of contracts and ensuing management of transactions whilst considering sales, promotions, credit levels, communication and advertising. As regards logistical channels, this regards the positioning of the product in terms of requisites of time and space of management of inventories, order processing and other services of a logistical value. From here, companies look to optimize services rendered at the lowest cost possible and pay attention to the characteristics of fulfilment, amplitude and multiplicity.

(d) Port promotion/communication has a considerable number of aims. Amongst the most significant, are the following: (1) facilitate information, ideas and activities on a certain theme to concrete recipients, (2) communication must increase the visibility of qualitative mid-long term factors, (3) it must promote the product, price and distribution in a regular way and in a qualitative and timely fashion, (4) seek differentiated communication strategies depending on addressees/payees, (5) communication integrates the co-ordination of advertising, promotion and direct action to consolidate functions, penetrate new markets and improve the port's reputation and notoriety, and (6) it must be suitable for financial resources and the desired benefits and be in line with the image it wishes to transmit. In short, it must be coherent.

This entire explanation is defined by various factors and variables seen in Tables 1–4.

In short, port marketing analyses reflect conditions of competitiveness and attractiveness. The former provide us with signs of efficiency around productive factors used and the latter include those elements which (when used in a specific way), allow one to capture new traffic and services for a concrete location. For this reason, specific resources contribute to determining port dynamics. Their attractiveness depends on resource specification and assets generated at the heart of the activity carried out by production factors. From this, the policies of territorial

Table 1 Variable of port marketing: product and port service (this refers to the location where the interface is established between maritime transport and land transport)

Basic elements	Factors or variables
1. Location and geo-strategic position	1.1. Opposite the markets, places of production, maritime routes and market niches
	1.2. Accessibility of the port to national and international routes, to logistical centres, industries and towns
	1.3. Physical characteristics of the port in terms of protection, safety, conditions of accessibility and anchorage
2. Technical factors of the port	2.1. Infrastructures of the port, quays, extension, equipment, buildings, maritime and land access
	2.2. Superstructure or software of the port, organization, IT systems, human resources, company network and services
	2.3. Logistical structure of the port, areas of storage, modal interface and logistical parks
3. Logistics of the port	3.1. Space
	3.2. Node to develop operations
	3.3. Areas and types of storage
	3.4. Management and control of logistics
4. Brand of the port	4.1. Image associated to the brand, logo and slogan
	4.2. Identification and differentiation of the port as compared to competitors
	4.3. Internal culture of the port
	4.4. The ports' attributes as perceived by the client

Table 2 Variable of port marketing: price (this covers the tariff system used by the Port Authority or Terminal for its clients for services offered and for the use/ enjoyment of its equipment and available infrastructure)

Basic elements	Factors or variables
1. Tariffs (may be set freely by the operator or fixed by the Port Authority)	1.1. Movement of ships 1.2. Logistics of merchandise 1.3. Intermodal operations 1.4. Storage and empty containers 1.5. Transhipment 1.6. Discounts
2. Structure of the tariff	2.1. Seek maximum profit 2.2. Reach minimum levels of quality 2.3. Size and power of clients 2.4. Market situation 2.5. Decision model 2.6. Shareholders' aims 2.7. Traffic profile
3. Complex tariffs	3.1. Commercial loyalty 3.2. Integration
4. Strategies	4.1. Focused on costs 4.2. Focused on clients 4.3. Focused on sharing

Table 3 Variable of port marketing: distribution (related to the use and facility of clients' access to vessels and merchandise in port surroundings for various services; also for its links to/integration in the railway network and in channels of intermodal transport)

Basic elements	Factors or variables
1. Interposition of products in logistical chains	1.1. Face major competition between various ports and in logistical areas 1.2. Analyse the phenomenon of delocalisation or delocalised production of components, development of transhipment 1.3. Links between production poles, consumption and distribution
2. Infrastructure of network and equipment	2.1. Integrated and intermodal transport system, responding to transport needs 2.2. Stimulate cost reduction and improve reliability 2.3. Favour consolidation of logistical chains 2.4. Presence of competition between port terminals
3. Adopt strategic decisions for the positioning of the port	3.1. Analysis of higher value markets 3.2. Integration of better logistic chains 3.3. Client relations

marketing insist on two strategies: a basic one based on the generic activities of the region (i.e. conditions of cost and profitability) and location strategies, i.e. those which force us to define more clearly the components of asset exploitation and the region's specific resources. These help us emphasise the essential components of resources and the agents' aptitude.

It is hardly surprising therefore that actions carried out by Port Authorities are directed through axes which combine various key elements, different complementary methods and multiple direct actions. In this sense, Table 5 shows the methods and direct actions linked to defining elements of the options of port marketing.

Table 4 Variable of port marketing: communication (focused on the promotion of gaining new clients, increasing market share and expanding the hinterland through the availability of services)

Basic elements	Factors or variables
1. Create value to the brand	1. Messages must be informative and named. Note concrete proposals or propose long lasting and permanent relations
2. Contribute to greater awareness and understanding of the services offered	2. Messages must respond immediately and react rapidly to the demand for information
3. Influence agents'/clients' preferences	3. Clients must possess up-to-date information on the functions carried out by the port, how to solve these problems, rival pricing structures and timing, knowledge of transport in the hinterland
4. Increase the use of certain services	4. Changes in beliefs, aptitudes, habits and scenario influence clients' perception on a port or terminal
5. Circulation of information and the message	5. One must reinforce the frequency and, above all, the identity of the message as well as evaluating its impacts and perception of these

Table 5 Methods and direct actions of marketing policies

Key elements	Methods	Direct actions
Locality	Communication mix	Direct-mail
Access	Selecting agencies	Mass e-mail
Offer	Definition of advertising aims and making decisions on publications	Telemarketing
Aims		Sales
International context	Internal communication messages and logos	Public relations
	Setting up campaign plans	Participation in trade fairs
	Setting up budgets realisation and evaluation	Launch of new products and services
		Conferences
		Interviews

3 Market Analysis

To set up marketing strategies more effectively and to implement necessary marketing policies, one must pay attention to the following types of market analysis: (a) external forces (via qualitative and quantitative analyses); (b) market surveys and market research; and (c) studying the competition.

The first concern is with external forces which are demographic (evolution of the population, occupation of geographical spaces, linguistic, cultural and/or ethnic composition and family additions). Thus, one needs to analyse the situations which characterise clients (consumers, consumption habits and typical and specific needs). Secondly, economic forces refer to economic indicators, whether they be regional, national or supranational. Depending on a particular moment's economic situation, there can be completely different customer reactions as regards preference for various products. Thirdly, political forces are taken into consideration as they define the higher or lower tendency of the market's interventionist capacity.

Markets and consumer patterns respond partly to traditions or to political and economic fluctuations. In this way, social and cultural forces have a significant influence on consumer patterns and on market behaviour. Finally, technological forces allow the development of high added value innovatory goods and services.

Market research and surveys are extremely useful for obtaining information to set up marketing strategies and to be able to implement suitable policies. Given that markets are not homogenous at world level, one can distinguish various groups of clients/consumers with common or differentiated characteristics. Market research provides us with very qualified tools to define the necessary marketing efforts.

Studying the competition involves analysing organisations depending on each competitor's possible scenario of actions. From the clients' perspective, a variety of products in a wide, valid and competitive market is desirable. From the company's perspective, competition forces it to make a constant effort to re-launch products using regular improvements, continuous innovation, optimisation of resources, minimisation of costs, and cutting down operational time.

Unfair competition, dumping, fraud, bribery and tax evasion, among others, introduce undesirable, perverse effects which lead to instability in markets and undesirable imbalances in the offer. At this stage, the State must intervene.

Market analysis considers the users' various actions. In this sense, one can emphasise the different aims and tools for users, producers and public institutions. Likewise, we can see distinct concepts concerning his/her acts and, above all, one must emphasise the fact that the purposes are being limited by dissimilar use of available tools. This framework is explained in Table 6 below.

4 Port Spaces and Its Positioning

The following tendencies form a backdrop to port spaces:

(a) Inter-port competitiveness increases due to growing overlapping hinterlands and greater ability to substitute each port in the frame of the productive process and the distribution of the transport cycle. Contracting by terminal companies thus has a greater influence.
(b) As the position of Port Authorities is reduced, the specific and strategic weight of large multinational groups increases. These are linked both to maritime companies as well as firms and terminals which aim to be partners of the Port Authorities.
(c) Port tariffs are reducing due to pressure from global carriers (although payment for port services increases given the new orders made by Port Authorities to improve the positioning of port enclaves).
(d) Modern port authorities move in "new competitive spaces", depending both on offers and on service providers to operators.
(e) Consequently, competition for the service and opportunity of costs are beginning to define a new maritime space.

Table 6 Aims, tools and indicators of port actions

Node of transport: port

	Long term aims	Short term aims	Tools	Indicators
Government	• Maximise social well-being • Improve competitiveness • Guarantee safety • Minimise environmental damage • Provide efficient activity of traffic	• Promote the use of maritime transport; promote familiarity with the port abroad • Internalise external costs (congestion, environmental, accidents, infrastructures) • Uniform and non discriminatory distribution; efficient and optimum use of infrastructures • Transparency of tariffs • Improve maritime access (depth, dredging) navigability and port infrastructures (berths, quays, mooring...) • Provide good and flexible connections with the hinterland and intermodal connectivity • Provide a high number of services (pilotage, towing...) • Guarantee optimum use of the land	• Quality Inspections • Socio economic negotiations • Suitable information flows • Standards of quality • Safety Regulations • Policy of infrastructures • Policy of land and concessions • Specialised terminals • Nautical access, land access	• Regulations and legislation • Principles of tariffs (indicators of prices and costs) • Proportion of damaged, lost (stolen) merchandise • Energetic equilibrium • Policy of infrastructures • Capacity of usage • "Frequency" of dredging activities • New investments in port infrastructures • Rates of mooring occupation
Producer	• Maximize benefits (private company) • Maximize production (public company)	• Increase market share • Guarantee safety and quality • Create value added • Increase productivity	• Control costs, minimise costs • Policy of investments, technical and technological improvements • Economies of scale • Activities of added value (consolidation and deconsolidation of containers, storage, inspections etc.)	• Average tonnage per boat mooring • Productivity of operations of loading/unloading, work productivity (e.g. containers manipulated by crane, manpower units, tonnage per ship/hour in the port, tonnage per metre of quay etc.) • Timing: waiting time, service time, time for loading/unloading.
User	• Maximise benefits for companies • Maximise use for homes	• Low prices, minimising logistical costs • Reliable and flexible services	• Power of negotiation • Quantity and size of ships	• Prices • Average size of consignment • Loyalty, customer satisfaction • Quota of services done in time (%)

Source: Collated by Blauvens et al. (2002). Nationale Maatschappij Der Belgische Spoorwegwn; Wobbe et al. (1999); Secretariat of UNCTAD (1976); Talley (1994); White (1995); Heavert et al. (2001), Best (2007) and Blauwers (2002).

Table 7 Elements of the marketing strategy

1. Characteristics of a modern port	2. Factors of competitiveness
Suitable infrastructure; efficient organisation; efficient management; specialisation; suitable levels of negotiation of companies linked to the port; suitable levels of service to territories close to the port; functional articulation with various modes of transport and logistical chains	Geographic localization and conditions of the port; size and type of the market which it covers and serves; infrastructure and existing equipment, logistical conditions and space available; access networks; articulation with various means of transport and services; organisational level of the port; relations with maritime companies and port operators
3. Analysis of the context of the port	4. Strategical analysis of marketing
Agents which influence the port; market segments; competitive ports; legislation; superstructures and equipment; quality of services; characteristics of the hinterland; strategic position; accessibility and integration; systems of transport and intermodality; technical staff	(a) Tools: products and port services; prices and tariffs; distribution and communication (b) Factors: competitiveness, differentiation; criteria

Source: my own

In short, the analysis is defined by two axes; one derived from the strong competition between ports (and maybe between territories), and the other from the growing vertical and horizontal integration of maritime agents.

In this sense, the methodologies for a port marketing strategy are limited to four elements:

This new conception (which we could call the "new port generation") was promoted and driven very clearly at the end of the 1990s and start of the twenty-first century, using the strong expansion of business and maritime transport. However, in the current period of recession, we are aware of certain port enclaves using a reduction in transport activity, new geographical relocation of production units and higher levels of decision making with which the Port Authorities and private agents seek goods and offer services (Table 7).

Under these parameters, a frame is built of the mission of a modern port Graph 2.

5 Conclusions on the Analysis of Port Marketing

There are many key questions related to the strategies of port marketing. Firstly, they highlight the actions which tend to improve infrastructures to promote productivity growth and improve the quality and efficiency of services offered.

Secondly, they emphasise aims leading to improve land connectivity and its insertion in global networks, i.e. those related to interchange and intermodality with other means of transport.

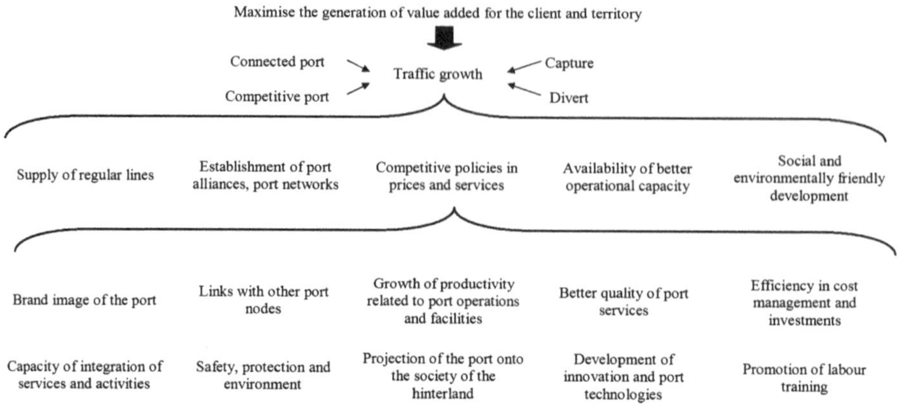

Graph 2 Modern task of the port

Thirdly, one sees the efforts in the fields of internal organisation and in management levels expressed in terms of costs like investment analysis and relations of public public/private partnership.

Finally, through marketing strategies, ports highlight those actions which refer to the continuous improvement of the professional qualification of its members, its pledge for technological innovations and to improve environmental indicators.

Most marketing strategies of main worldwide ports are built on these actions.

References

Best R (2007) Marketing Estratégico. Pearson Prentice Hall, Madrid

Blauwens G, De Baere P, Van de Voorde F (2002) Transport economics. De Boeck, Antwerp

Heavert T, Meersman H, Vande Voorde E (2001) Co-opetition and competition in international container transport: strategies for ports. Marit Pol Manag 28(3):293–305

Talley WK (1994) Performance indicators and port performance evaluation. Logist Transport Rev 30:339–352

UNCTAD (1976) Port performance indicators. Conference on trade and development, Geneva

White P (1995) Public transport, its planning, management and operation. Biddles, Westminster

Wobbe W et al (1999) Transport benchmarking. OCDE, Paris

Contextual Port Development: A Theoretical Approach

Ricardo J. Sánchez and Gordon Wilmsmeier

Abstract Ports play a critical role as gateways and facilitators of trade. In the last 20 years, ports have undergone an intensive evolution in trying to adapt to a changing environment (change in demand, etc.). The results and models from this evolution process vary by regions and economic contexts, particularly in developing countries. While ports have developed in scale and have consequently taken on the challenges of growing trade flows, access infrastructure to ports or port delivery corridors and institutional developments have lagged behind. The resulting bottlenecks in some way reflect deficits and insufficiencies in the interplay of the economic system and factors defining port development: transport demand, the structure of trade, transport services, institutional capacities etc.

A time lag in the resolving of infrastructural bottlenecks, which to a great extent depends on the efficiency and effectiveness of institutions, can cause significant impacts for regional economies. The probability of time lags is especially prevalent in the interaction between the port and the maritime system as the port system has longer, 'discrete' development cycles in comparison to the maritime system.

This chapter investigates and evaluates port development as the consequence (result) of the interaction of three systems: the economic system, the maritime system, and the port system; and develops a relational approach to port development.

R.J. Sánchez (✉)
United Nations Economic Commission for Latin America and the Caribbean (UNECLAC), Santiago, Chile
e-mail: Ricardo.Sanchez@cepal.org

G. Wilmsmeier
Transport Research Institute (TRI), Edinburgh Napier University, Merchiston Campus, EH10 5DT, Edinburgh, UK

P. Coto-Millán et al. (eds.), *Essays on Port Economics*, Contributions to Economics,
DOI 10.1007/978-3-7908-2425-4_3,

1 Introduction

Ports facilitate trade and are 'gateways to globalisation'. Exponential growth in trade volumes paired with continuous increases in ship size and advances in the technological evolution of handling have constantly altered the 'environment' for port development. Rodrigue et al. (1997) argue that the transposition of an economic system into space significantly increases the reliance on transportation. Ports have responded to these changes through investment in infra- and superstructures, but also through devolution. Under this expanding environment, the limitations of transport infrastructure for spatial diffusion in certain regions have become obvious.

The results and effectiveness of these responses differ throughout the global port system. Ports, particularly in developing regions, have struggled in reacting timely and adequately to successfully meet the challenges of growing trade flows. Limitations surged into public awareness firstly in the port industry, revealing the inefficiencies of these proclaimed facilitators of trade and an absence of contextual development.

The main focus in the literature and studies on port development in developing countries, including Latin America (LA), has been on the effects of port devolution on infrastructure investment, port productivity and efficiency. We argue that, particularly for developing countries, the following issues need to be added to the discussion:

(a) The institutional framework for port operations and the changing role of port authorities under a competitive and privatised port environment
(b) The changing relation and play of power between port authorities, port operators and shipping lines
(c) The conceptualisation of the hinterland: regionalisation of maritime and land hinterlands

The authors argue that the prevalent lagging in port development results from defects and insufficiencies in the interplay of the economic system, specifically, factors that determine port development: transport demand, the structure of trade, transport services, port capacities and development within the maritime system, etc. This approach allows for the consideration of the three issues and a discussion of port development from a relational and contextual perspective.

2 Introduction to Challenges in Port Development

Ports play a critical role as interfaces and facilitators of trade. As the economies of developing countries´ become increasingly integrated with the global economy, their ports must facilitate such integration in order to help reach the countries' development objectives. Ports in developing regions, i.e. LA, underwent an intensive evolution in the last 20 years with their attempts at adapting to the changing environment (changes in demand, etc.). This evolution process has brought different results and models within the port system. While ports have developed and in some way took on the challenges of growing trade flows, access infrastructure to ports or port delivery corridors and institutional structures have remained in the early stages of development. Current infrastructural bottlenecks arise especially in relation to port infrastructure. These bottlenecks in some way reflect deficits and insufficiencies in the interplay of the economic system and factors defining port development: transport demand, the structure of trade, transport services, port capacities and institutional frameworks, etc.

In the last decade a significant number of countries – particularly those in LA – have implemented policies aimed at reforming their port industries. In the attempt to support such a capital-intensive industry, privatisation has often formed an important strand of such policies. A key claim in favour of privatisation is that the transfer of ownership from public to private hands will ultimately lead to an improvement in economic efficiency and, hence, financial and operational performance. However, the success of such strategy depends on the appropriateness and parallelism of the evolution of institutions (e.g. port authorities, regulatory bodies etc).

Port development in developing regions has mainly been driven by external factors and the port system significantly influenced by the need to satisfy the requirements of the maritime system. But privatisation efforts in the region have only partly been a cure for ports as the role of ports continues to change. Today, ports can no longer expect to attract cargo simply because they are natural gateways to rich hinterlands. The ports' role has changed from a monopoly to a dynamic interlinkage and on to a subsystem in the logistics chain. We argue that privatisation is only a partial cure for what ails ports in developing countries, i.e. LA, and that, if implemented in isolation it simply lacks a relational view of port development.

Based on these observations we argue that the interplay between three systems particularly in developing countries, is constrained. The ability of ports to deal with their changing relationships with the maritime sector and the new requirements from the expanding economic system is only partial.

Reform in the maritime sector, especially in Latin America and other developing countries around the world, has principally taken place under Fordist principles, based on the economies of scale and efficiency gains, driven by standardisation of products and services. However, Fordism has structural boundaries because these economies eventually reach their limit.

New developments towards a post-Fordist economic environment change the source of competitiveness for ports from economics of scale based on basic production factors (capital, land, labour), to economies of scope based on advanced production (service) factors, know-how, and procedures.

Moreover, the nature of required services is changing from standard services, with long life cycles, to large differentiated service requirements, with short life-cycles.

The environment for ports has evolved in a highly dynamic manner with more uncertainty and risk.

Finally, the forms of organisation have changed from integrated structures based on standard procedures and processes, to flexible, decentralised structures with incident management needs.[1]

The maritime industry is in a period of unprecedented change. New forces are requiring adjustments and adaptations: ports are particularly vulnerable because intermediaries are in complex and competitive transport chains. Ports are required to act and react to developments in both land and water spheres. Ports have lost the means to influence events to the degree they used to be able to and are being forced to react to changes. A variety of issues are presented that suggest that there are opportunities for port authorities to intervene and better influence the future. These are opportunities, however, that require reappraisal of the role of ports in logistical chains on land and water.

This change of power is also driven by the process of vertical and horizontal integration in the maritime sector. This integration has lead to an increasing involvement of global port operating groups and shipping lines in port management and operations.

Given these considerations, it is important to take into account that next to its own dynamics the maritime sector underlies a cyclic pattern that is dependent on economic development. The denominated shipping cycle influences maritime industry development and thus also conditions port development. Consequently port development, conceived in a relational and contextual manner, results from the interaction of the economic system, the transport system, with particular emphasis on the maritime and port subsystems and the social system.

Ports, particularly in developing countries, must determine strategies for long-term sustainability which is not necessarily at a maximum efficiency level, but at an optimum efficiency and efficacy, respecting the contextuality of port development under specific conditions, considering the game of crossed-pressures that acts over ports and their development.

3 A Systematic Approach to the Theoretic Complexity

In order to understand the relational perspective and contextuality of the economic, maritime and port system, an approach to space and transport theory is required. Transport is characterised as the interconnecting process of fixed elements within space. Therefore, transport is the instrument leading to the development of spatial order and organisation as well as spatial perception by the individual. The historical spatial development of transport defined as interconnecting flows between points in space is described below.

---

[1]For further details about these matters, see Baltazar, Ramon and Mary Brooks: "Port Governance, Devolution and the Matching Framework: A Configuration Theory Approach", in Brooks and Cullinane (2008).

In a landscape system where locations are geographically separated by distance, the transfer of goods, material and individuals is an inevitable necessity of spatial organisation, since both producers and consumers are located at different points in geographic space.

What is stressed is that the structure of the transportation networks of any area cannot be divorced from the geographic characteristics and the appearance of space to its users. The goal is to analyse the interdependency of nodes (i.e. ports) with economic and maritime industry development. What are the reasons for lagging port development and how can the responsiveness and effectiveness of port development to changes in the economic and transport system be improved? How can the theoretical approach of contextual port development be transposed from theory to reality?

Space and flows are the basis for this section. While space is vital and strategic, flows are spatial and temporal, but above all material. The economic system defines the materiality and structure of flows, while anthropocentric variables influence flows by perception and cognition. The flow of freight in geographic space underlies a system with internal complexity. 'Transport is an epitome of the complex relationships that exist between the physical and political activity and levels of economic development' (Hoyle and Knowles 1998).

> Complexity in this sense is a measure of vagueness or lack of information. Complexity is the information, which a system is lacking, to describe and register its environment or respectively itself entirely (Luhmann 1984a, b).

Transport constitutes itself as non-trivial cyclic and self-referring system being principally multi-dimensionally linked with other systems. One system is the physical environment, which includes physical characteristics, space, and natural resources. The transport system constitutes itself from physical, economic and social characteristics. Transport flows are always reflected in the physical environment as transport requires physical infrastructure. The transport flows themselves are initiated from the economic system, which includes all monetary factors (consumers, industries etc.), while the decision on the direction of transport flows is realised in the system of society, which refers to non monetary and cultural dimensions (Fig. 1).

The transport system makes use of all of the three other systems to be able to realise its functions, but does not overlap totally with these other systems. The figure above portrays the interconnecting function of the transport system.

The entrepreneur trying to act in such a system will encounter certain stabilising feedback effects, which will reduce the desired effects of his or her intervention. The same outcome is not necessarily the case with regard to negative effects, but self strengthening feedback effects can be over-directed and can consequently lead to more negative than positive effects on the system. As the transport system contains several subsystems (modal interfaces, e.g. ports), these lead through delaying or accelerating effects to the creation of a complex time dynamic in the system. The behaviour of the transport system composes from characteristic regularities and structures, even though it is not predictable. It can be said that the

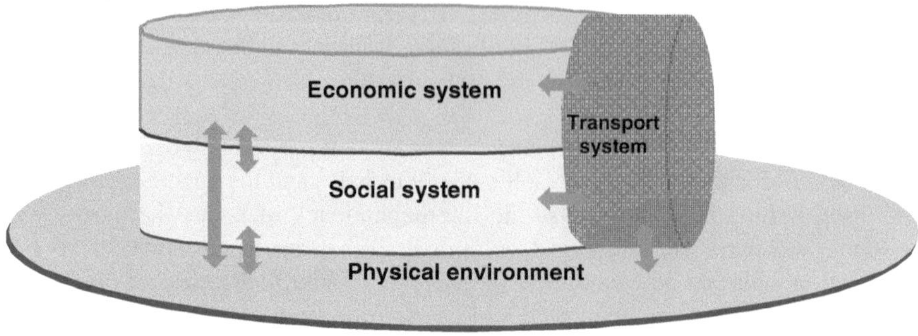

Fig. 1 Interconnection of the transport system with other systems

system has chaotic characteristics. Chaotic in this context does not refer to eventual processes: they are highly organised. But through the impedance of the smallest variables (factors), the transformational behaviour of the system can jump from one characteristic to another. The feedback strength of minimal changes in the system is also known as the 'butterfly effect', saying that a butterfly can generate a hurricane by moving his wings. The changing factors also can be so minimal that any differences may not be felt at all.

With each transformation of the input, the system changes its state, and there with its transformation rules (Schober 1991, p. 3520). The transport system has a self-organising structure which is defined as 'Autopoiesis'. By virtue of the transport system's own function, it is able to adjust itself while obtaining its identity and defining its limits. Looking at transport 'Autopoiesis', it seems to have an especially high inertia when it comes to changing system variables (see Maturana 1994, p. 77; Jantsch 1982, p. 64).

It can be observed that transport under the pressure from an uncertain environment takes actions itself in order to tackle existential situations (otherwise market forces will deconstruct the organisation of the transport system). In case feedback loops are missing, parts of system can grow in an uncontrollable manner, and through the limitations of its physical characteristics it will lead to overshooting and collapse of the system. In developing countries we argue that Autopoieses cannot 'act'.

The factors leading to the transport system's organisational characteristics are identical to the factors which lead to development and competitiveness. Therefore transport can manoeuvre its epochs and processes. But even though the transport system steers and organises itself, the global tendencies of the system are defined by its environment and not itself.

The authors set nodes (ports), as a subsystem of the transport system, in the centre of investigation. What are the characteristics of nodes (ports)? Their development has a tempo and rhythm as well as direction and affinities. The significance of the material quality of nodes (ports) is that they have structure, beyond facilitating processes. Nodes (ports), as the facilitators of flows, in this study indicate a 'bridge' between the outputs of the economic system and the movement of these outputs within globalised trade. Ports have grown to be a key component of competitiveness.

This chapter focuses on the contextual analysis of node development. Topographic settings are external limits, which give a level of predictability to flows; nodes (ports) speed up flows by enabling mobility.

Following the arguments from Hoyle and Knowles (1998), the understanding of port development incorporates the following five essential ideas for a theoretic framework of transport and development.

Historical Perspectives: The appearance of transport systems today evolved from theoretic ideas and concepts in the past and the current recognition of these systems is an appreciation of the ideas from the past. Besides this, all current transport networks are our heritage from the past. Attempts at revisiting and analysing transport networks today should bear in mind that these networks were designed to serve certain purposes in the past, which can differ from those they do our should fulfil today.

Nodes, Networks and Systems: Transport nodes are a critical measure. Nodes define the starting and endpoint of relationships and are connected by links. The pattern of nodes and linkages forms the transportation network. Transport systems develop from technological capability and economic resources.

Modes, Choices, Intermodalism and Flexibility: The third idea deals with the interconnection between the conventional transport modes: roads, railways, air, waterborne, pipelines. Hoyle and Knowles (1998) define three relevant dimensions:

- The relative significance of different transport modes
- The degree of inter-modal choice
- Dependence of modern transport systems on the concept of intermodalism

Containerisation nourishes these three dimensions, bridging the gap between modes in freight transport, making transport modes complementary and not competitive.

Deregulation and Privatisation: Transport services have undergone a change from regulated patterns and charges towards deregulated and liberalised Deregulation, which has affected all modes of transport, has been underpinned by the theory of contestable markets, though the transition to deregulated markets has not always been contestable since transport industries have a natural tendency to strive towards monopolistic industry structures.

Another motive for transport deregulation and privatisation is reducing the cost of transport subsidies and raising enormous capital receipts from selling publicly owned companies.

Holistic Approaches: Transport systems are dynamic wholes and their evolution and operation should ultimately be perceived in this context. Their origin, development and current operation should be taken into consideration in order to approach transport as a whole. Transport modes should not be seen as individual transport modes, but should be analysed jointly and in an integrating manner.

4 Understanding System Relationships: The Maritime and Port System

The economic system, which we consider as exogenous, initiates transport flows and, in consequence, acts on the maritime transport system – both shipping and port services. In general terms, it is the driver for commercial activity and, in particular, for the provision of transport services as a derived demand from the direct commercial demand for internationally traded products. Dynamic changes in the nature of this commercial demand have required structural changes in the composition of the transport fleet, in transport network configurations and in the organisational structure of transport services. The maritime system represents a subsystem (although a crucially important one within the context of international trade) of what is a wider multi-dimensionally linked transport system. The maritime system encompasses both shipping and port subsystems which, although always irrevocably interlinked, have become increasingly overlapping and less dichotomous in recent years as shipping companies have diversified into container handling in ports, initially through the promulgation of dedicated terminals and more lately as newly emergent global terminal operators in their own right (Fig. 2).

One basis for distinguishing between the shipping and port subsystems remains the fact that the constituent elements of the latter are composed of physical characteristics in space, while the former comprises mobile elements. The economic and the shipping system together generate pressure on the port system in the form of ever-evolving specific requirements with respect to infrastructure, superstructure, equipment, efficiency, organisation etc. This prompts a process of time-lagged reaction within the port system to satisfy this changing demand and it is this reactive process which actually constitutes the port development process.

Changes in the port system occur in an almost completely discrete manner, since variations in port infrastructure and superstructure, as well as organisational changes, appear to be rather abrupt and are neither implemented nor do they 'grow' in a continuous fashion; investment in the port sector is often characterised as being

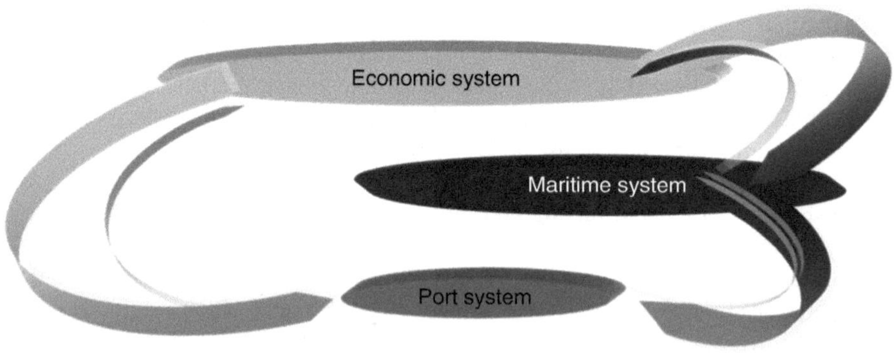

Fig. 2 Port development – the interaction of systems
Sources: Authors elaboration

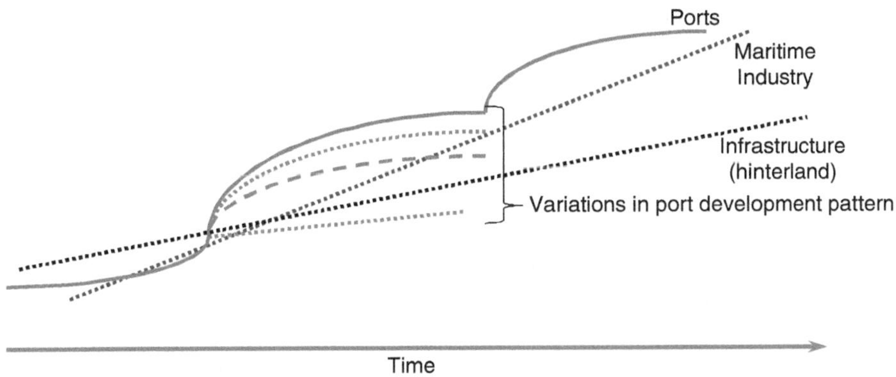

Fig. 3 Development pattern
Source: Authors

'lumpy'. Moreover, port development is very often dependent upon and determined by the degree to which a specific port in question is embedded within local and regional institutional considerations. It is certainly the case, for example, that the conditions under which any port system interacts with other subsystems of the wider transport system – in particular, the local port access infrastructure – are very often locally and regionally defined and, therefore, beyond the direct sphere of influence of the port system itself. This is critically important not only to the port but also to the economy it serves as it is this which ultimately defines the degree of connectivity enjoyed by the economic system that prevails within a port's hinterland (Fig. 3).

Due to the fact that the port system development cycle advances in a discrete manner, its adjustment to the continuous evolution of freight transport demand will inevitably lead to alternating situations of either infrastructural insufficiency and scarcity of supply on the one hand (i.e. excess demand), or to a surfeit of port infrastructure (i.e. surplus supply). This somewhat natural characteristic of a virtually constant harmonic mismatch of port infrastructure supply and demand can be dramatically exacerbated by failures in local and regional decisions which impact upon the port system. In either case, the effect on the efficiency and performance of a port will be negative.

The major factors determining and reflecting port system development are: (a) physical (infrastructure and superstructure), (b) economic, (c) social/environmental, and (d) institutional arrangements. Quite apart from the scale economies that may be derived from port system development, there are substantial impacts on port facilities as well as capacities in terms of water depth and handling equipment.

5 A Theoretic Review on Port Development

What is port development? Goss (1990a) defines ports as a gateway through which goods and passengers are transferred between ship and shore. Following

Wang et al. (2005) port functions of today's ports have changed and ports have undergone important processes of evolution.

Traditional port development models (Taaffe et al. 1963; Bird 1980) focus on the explanation of different stages of port development over time based on economic, political and technological factors. Hayuth (1981, 1988) points out the importance of technological change and its impact on a competitive structure. Robinson (2002) and Notteboom and Rodrigue (2005, 2008), among others, add logistical integration from a functional and organisational perspective as a subsequent development stage to the models postulated from Taaffe et al. (1963) and Bird (1980).

These models have been useful in explaining port development as a function of technological, organisational and functional aspects, but fall short in that they do not take into account knowledge management and organisational capacity.

These studies point out the important relationship between port development and spatial distribution of port activity, which particularly develops through technological development i.e. containerisation.

From this point of view, port development and the spatial distribution of port activity is characterised by a gateway facing competition from developing smaller ports at the periphery (Hayuth 1981 and 1988). Complementary to these ideas, Notteboom and Rodrigue (2005) introduce a new phase of regionalisation where logistical integration and network orientation explain the emergence of the so-called 'offshore hub ports' and the geographic and functional expansion of load centres to become 'regional load centre networks'. Here, the concept of centrality, that explains to some extent the formation of gateways, is replaced by the concept of 'intermediacy' (Fleming and Hayuth 1994), where a large direct hinterland market is not a necessary condition for concentrating large traffic volumes. Instead, discontinuous hinterlands are supported by logistic zones and inland distribution centres, which at the same time reflects the degree of logistic integration among carriers and the new 'mega carriers'. In this sense the adjusted definition of hinterland operates, one that considers core, congruent and extended hinterlands, which adjusts to the stretches or functions of the aforementioned port service demand (Sánchez and Wilmsmeier 2007) (Fig. 4).

It is difficult to define port development universally and to set up a general development classification due to the complex nature of port activities. As a basis for port development, UNCTAD (1992) defines the basic provisions necessary for a port (see Table 1).

The layout, organisation and performance of these basic provisions changes and develops depending on the demands and requirements of port users. UNCTAD (1992) Tries to classify this evolution based on the strategy of the port, the scope of activities and organisational and production characteristics.

However, one has to notice that the development time frames from one generation to the next differ throughout the world and that especially in LAC certain ports cannot be defined as 'third generation' ports. Bichou et al. (2004) add that the generation-type taxonomy falls short in more than one aspect. First, it identifies port generations through sea/shore interface developments with little interest in port potential for shore/land-side expansion, as in the case of dry ports and distriparks. Second, it applies a rigid categorisation far from reflecting the composite reality of ports.

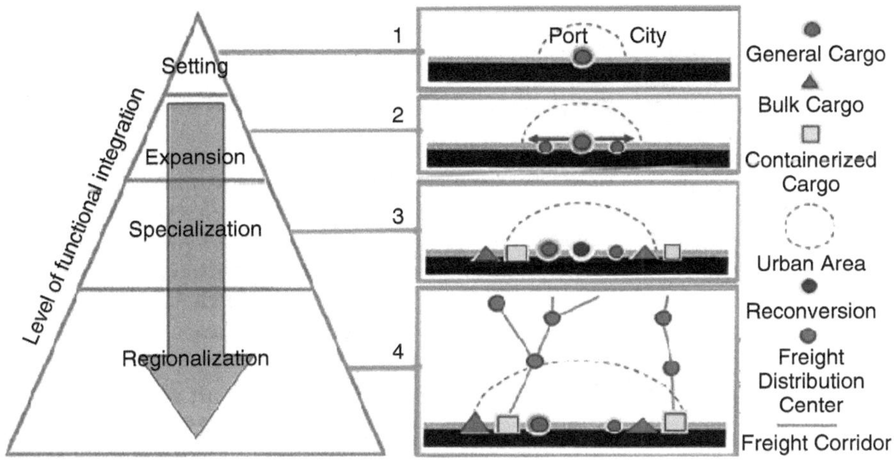

Fig. 4 From setting to regionalisation
Source: Notteboom and Rodrigue (2006)

Table 1 Port facilities and services

Infrastructure	Approach channel, breakwater, locks and berths
Superstructure	Surfacing, storage (transit sheds, silos, warehouses), workshops, offices
Service to ships	Harbour Master's office, navigational aids, pilotage, towage, berthing/unberthing, supplies, waste reception and disposal, security
Service to cargo	Handling, storage, delivery/reception, cargo processing, security

Source: UNCTAD (1995, p. 27)

Many ports in the world still perform first or second generation-type functions, and even within a single port, there may be a variety of operational and management systems intersecting across different generation categories. Third, it hypothetically equates all cargo/ship type operations and functions under the same generation. In practice, though, many ports declared to be fourth generation still carry out activities of first or second generation-types through, for instance, handling first generation-type cargo and ships.

Bichou and Gray (2005) summarise the variations in institutional and organisational management models across major port assets, facilities and services. The divisions between private and public ownership are hypothetical but typical and based on a thorough literature review (Illustration 1).

It is likely that the major obstacle to adopting a single valid taxonomy for port management comes from to the complexity and diversity of the port business at more than one level, inter alia:

(a) Organisational differences: issues of ownership (public versus private), institutional status (landlord/tool versus service), social arrangements (labour and manpower), etc

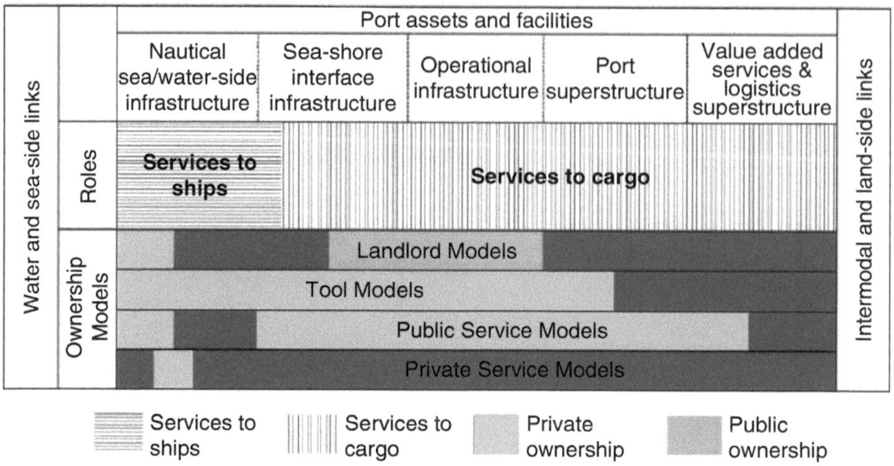

Illustration 1 Variations of functional roles and institutional models across different port services and facilities
Source: Bichou et al. 2004

(b) Operational differences: types of cargo handled, ships serviced, terminals operated, etc
(c) Physical and spatial differences: location, access, connectivity, available capacity, etc
(d) Legal and regulatory differences: trade and transport policy, administrative procedures, safety and security regulations, environment, etc

Port development throughout time, for the following analysis, is defined as the process of creation and adaptation to satisfy changing demands of clients. With shifting requirements from basic port facilities to logistics facilities, the needs in the provision of port services are geared towards logistics and can develop in four directions:

(a) The geographic scale of port networks ranging from local to transnational presence
(b) The complexity of interfaces (referring to the potential of inter- and multi-modality in the port)
(c) The number of activities in the port (ranging from general haulage to high value-added services)
(d) The degree of specialisation (type of products, shipment sizes etc.)

Hence as far all future developments are concerned, there are different options of the port's service providers. Since the port represents a physical and functional link between logistic and transport networks, ports need to meet certain requirements in the future and these are influenced by a number of restrictions and external drivers.

Leal et al. (2009) describe port development as the common interaction of three groups of variables: accessibility, the kind of formal and informal industry relationships, and the institutional framework:

1. Accessibility profile[2] of the port, which makes market players decide to use the port in their logistic network, but at the same time makes terminal operators decide to allocate resources to the port
2. Proximity – formal and informal industry relationships,[3] whereby certain kinds of know-how prevail and others disappear. This happens through a process of knowledge spill-over coming from the industry (market players) itself or informal networks to which the terminal operators belong
3. The institutional framework[4] has a decisive influence upon port performance and thus upon the spatial distribution of activity. At the same time, differences in spatial distribution of port activity could imply different levels of competition

The interaction and relationship between this group of variables end up as determinants of port development. However, all parts must work in a coordinated and appropriate way; otherwise, port development might not meet its potential or might not eventuate at all. The following figure attempts to illustrate the relationships between the three main components (Fig. 5).

For instance, within a fragile (weak) institutional framework, even industry relationships and accessibility are strong; port development will be weak, because investment will be very expensive (and inefficient) due to an inappropriate institutional/economic context.

Sánchez and Tuchel (2005) and Sánchez et al. (2006) addressed port development under a systems approach. Their explanation is constructed from the interaction of the different components as shown in the figure below (Fig. 6).

All four components are interrelated and have varying impacts on port development. The authors differentiate the type of environment which acts upon each of the components, while components are influenced by different 'levels' of environment. For certain components the local or global influence is predominant.

The model uses similar components as are used for the definition of sustainability. Therefore it can be assumed that if all components reach a state of equilibrium, sustainable port development is possible. Port development can thus be defined as an accumulation process which is formed and directed by the four identified components that constitute themselves from numerous factors that influence this development directly or externally. The following figure describes the horizontal accumulation process (Fig. 7).

The figure describes an interaction of components that results in 'ideal' port development, where a variation in one or more components has an equilibrated positive effect on port development and the interrelated components themselves. Such a scenario allows for continued positive port development.

The interrelation between the four components and their conditioning of one another is a prerequisite for the vertical accumulation process: in other words, the development of a port to a different level. Consequently, two directions, horizontal

[2]Accessibility includes location, infrastructure, transport layer and logistical layer.

[3]For vertical and horizontal industry relationships.

[4]Public, social and regulatory institutions and involved.

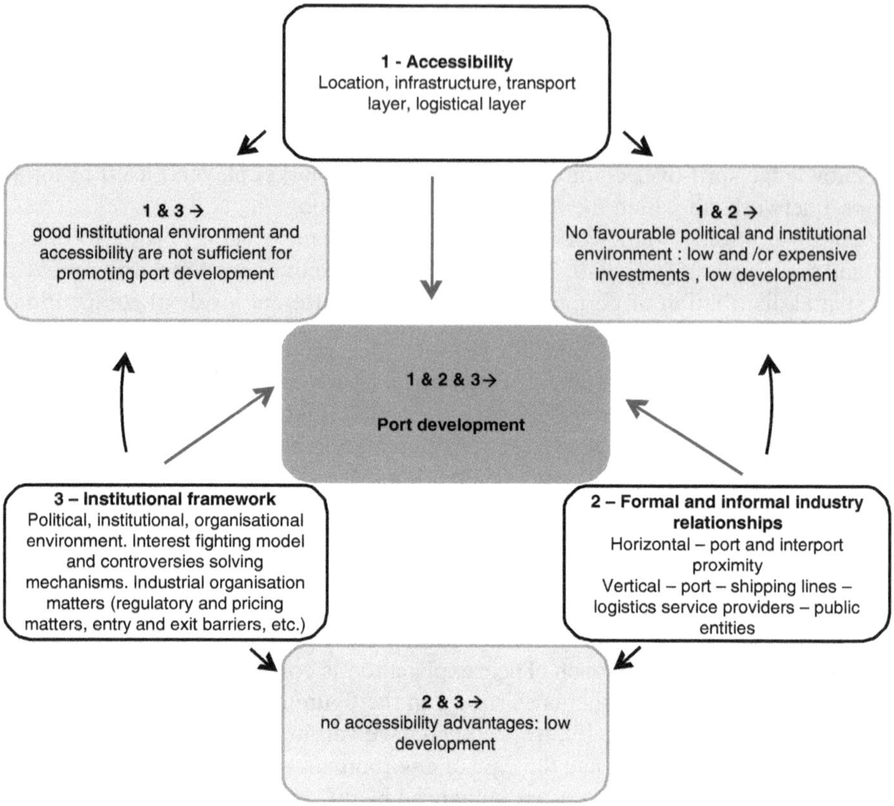

Fig. 5 Port development as a function of three main components
Source: Authors based on Leal et al. (2009)

and vertical are identified. The first describes the change in a port towards a higher level of development i.e. first, second and third generation of ports. However, in order for the 'movement' in this direction to take effect it is necessary that the factors of each component have reached similar levels/stages, because only then can a higher level of development be reached, as developed through the horizontal direction. The following figure depicts port development if the two directions act upon the port in an equilibrated way (Fig. 8).

Under the abovementioned conditions it is possible to argue that port development derives from horizontal movements (circular): that through externally imposed inequity between the four components the port is allowed to develop to a different level or generation. Is port development thus an induced process through those four components? Does the influence of factors within the components change as ports progress in the stages/generations of port development? (Fig. 9)

The notion of port development cycles needs to be discussed. It can be argued that it is necessary that ports develop a certain 'rhythm' and flexibility to be able to adapt to changing environments. We argue that port development has intervals in

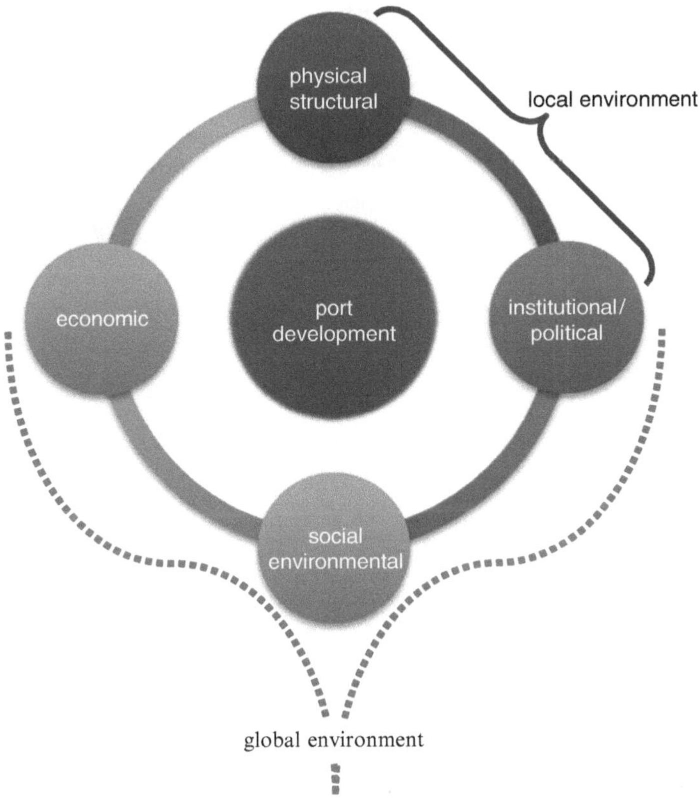

Fig. 6 Components and influences of port development
Source: Sánchez and Tuchel (2005)

which a recurring sequence of events take place and, with their occurrence, create a progressing port life-cycle.

However, an imbalance in the interaction of components or a significant variation in the different factors can have negative impacts if 'gaps' between components get too big. One example would be a significant progress being made in economic development that is not followed by the institutional/political component as it cannot overcome its 'sclerosis'. Such a situation will customarily lead to a fracture in the port development cycle and, if not addressed properly, will create lagged port development as can be seen in a number of developing countries.

6 Towards a Relational Perspective on Port Development

Global, regional and local factors can be defined in each of the three systems. The development of the maritime system is mainly influenced by global institutional and organisational factors, factors which are not in reach of the 'locally and regionally' embedded port system.

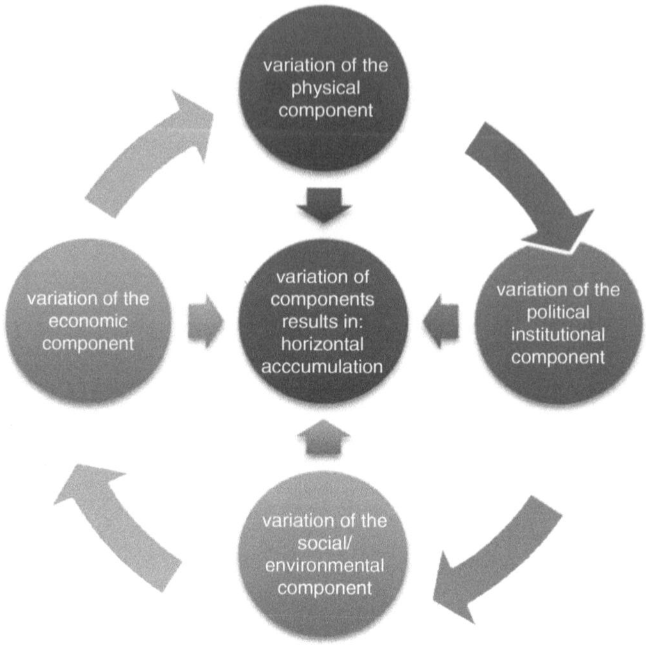

Fig. 7 Horizontal accumulation process
Source: Sánchez and Tuchel (2005)

Changes in the port system occur in an almost completely discrete manner, since variations in infrastructure and superstructure of ports as well as organisational changes appear almost abrupt and do not grow in a continuous way. Port development is dependent on its embeddedness and determination by local and regional institutional factors, which are not directly related to the global factors. This stands in high contrast to the upsurge of global players in the operation of ports. However, certain locally and regionally defined conditions of interaction between other subsystems of the transport system, such as local port access infrastructure, which define the connectivity with the economic systems of the port's hinterland, abscond from the port's influence.

As the port system development 'cycle' advances in a discrete manner, its adjustments to the evolution of freight transport demand (which are continuous) can provoke situations of infrastructural insufficiencies and scarcity of supply on the one hand, and on the other hand failures in local and regional decisions on ports can lead to a superavit of port infrastructure, resulting in an excess of port infrastructure. In both cases the effects on the efficiency and performance of ports is negative.

We argue that a port development cycle based on the spatial reach of a port is not an adequate measure for the ports participating in the global container market and that this measure has to be extended to include the type of operation of a port.

These development cycles vary in their development time and certain activities can lead to changes in the normal development curve. One development cycle can

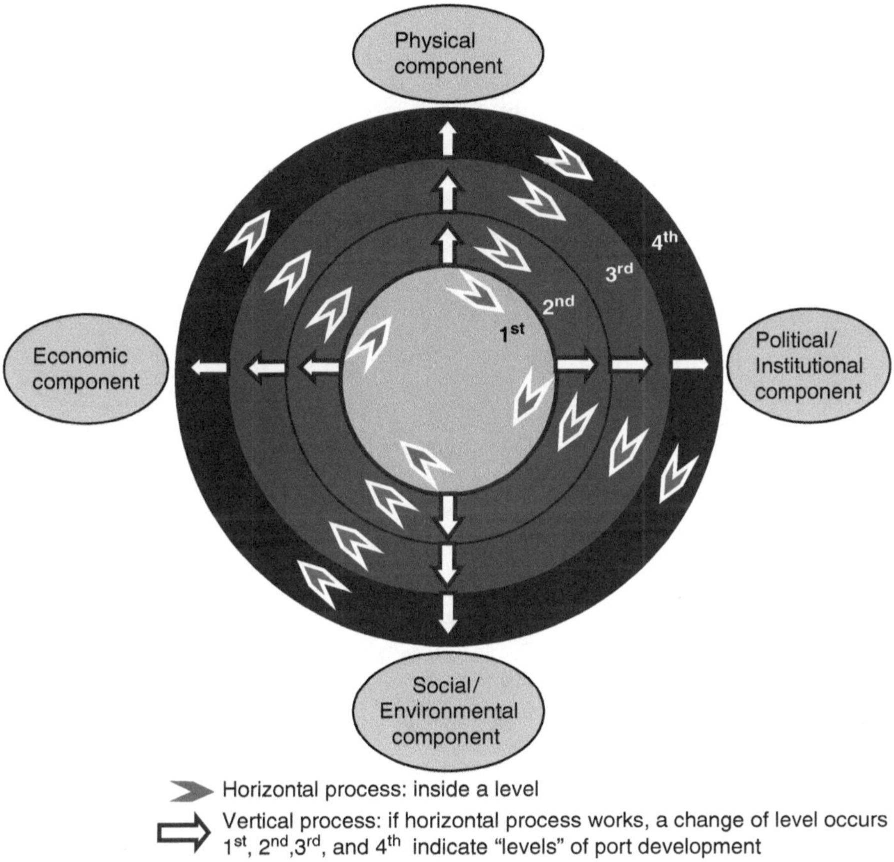

Fig. 8 Horizontal and vertical accumulation processes and their components acting on port development
Source: Sánchez and Tuchel (2005)

be substituted with the next one before the first cycle has ended. But port development cycles can also be extended in their duration by technological changes or extensions of the standardisation phase. The development cycle is driven internally by processes (handling) and product (service) innovations. The pressure of change from a Fordist regime to a post-Fordist regime is also present in the port system, for instance in changing the pattern in products (services), governance and spatial organisation.

The duration of port cycles depends significantly on external pressures (economic, maritime system) and the effectiveness of port organisation and the institutional framework to respond to these pressures. Port development is based on three columns: technology, organisations, and territory, with intense interaction between them. This concept has been introduced by Storper (1997). He argues that untraded interdependencies realise organisational, communication and learning processes

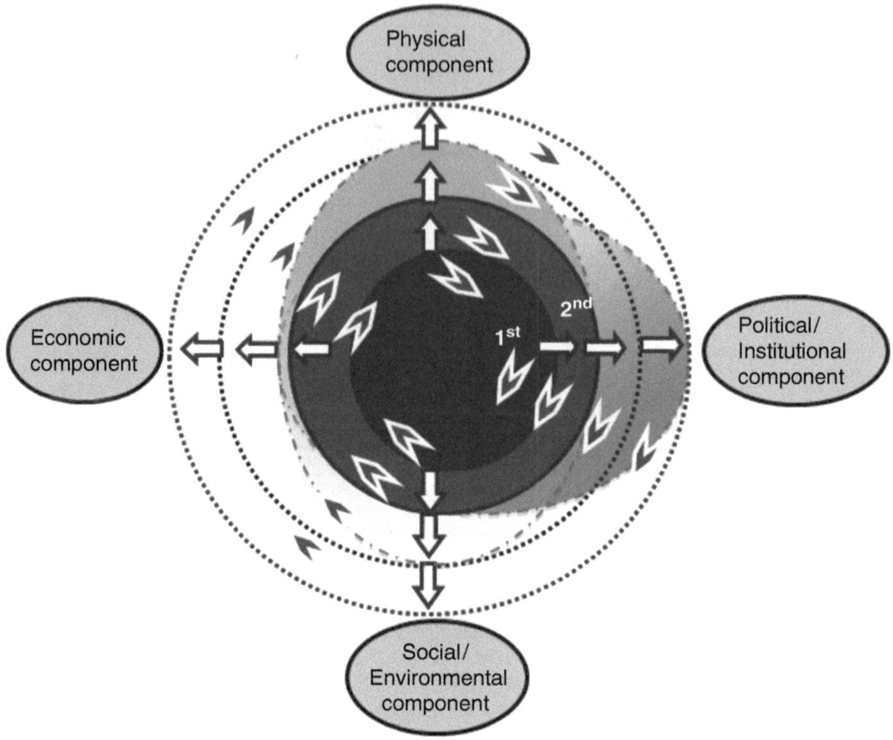

Fig. 9 Unbalanced port development
Source: Sánchez and Tuchel (2005)

and lead to an advantage in competition. A main argument of Storper (1997) is that the development of organisation is influenced by technological development and the territorial context (cultural).

Port development can be defined as a discontinuous, cumulative process, which develops and appears as a series of innovations. Development in this context leads to a structural transformation of the port. To this end, it is necessary to differentiate the term 'growth', which occurs without structural change, from the idea of 'structural transformation'. We argue that beyond the spatial approach an organisational and relational perspective is required for the analysis of port development because the maintenance of a port's competitiveness requires structural transformation.

In what follows, the effect of growth on ports and the ensuing required structural transformations are analysed from a micro-economic perspective through the application of the product life cycle. The product life cycle (Schaetzl 1996) assumes that each product has a certain economic life-time which is characterised by a set of common phases. During its economic lifetime, a product is subject to changes in product design, market conditions and conditions of production. Generally, the product life cycle suggests that the lifetime of a product, service or branch can be divided into four or five stages as shown in Fig. 10 below.

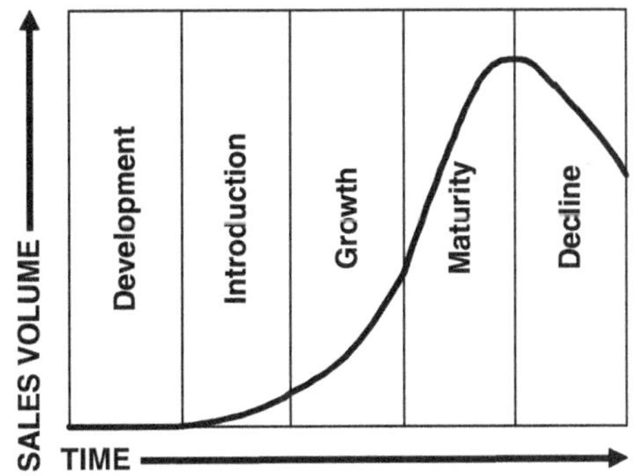

Fig. 10 The product life cycle
Source: Derived from Kotler and Armstrong (2004)

Within the context of the container port sector, a port can be considered to be analogous to a service (product) in that it can be perceived as exhibiting the same stages in its life-cycle. Therefore it is logical to fit port development into the theory of a Product Life Cycle in order to discuss the interrelation of port development with the other four identified systems. Applying the generic Product Life Cycle theory to the port context implies a definition of its five different stages as follows:

6.1 Development and Introduction

The introduction of a port site with related services allows for direct trade with other non-adjacent regions. Services are basic and not standardised as being cargo-based. The port services provided during this stage of a port's development cycle are commonly from a monopolistic public supplier, with human capital as the main factor of production. In addition, the geographic reach of the port's hinterland during this stage of development is typically restricted to the adjacent city.

6.2 Growth

Mirroring the seemingly inevitable long-term growth of international trade, activity in most newly created ports will also inevitably grow from the initial development and introduction stages. As this happens, economies of scale will be realised that will fuel a quickening pace of development. Standardisation and process innovation are addressed and implemented, while capital equipment gains in importance over human capital. Capital intensity of development will require transformations in port operations from public to the private sector; thus, growth becomes a principal driver of port devolution. The geographic reach of the hinterland expands driven by land infrastructure development, and the required port area for storage and port related activities increases.

6.3 Maturity

Port activity grows at a slower rate, standardisation (usually in the form of containerisation) becomes fully implemented and competition in the market increases. This latter characteristic is true both internally and externally. As the number of terminals increases within a port, by promoting greater private sector involvement in container handling activities, port authorities will typically move towards stimulating internal competition through the creation of an internal market structure. At the same time and commensurate with greater maturity, external competition increases as the geographic reach of a port's hinterland expands even further and potentially starts to overlap with the hinterlands of other ports. Also during this stage, the port area required for container storage and other port-related activities increases still further, but approaches and sometimes reaches either a physical constraint on further expansion, or possibly a competitive constraint from other activities and land use in areas adjacent to the port. In consequence, investment during the maturity stage of the port development cycle focuses on the rationalisation of port services, particularly as land becomes a scarce commodity and commands premium prices or rents.

6.4 Decline

This occurs once the point has been reached where the limitations in feasible rationalisation, investment and access are reached. Port activity reduces. As no further expansion of the port area or no other efficiency gains are possible, the supply of port capacity becomes fixed. As land access becomes increasingly congested, market share is lost to competing ports with overlapping hinterlands and this falling market share will soon manifest itself as declining throughput and sales volume.

During the course of any product life cycle, key factors will shift and change. For example: production shifts from being human capital intensive to becoming capital intensive; innovation transfers from being product-based innovation (e.g. the container in port cargo handling) to process innovation; investment is proportionately reduced in R&D and increased in rationalisation; production runs shift from small batches (general cargo in the port context) to mass production (containers); and the market develops from a seller's market to a buyer's market (Schaetzl 1996).

During the transitional process by which a product moves from the development and introduction phase through the growth, maturity and decline phases of its life cycle, the conditions for production and of the market (consumption) will change. One observable consequence of this is that the optimal location for production will typically change from a central location to the periphery. It is this inherent characteristic of a product's evolution through its life cycle which helps explain the way in which the operational scale and scope of freight distribution has become extended over time. Indeed, the five standard stages of the product life cycle are likely to relate, with a high degree of correlation, to the four stages in 'the extension of the operational scale of freight distribution' identified by Rodrigue (2006). Ports are basically responsive (or reactive) to the demands of their customers (primarily the maritime carriers) and, at a fundamental level, the shipping industry's evolution simultaneously reflects and has facilitated the extension of freight distribution operations to a global scale.

It is well known that the globalisation of economic activity has seen the emergence of global supply chains with their space of flows. The operational scale of the actors involved has consequently expanded substantially – processes well understood by contemporary economic geography. The economic system today is based on mobile production factors, comparative advantages, and intra-industrial linkages. For the transport sector, multi-scale transport systems have emerged alongside, linking global processes with local realities. In general this development is referred to as Postfordism or Neofordism. The postfordist structure is more directed towards economies of scope, flexible organisations and co-operation in economic networks. This change has brought a territorial transformation of the global economy.

For the economic system, one result is that the markets produced for and transported to are becoming more and more diversified and are covering a growing geographic scale. The development tends towards a buyer's market based on less-standardised production, entailing smaller, varied product series and hence smaller batches in transportation. These economies of scope tend to become as important as economies of scale.

It is not unreasonable to assert, therefore, that the port life cycle (and where any individual port is positioned within it) is very much (functionally) dependent upon the level and nature of engagement of the economic system within a port's hinterland with the wider international trade arena. This is a factor which itself is heavily influenced by the aggregate (or average) stage of product life cycles within that economic system as reflected in, and facilitated by, the level of accessibility of that hinterland and the subsequent geographic scope of freight distribution.

7 A Need for Contextuality in Port Development

> Globalisation of sea shipping operations is a reality, but the port and land operations are subject to regional and local variations.

This chapter attempts to re-conceptualise port development and to perceive and understand it as a historically contingent outcome of complex and multiple bounded and unbounded, economic, institutional and political processes.

What are the main characteristics to model port development in a contextual form? Different characteristics of port models can be found, but none integrates the economic, port and sub-systems of the transport system. In general the UNCTAD classification of ports is accepted and the applied classification defines different 'generations' of ports that can be found throughout the world. However, the classification of port generations is primarily based on the port operator model.

We argue that this classification fails to describe 'generations' of ports, because it neglects the relational perspective in a port's development with the environment, potential parallelism of functions and their evolution in time.

As outlined in the previous section, port development is driven by technologies and (re)organisation which lead to the rise of new, and the fall of old, products (services), processes and locations. At the time of their initiation, ports operate at a local level. They then extend their influence on the seaward side to the regional and later to the national and continental level. The final stage is the global level. Because of the latest developments in the structure of maritime networks, in this final stage, one also has to differentiate between hub and gateway ports that operate at the global level.

These port development cycles vary in their duration; their length depends significantly upon the external pressure emanating from the economic and maritime systems and the degree of effectiveness of port organisation and the institutional framework within which the port works to respond to this pressure. Storper (1997) avows that port development is based on three pillars: technology, organisation, and territory, with intense interaction between the three. He argues that the development of port organisation is influenced by technological development and the spatial development context, and that it is this which leads to the realisation of organisational, communication and learning processes that produce a competitive advantage for a port.

Certain activities or actions can lead to changes or shifts in the normal port development curve. Not only can one development cycle be substituted by an alternative prior to the full cycle having completed, but also phases within the port development cycle and, therefore, the overall port development cycle itself can be extended in duration through technological or organisational changes or, more specifically, through the extension of the phase where increasing standardisation takes place. This renewal of the port development cycle is driven internally by process (handling) and product (service) innovations.

Examples of challenges during the growth phase can particularly be observed in the seaport sector in developing countries.

Continuing growth in container traffic and changes in the nature of container shipping operations have led to increased pressure on ports in a number of ways. The deployment of ever-larger containerships has resulted in increased draft requirements to allow better access from the sea. Growing traffic levels have also led to significant pressure on investment requirements for landside and hinterland access. In addition, growth has required process innovation and the rationalisation of container handling, as well as maximising the use of storage areas in ports; aspects that have now attained a high level of sophistication.

Continued growth in container traffic leads to a lack of space at seaport terminals and growing congestion on the access routes (land and seaside) that serve them. Port management will be very much aware that underinvestment and any persistent lack of capacity will mean that, where a choice exists, customers will eventually divert their business to competitor ports. In any case, even solely in terms of opportunity cost, a significant loss in container traffic will result and this possibility puts pressure on ports to invest in infrastructure and superstructure. However, such investments constitute a significant financial burden and lead to the requirement for further expansion of other port facilities. Due to a lack of available sources of finance, legislative barriers, restricted land availability and other difficulties, the actual development of ports or terminals particular in LAC has been rather slow and sparse.

Ducruet et al. (2009) 'Concentration stems from the path-dependency of large agglomerations'. The assumption, therefore, is that seaports in developing countries now find themselves having not progressed sufficiently throughout the growth stage of the product life cycle. In accordance with theory, a large proportion of ports in developing countries face inevitable and imminent bottlenecks and inefficiencies. If this is indeed the case, the concern of port authorities and governments, in particular, is to determine how to effectively counteract this situation.

A time lag in the realisation of resolving infrastructural bottlenecks can provoke significant impacts on regional economies. These impacts are especially prevalent in the interaction between the port and the maritime system, because the port system has longer development cycles than the maritime system. For the case of LA and other developing regions, these two cycles are decoupled; while the maritime system cycle until recently underwent a strong growth period the port system cycle developed in a constricting manner with an almost stagnant infrastructure provision. As the expansion of the maritime system is particularly driven by economies of scale, which is predominantly reflected in increasing ship sizes, stagnant port infrastructure endowment does inevitably create bottlenecks and system inefficiencies.

The interplay between the three systems and their development cycles in the case of developing countries and particularly LA is crucial, as change itself can be considered a fundamental part of all subsystems within the transport system. The discussion develops further the identification of components which allow a certain monitoring of the interplay and impacts.

In the analysis of the current state of port development the institutional set up is decisive as it constructs the tie (interface) between the different systems and sets the roles for interaction. Existing institutional constraints together with the prevalent lack of adequate hinterland infrastructure, particularly in LA have a large negative impact on ports and hinder them from adopting innovative strategies to construct a long term development path of interaction and coordination and maximum efficiency and efficacy.

We argue that analyse from a relation perspective includes four fields: Organisation-Structure, Evolution, Innovation-Strategy, and Interaction. The configuration of these fields interacts with the relational perspective drivers of contextuality, contingency and path-dependency.

The interaction of the relational perspective and the ions is influenced by the interference of other systems, namely the maritime system and the economic system. The following figure shows the interaction between the systems and the relational approach (Fig. 11).

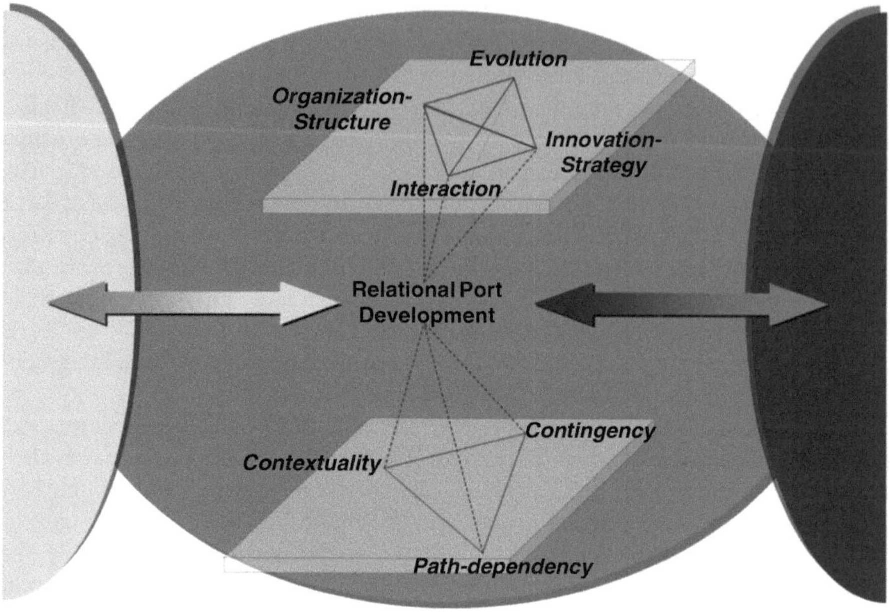

Fig. 11 Relational port development

8 Conclusions

Developing countries have to adjust their port development strategies in order to face current challenges induced though increased demand for port services from trade. However, it seems that current strategies lack understanding that port development is the result of the interaction of three systems the economic, the transport and the port system. Each one of these responds to cyclic developments, which has significant impact on the interaction between the three systems.

Therefore it is important to understand the theoretic discussion on port development in order to understand the contextuality and embeddedness of port development that if ignored can lead to a lack of port development or incoherent strategies that will in result impact on a country's competitiveness in trade. This approach also allows taking into account the notion of sustainability.

A relational view of port development has been developed, which rests on the following propositions. Ports are structurally situated in contexts of economic and relations, which leaves them directly exposed and vulnerable to changes in their environment

Further, port development is path-dependent to the extent that future action is dependent on past decisions, structures and processes. Finally port development is contingent and open-ended as decisions they might deviate from an existing development path.

This conceptualisation of port development underlines the necessity for decision makers to further develop a clear understanding of the complexity of port development as such knowledge can potentially reduce risks and allows to see port development and the wider impacts on other systems. At the same time this conceptualisation decision makers to critically reflect their own role as a factor for port development

Our discussion does not attempt to develop a comprehensive theory, which allows explaining and predicting port development. Rather, the framework developed in this paper presents a more multidimensional relational view, which can be applied to analyse port development outcomes and challenges in a variety of contexts and is particularly useful when dealing with port development problems as encountered and analysed by us over the last years.

When analysing port development issues it is important to be aware of contextuality, path-dependence and the contingency of port development. Our approach allows following research questions analysing how global changes in demand i.e. the current economic downturn, competition and technological development impact on port development and how these effects vary in different regions or countries.

References

Bichou K, Gray R (2005) A critical review of conventional terminology for classifying seaports. Transportation Research A, Berkeley, California 39(1):75–92

Bird J (1980) Seaports and seaport terminals. Hutchinson University Library, London

Brooks M, Cullinane K (eds) (2008) Devolution, port governance and port devolution. Elsevier, Amsterdam

Fleming DK, Hayuth Y (1994) Spatial characteristics of transportation hubs: centrality and intermediacy. J Transport Geogr 2:3–18

Hayuth Y (1981) Containerisation and the Load Center Concept. Econ Geogr 57:160–176

Hayuth Y (1988) Rationalization and deconcentration of the US container port System. Prof Geogr 40:279–288

Hoyle B, Knowles R (eds) (1998) Modern transport geography, 2nd edn. Wiley, London, p 13

Jantsch E (1982) Selbstorganisation des Universums. Munich, Germany

Kotler P, Armstrong G (2004) Principles of marketing, 10th edn. Pearson Education, Upper Saddle River, NJ

Luhmann N (1984a) Soziale Systeme. Frankfurt a. M, Germany

Luhmann N (1984) Soziale Systeme, Suhrkamp, Frankfurt a. M., pp 50ff

Maturana HR (1994) Was ist Erkennen? Munich, Germany

Notteboom TE, Rodrigue JP (2005) Port regionalization: towards a new phase in port development. Marit Pol Manag 32:297–313

Notteboom TE, Rodrigue JP (2006) Re-assessing port-hinterland relationships in the context of a global commodity chains. In: Port-cities in global supply chain. Ashgate, London

Notteboom TE, Rodrigue JP (2008) Containerisation, box logistics and global supply chains: the integration of ports and liner shipping networks. Marit Econ Logist 10:152–174

Robinson R (2002) Ports as elements in value-driven chain systems: the new paradigm. Marit Pol Manag 29(3):241–255

Rodrigue JP (2006) Transportation Modes. (www-pages). Available at URL: http://people.hofstra.edu/geotrans/eng/ch3en/ch3menu.html

Sánchez J, Tuchel N (2005) El desarrollo portuario, un modelo de acumulación circular (Port Development – A circular accumulation process). UNECLAC-DRNI Working Paper, September

Sánchez RJ, Wilmsmeier G (2007) Governance and port devolution: the case of the River Plate basin. In: Research in transportation economics, vol 17, Devolution, port governance and port performance. Elsevier, Amsterdam, The Netherlands

Sánchez RJ, Tuchel N, Wilmsmeier G (2006) Port development cycles – a theoretical approach. 2006 Conference of the Association of American Geographers, Chicago, Illinois, USA

Schaetzl L (1996) Wirtschaftsgeographie 1 – Theorie, 6th edn. UTB, Paderborn

Schober H (1991) Irritation und Bestätigung – Die Provokation der systemischen Beratung oder: Wer macht eigentlich die Veränderung? In: Hofmann M (ed) Theorie und Praxis der Unternehmensberatung. Physica, Heidelberg, p 352

Storper M (1997) The regional world: territorial development in a global economy. Guilford Press, New York

Taaffe EJ, Morrill RL, Gould PR (1963) Transport expansion in underdeveloped countries: a comparative analysis. Geogr Rev 53:503–529

Wang TF, Cullinane K, Song DW (2005) Container Port Production and Economic Efficiency, Palgrave-MacMillan, Hampshire, UK

The Conditioned Demands of "General Merchandise", "Dry Bulk" and "Liquid Bulk" Sea Transport

Pablo Coto-Millán, José Baños-Pino, Miguel Ángel Pesquera, Juan Castanedo Galán, and Lucía Inglada-Pérez

Abstract In this paper we use a theoretical model for sea transport demand services in Spain for the period 1975–1990. Using quarterly data, we estimate separate equations for the different of sea traffic, focus in Liquid Bulk traffic.

1 Introduction

The evolution of transport economic analysis is not constant through time. At the beginning of this century, transport economics was the aim of many figurative works. However, positive studies on the factors of transport service demand and costs, have prevailed throughout recent decades (Winston 1985). Research on transport demand has developed from the primitive models of flow and tariff engineering, to more modern microeconomic models which have based on the understanding the behaviour of individual agents. Passengers demand models assume that passengers optimize their utility (Varian 1978 and Oum 1979). Goods demand models assume that companies try to minimize transport costs. In these models, goods transport demand is derived from a neoclassical cost function for a particular company. Such demand is dealt with such as like input demands in the production process, according to the well known Shephard's Lemma (Winston 1983). The most outstanding works on goods transport demand are as follows: Levin (1978), Friedlaender and Spady (1980) and Winston (1981a) for railway and road transport. There has not yet been any known research on air goods

P. Coto-Millán (✉), M. Ángel Pesquera, and J. Castanedo Galán
University of Cantabria, Santander, Spain
e-mail: cotop@unican.es

J. Baños-Pino
University of Oviedo, Oviedo, Spain

L. Inglada-Pérez
UNED, Madrid, Spain

P. Coto-Millán et al. (eds.), *Essays on Port Economics*, Contributions to Economics,
DOI 10.1007/978-3-7908-2425-4_4,

transport. However, there have been many published and non published works on sea goods transport demand. Winston (1981b), Winston (1983) and Coto-Millán (1991a, b), are among those works which have been published. There are some works on the demand which are known within the field but which have not been published, some of them being, the reports carried out mainly by Lloyd's Shipping Economist (London), Institute of Shipping Economics and Logistics (Bremen), and Norwegian Shipping News (Oslo). We must also mention that since being settled, the OECD and the UNCTAD have published reports on international sea transport. Spanish research on the estimation of sea transport demand dates from the Development Plans of the Sixties. A more recent work was carried out titled Future Needs of the Spanish Merchant Fleet 1983), carried out by the Institute of Communication and Transport Studies to work on the Fleet Plan. There have been more recent works such as the Fleet Plan (1985), as well as the Instituto de Estudios de Transportes y Comunicaciones (Institute of Transport and Communication Studies), and the works by Coto-Millán (1986), Coto-Millán (1988a, b), and by Coto-Millán and Sarabia (1993). These works highlight the study of sea transport aggregate demand of three groups of services: "General merchandise", "Solid bulk" and "Oil products".

The quantity of products which has the customs duties of "General merchandise", is included in this kind of service, as well as the transport of small and non homogeneous package units, which are carried in different packings or even unpackaged. Goods transported in containers are also included in this group. In "Solid bulk" service there is an estimation of the quantity of products transported in bulk such as the following: coal, iron, cereal, cement, fertilizers, and so on. Such kind of transport is generally carried out in special ships, although occasionally, non specialized ships can be used, allowing a variety of different loads. Finally, "Oil Products" service includes oil and all its derivatives which are carried by sea.

In addition to this, if we exclude "Oil Product" service, the sea transport services which are left, are distinguished by the quantity of goods carried rather than by their nature.

Sea transport demand research can be interesting for outfitter companies, shipping companies, consignees, stowage companies, port authorities and regulation bodies for the following reasons: income predictability, size and structure of the Spanish fleet and port infrastructure, predictability of regular line demand, fixing of tariffs and traffic arrangement.

2 A Theoretical Model of Sea Transport Demand

The following model has been based on research done by Winston (1981b, 1983) and Coto-Millán (1986).

Assume an aggregate production function of economy

$$Y = Y(\mathbf{F}; \mathbf{STM}) \tag{1}$$

where Y, represents the total product, F is the vector of the production factors which are different from sea transport services, and **STM** is the vector of sea transport services which have been lent to produce Y. The vectors of the respective factor prices **f, pf,** and the output price, p are also considered to be influenced by external factors.

Assume that companies chose **F** and **STM** levels which maximise their profits,. Profits are defined as:

$$\pi = \mathrm{p\,Y} - \mathrm{C}$$

where C represents the total costs of the production of Y:

$$\mathrm{C} = \mathbf{f} \cdot \mathbf{F} + \mathbf{pf} \cdot \mathbf{STM} \tag{2}$$

If companies maximise their profits, it must be verified that:

$$\frac{\partial Y/\partial F}{\partial Y/\partial SMT} = \frac{f}{pf} \tag{3}$$

Solving the system made up by (2) and (3), and substituting in (1), we can obtain the factor conditioned demands which depend on the factor prices and the output amount produced. i.e.:

$$\mathbf{F} = \mathrm{F}(\mathbf{f}, \mathbf{pf}, \mathrm{Y}) \tag{4}$$

$$\mathbf{STM} = \mathrm{STM}(\mathbf{f}; \mathbf{pf}; \mathrm{Y}) \tag{5}$$

Expression (4) is the condition demand function of the classical factor production. Expression (5) must be explained more in detail. Vector **STM** has three components. Using subindices, we can say that STM_i represents the amount of transport service demand of group i (i = 1,2,3), where subindices refer to the following group of services: "General merchandise", "Solid bulk" and "Oil Products".

We assume that such quantities can be estimated taking into account the number of tons of goods transported , and leaving aside the distance travelled by each mode of transport. There are no statistics or any approximation of the number of tons per mile which are really demanded in each transport service. However, It can be assumed that, given the variety of origins and destinies of goods, the growth rate of transported tonnes is similar to the growth rate of tonnes per mile. Finally, it is assumed that changes of these growth rates are insignificant and the increase in the number of sea miles travelled by each group of sea transport service, is compensated by the decrease in other goods of the same group.

Expression (5) can be expressed in more detail assuming that STM_i, is a function of the total output of the Spanish economy Y, and of the price vector **P**, which changes depending on the case, where both the price of the specific sea transport services, and the prices of different substitutory and complementary sea transport services, as well as other factors linked to goods and service production, are included for each time period t.

Therefore, we can write:

$$\mathrm{STMi,t} = \mathrm{o(Yt, Pt)} \tag{6}$$

where,

$$\frac{\partial \mathrm{STM_{i,t}}}{\partial \mathrm{Y_t}} > 0 \text{ y } \frac{\partial \mathrm{STM_{i,t}}}{\partial \mathrm{P_{i,t}}} < 0$$

That is to say, it is assumed that sea transport services behaves as normal factors with respect to product and their own prices. It is also assumed that substitutability and complementarity relationships between factors are usual relationships in conventional economic theory.

This model is general enough to analyse national and international sea transport demand, in each of the different sea transport services which have been highlighted.

3 Data

Data on sea goods transport refers to tonnes transported between the Spanish ports, whatever their type of navigation. The starting point is monthly data on the goods carried in each Spanish port. We have calculated the tonnes transported between the following ports: Algeciras-La línea, Alicante, Almería, Avilés, Barcelona, Bilbao, Cádiz, Cartagena, Castellón, Ceuta, El Ferrol, Gijón, Huelva, La Coruña, La Luz and Las Palmas, Malaga, Melilla, Palma de Mallorca, Pasajes, Pontevedra, Puerto de Santa María, San Esteban de Pravia, Santa Cruz de Tenerife, Santander, Seville, Tarragona, Valencia, Vigo and Villagarcía de Arosa. To the amount obtained, we have added the transport of goods in ports belonging to the Comisión Administrativa de Grupo de Puertos (C.A.G.P.) which accounts for the goods transport carried out between Spanish ports which are not included within the group of 29 ports previously mentioned. However, the series used are not quarterly due to the availability of the other variables.

The data on the product variable used (the real quarterly GDP) has been obtained from Banco de España and Contabilidad Nacional Española (Spanish National Account).

Sea transport services price variables have been obtained from different international bodies. Dry cargo tramp trip charter freight (FPV) variable, has been obtained from the General Council of British Shipping (London). Dry cargo tramp time charter freight (FPT) variable, has been obtained from Norwegian Shipping News (Oslo). Liner freight rates (FBL), variable has been obtained from the Institute of

Shipping Economics and Logistics (Bremen). PRF_t is a variable which has been especially drawn up and estimated as follows. We have taken an index of the maximum freight regulated in Spain, and information which appears in the B.O.E. (State Official Report), when the average international index of the tanker freight is higher than the index which has been regulated in Spain. Moreover, we consider the average international index of the tanker freight when this is lower than the maximum price regulated in Spain. Such an average international index of the tanker freight has been estimated from the VLCC freight (very large crude carrier), and from the SCPC (small crude and products carrier), both having been obtained from the Lloyd's Ship Manager Freight Indices(London).

The main variables, according to the theoretical model, which could explain the different variety of sea transport, are as follows: GDP evolution (in order to approach the final product), national and international regulated freight evolution, and prices of other complementary or substitutory productive factors.

4 Traffic Equations of "General Merchandise", "Dry Bulk" and "Liquid Bulk"

The estimations have been carried out from the different specifications which correspond to model (6), fitting the variables to each mode of transport. We have used a cointegration point of view which complements other opinions found in Coto-Millán (1988a, b) and Coto-Millán and Sarabia (1993). The results were presented as in the works by Pérez (1993) and Pérez et al. (1993).

All the variables appear in natural logs. The standard deviation of each coefficient appears within brackets. With cointegration techniques, the model which has been carried out presents one long-run and two short-run analysis estimations. One of the short-run estimations corresponds to a non lineal equation joining in one single stage, both long and short-run equations; see Engle and Granger (1987), and Johansen (1988). Each equation has been widely checked.

4.1 "General Merchandise" Traffic

As far as this demand function estimation is concerned, two different problems arise. The first is one of the identification, having taken into account that not only supply but also demand usually develop simultaneously through time, as a response to very similar variables. With regard to the group of services studied here, supply comes from both the Spanish and the international fleet, the latter being higher than the Spanish supply. This fact means that Spanish supply exerts an insignificant influence over the world prices. Since the international sector is extremely competitive, international freight prices can be considered as a parameter.

The second problem is the Spanish regulation. Such regulation can be considered as virtually non existent. That is to say, freights are essentially international.

The long-run variables which have been considered are the following:

LMG: "General merchandise" log in tonnes.
LPIB: Gross Domestic Product log in real terms.
LFPT: log of dry cargo tramp time charter freight index.
LFPV: log of dry cargo tramp trip freight index.
LFBL: log of regular liner freight index.
MCE: long-run equation remainders which are to be used as error correction mechanisms in the short-run equation.

Time freight represents the prices of contracting ship services for a limited period of time, and refer to changeable trips which can be shifted to regular lines, either as outsiders or as freighters, or they can follow various routes ("tramp" navigation).

Travel freight refers to the prices of contracting ships for a particular trip. Again, cargoes can follow either a regular route or any other route agreed on by both parties.

Regular line freight refers to regular trips between ports. Moreover, such freight is stipulated by the shippers (outfitter or shipping companies) in the so called liner conferences. Such conferences are shipper assemblies made up to start a regular line, to stipulate its freight, to agree on systems of price reduction for the user of the line, and finally, to lend joint services, when they agree to do so. The conditions are usually named "liner terms".

4.1.1 Long-Run Analysis

$$LMG_t = \underset{(0.77)}{-2.319} + \underset{(0.133)}{1.074}\ LPIB_t \ \underset{(0.053)}{-0.193} LFPT_t + \underset{(0.149)}{0.607}\ LFPV_t \tag{7}$$

R^2 adjusted = 0.710; F = 52.52
S.E. = 0.094; DW = 2.29
ADF = −4.670; DW = 1.94
Estimation period: 1975.I; 1990.IV
Estimation method: Ordinary linear square.

The first statistic DW refers to (7), while the second is the statistic of the regression used to compute Dickey-Fuller statistic.

In addition to this, we have used the method presented by Johansen (1988) for the long-run analysis.

For a self-regressive VAR (1) vector with constant restricted so that it appears in the long term relationship, the test on the quantity of cointegration vectors has led to the following result:

Null hypothesis	Trace test value	Critical value at 5%
r = 0	48.13	47.21
r = 1	22.51	29.68
r = 2	6.5	15.41
r = 3	0.90	3.76

where r is the number of cointegration vectors.

Therefore, we conclude that there is only one cointegration vector which can be written as:

Proper vector	Proper vector				
	LMG_t	Constant	$LPIB_t$	$LFPV_t$	$LFPT_t$
0.3217	−1.87	5.03	2.08	1.28	−0.46

Normalizing the proper vector corresponding to the highest proper value, we can obtain the following long-run relationship:

$$MGt = 2.69 + 1.11 \text{ LPIBt} + 0.68 \text{ LFPVt} - 0.24 \text{ LFPTt}$$

Note that we obtain coefficients similar to those estimated in (7).

The long-run elasticities which have been obtained for this model are as follows: a unitary revenue elasticity of 1.074, a price elasticity of −0.193 and a cross price elasticity of 0.607. "General merchandise" traffic has the highest added value and at the same time, from the point of view of port industry, it generates higher added value. The reason for this is that such goods require more port manipulation and at the same time, generate more gross operating surplus not only for the cargo but also for the port industry (Port Authorities, consignees, stowage companies, customs agents, suppliers and so on) and for the user of these services. Such a situation guarantees a unitary product elasticity. Price elasticities which are lower than the unit, show that traffic demand is not very liable to freight changes. If sea transport services could be disaggregated into smaller categories, it would be possible to find out a higher reaction to both price changes and sustitutability and complementarity relationships. Such relationships can appear unclear in such aggregated terms.

According to the long-run equation, we have calculated the short-run equation. We have defined the variables as follows:

DLMG: growth ratio of "General merchandise" traffic.
DLFPT: growth ratio of dry cargo tramp time freight index.
DLFPV: growth ratio of dry tramp trip freight index.
D87.III: dummy variable for the third quarter of 1987.
D1: dummy variable for the first quarter.
D4: dummy variable for the fourth quarter.

4.1.2 Short-Run Analysis

$$\mathrm{DLMG_t} = -\underset{(0.088)}{0.166}\ \mathrm{DLFPT_t} + \underset{(0.259)}{0.628}\ \mathrm{DLFPV_t} - \underset{(0.076)}{0.176}\ \mathrm{D87.III} - \underset{(0.021)}{0.045}\ \mathrm{D1} + \underset{(0.020)}{0.072}\mathrm{D4} - \underset{(0.118)}{1.082}\ \mathrm{MCE_{t-1}} \quad (8)$$

R^2 adjusted = 0.70; F = 31.35
S.E. = 0.07; DW = 1.90
Ljung-Box Q(4) = 4.83
Ljung-Box Q(6) = 8.04
Ljung-Box Q(8) = 14.32
Test Jarque-Bera = 2.97
Test ARCH(4) = 5.79
Estimation period: 1975.II; 1990.IV
Estimation method: Ordinary linear square.

Short-run price elasticities can be obtained by this estimation: −0.166 for time freight and 0.628 for travel freight. Short-run elasticities corresponds essentially to long-run elasticities.

Joint non-linear estimation has led to the following results:

$$\mathrm{DLMG_t} = -\underset{(0.096)}{0.173}\ \mathrm{DLFPTt} + \underset{(0.274)}{0.754}\ \mathrm{DLFPVt} - \underset{(0.081)}{0.194}\ \mathrm{D87.III} - \underset{(0.026)}{0.069}\ \mathrm{D1} + \underset{(0.025)}{0.054}\mathrm{D4} - 1.021\left[\underset{(0.122)}{\mathrm{LMG_{t-1}}} + \underset{(0.756)}{2.362} - \underset{(0.174)}{1.100}\ \mathrm{LPIB_{t-1}} + \underset{(0.054)}{0.206}\ \mathrm{LFPT_{t-1}} - \underset{(0.154)}{0.590}\ \mathrm{LFPV_{t-1}}\right] no \quad (9)$$

R^2 adjusted = 0.68; F = 16.07
S.E. = 0.07; DW = 2.04
Ljung-Box Q(4) = 5.33
Ljung-Box Q(6) = 8.21
Ljung-Box Q(8) = 13.75
Test Jarque-Bera = 2.22
Test ARCH(4) = 5.76
Estimation period: 1975.I; 1990.IV

Given that $\mathrm{MCE_{t-1}}$ coefficient is virtually the unit, the previous equation has been again calculated at level according to Andrés et al. (1989):

$$LMG_t = -\underset{(0.091)}{0.168}\ DLFPTt + \underset{0.256}{0.737}\ DLFPVt - \underset{(0.080)}{0.193}\ D87.III - \underset{(0.024)}{0.071}\ D1$$
$$+\underset{(0.025)}{0.059}\ D4 - \underset{(0.64)}{2.353} + \underset{(0.118)}{1.099}\ LPIB_{t-1} - \underset{(0.046)}{0.250}\ LFPT_{t-1} + \underset{(0.130)}{0.588}\ LFPV_{t-1}$$

R^2 adjusted = 0.79; F = 32.05
S.E. = 0.07; DW = 2.09
Ljung-Box Q(4) = 5.37
Ljung-Box Q(6) = 8.26
Ljung-Box Q(8) = 13.76
Test Jarque-Bera = 2.10
Test ARCH(4) = 5.62
Estimation period: 1975.I; 1990.IV
The elasticities derived from (9) are as follows:

Unitary product elasticity of 1.1, price elasticity of −0.206 for time freight and cross price elasticity of 0.59 for trip freight. The elasticities which have been obtained from the estimation are robust with respect to the choice of the initial conditions. The results obtained in model (9) and (7) have not changed very much.

4.2 "Solid Bulk" Traffic

"Solid bulk" sector is dealt with similarly to "General merchandise" sector, in spite of the oddities which have arisen, which we will try to explain in the following terms. Firstly, we must say that the Spanish fleet of "Solid bulk" is insignificant with respect to the world fleet. Secondly, these services are regulated as regards freight and fleet. Spanish government has fixed, during this research, a maximum freight for some goods such as: cereal, coal, coffee, frozen meat and so on. Regulation has been carried out along with the requirement that the transport of goods is performed in ships belonging to the Spanish fleet. It is possible that such regulation seems to alter this traffic, however, the government grants the outfitter or shipper an amount of money when the maximum freight is lower than the international freight. On the contrary, when the freight paid to the outfitter and/or shipper is higher than the international freight, the government grants money to the user (Coto-Millán 1986). The reason for this is to offer an international competitive price to the user.

Due to all this, the problems which have arisen as regards identification and regulation in this kind of transport, can be easily solve.

Long-run equation variables are as follows:

LGS: log of "Solid bulk" in tonnes.
LPIB: log of Gross Domestic Product in real terms.
LFPT: log of dry cargo tramp time freight index.

LFPV: log of dry cargo tramp trip freight index.
D82I-IV: dummy variable with value 1, from the first to the last quarter of 1982.

4.2.1 Long-Run Analysis

$$\begin{aligned} LGS_t = \underset{(0.574)}{1.640} + \underset{(0.099)}{0.323}\ LPIB - \underset{(0.040)}{0.34}\ LFPT_t + \underset{(0.112)}{1.049}\ LFPV_t \\ + \underset{(0.042)}{0.144}\ D82I - IV \end{aligned} \tag{10}$$

R^2 adjusted = 0.74; F = 46.47
S.E. = 0.07; DW = 1.65
DF = −6.99; DW = 2.05
Estimation period: 1975.I; 1990.IV
Estimation method: Ordinary linear square.

The application of Johansen method (1988) to VAR(1) with a restricted constant provides the following result:

Null hypothesis	Trace test value	Critical value at 5%
r = 0	74.18	68.52
r = 1	43.31	47.21
r = 2	18.81	29.68
r = 3	5.33	15.41
r = 4	0.86	3.76

Next, we accept the existence of a single cointegration vector.

Proper value	Proper vector					
	LGS_t	Constant	LPIB	$LFPT_t$	$LFPV_t$	D82I-IV
0.3735	−1.78	−1.03	0.92	−0.58	1.81	0.56

The normalized proper vector leads to the following long-run relationship:

$$LGS_t = -0.581 + 0.517\ LPIB_t - 0.328\ LFPT_t + 1.013\ LFPV_t + 0.513\ D82I\text{-}IV$$

Product variable long-run elasticity is 1.049, price elasticity is −0.297 and cross elasticity is 1.049. Time freight index, FPT, can be identified with the price of the service itself. While trip freight index, refers to substituting services. Unlike "General merchandise" sea transport, in "Solid bulk" transport, elasticity is lower than the unit. The reason for this is that the goods carried have low added value and, given that they are carried in bulk, also generate low added value. They are both loaded and unloaded automatically. Price elasticity again shows that this transport is not very liable to freight changes. Again, the results are conditioned by the aggregation which provided by the availability of the data.

From (10), we can obtain the short-run relationship which uses the following variables:

DLGS: growth ratio of solid bulk.
DLFPT: growth ratio of dry cargo tramp time freight index.
DLFPV: growth ratio of dry cargo tramp trip freight index.
D78I: dummy variable which takes value 1 in the first quarter of 1978.
D82I-IV: dummy variable which takes value 1 during the first and to the last quarter of 1982.

4.2.2 Short-Run Analysis

$$\begin{aligned} DLGS_t = &\underset{(0.155)}{0.645}\, DLFPV_t - \underset{(0.034)}{0.091}\,(DLFPT_{t-1} + DLFPT_{t-2}) - \underset{(0.056)}{0.220}\, D78I \\ &- \underset{(0.057)}{0.114}\; D82I - IV - \underset{(0.105)}{0.625}\; MCE_{t-1} \end{aligned} \tag{11}$$

R^2 adjusted = 0.54; F = 19.38
S.E. = 0.056; DW = 2.13
Ljung-Box Q(4) = 3.75
Ljung-Box Q(6) = 7.17
Ljung-Box Q(8) = 14.89
Test Jarque-Bera = 1.47
Test ARCH(4) = 1.52
Estimation period: 1975.II; 1990.IV
Estimation method: Ordinary linear square.
Joint non-lineal estimation of (10) y (11) is as follows:

$$\begin{aligned} DLGS_t = &\underset{(0.149)}{0.708}\; DLFPV_t - \underset{(0.036)}{0.095}\,(DLFPT_{t-1} + DLFPT_{t-2}) + \underset{(0.060)}{0.133}\, D81II \\ &- \underset{0.098}{0.557}\,[LGS_{t-1} - \underset{(0.484)}{1.771} - \underset{(0.093)}{0.375}\; LPIB_{t-1} + \underset{(0.045)}{0.285}\; LFPT_{t-1} \\ &- \underset{(0.132)}{0.929}\; LFPV_{t-1} - \underset{(0.038)}{0.107}\; D82I - IV] \end{aligned} \tag{12}$$

R^2 adjusted = 0.52; F = 8.79
S.E. = 0.058; DW = 2.20
Ljung-Box Q(4) = 3.82
Ljung-B = 6.42
Ljung-Box Q(8) = 16.03
Test Jarque-Bera = 2.46
Test ARCH(3) = 1.67
Estimation period: 1975.II; 1990.IV

The elasticities derived from (12) are as follows: product elasticity lower than the unit of 0.375, price elasticity of −0.285 and unitary cross price elasticity of 0.929.

Estimation (12) is robust with respect to the initial conditions and the estimated parameters are similar to estimations (10) y (11).

4.3 "Liquid Bulk" Traffic

The traffic of "Liquid Bulk" or "Oil products" has special features with respect to "General merchandise" and "Solid bulk". During this research, it has been completely regulated. Regulation fixes maximum freight for each transport ship as regards its size and route. It is also necessary to take into account the regulation of the price of the goods which have been carried. Moreover, this additional condition forces any transport activity of this kind to be carried out by the national fleet, as long as it is available. Unlike "General merchandise" and "Solid bulk", in the transport of "Oil products", the user pays the regulated freight instead of the international freight.

The identification problem can again be solved since there is a possibility to resort to the international supply, which is much more relevant than the Spanish fleet, when this fleet cannot satisfy the demand.

"Oil products" transport has been estimated with models similar to those used above. Long-run equation has the following variables:

LPP: "Oil products" log in tonnes.
LPIB: Gross Domestic Product log in real terms.
LPRFP: regulated oil freight index log.

4.3.1 Long-Run Analysis

$$\text{LPPt} = \underset{(0.702)}{5.068} + \underset{(0.103)}{0.403}\ \text{LPIBt} - \underset{(0.018)}{0.043}\ \text{LPRFPt} \tag{13}$$

R^2 adjusted = 0.205; F = 9.16*
S.E. = 0.089; DW = 2.42
ADF = −5.50; DW = 1.93
Estimation period: 1975.I–1990.IV.
Estimation method: Ordinary linear square.

The application of Johansen method (1988) to a VAR(1) with restricted constant, has led to the following result:

Null hypothesis	Trace test value	Critical value at 5%
r = 0	39.13	29.68
r = 1	7.18	15.41
r = 2	1.24	3.76

Then, we accept the existence of one single cointegration vector.

Proper value	Proper vector			
	LPP_t	Constant	LPIB	$LPRFP_t$
0.4026	2.35	2.03	−0.95	0.08

The normalized proper vector has led to the following long-run relationship:

$$LPPt = -5.057 + 0.405\ LPIBt - 0.036\ LPRFPt$$

The product variable long-run elasticity is 0.403, price elasticity is −0.043. The low elasticity product can be explained, as in the case of "Solid bulk", by the low added value of the goods which have been carried (especially import oil). The regulated price of these sea transport services, PRFP, represents the expected relationship: transport demand is not very liable to freight changes. With the aim to capture a cross effect, petrol and fuel prices have been included, but the results obtained have been insignificant.

From (13), we can obtain the short-run relationship which uses the following variables:

DLPP: growth ratio of "Oil products".
PRFP: growth ratio of the freight index of cargoes.
D85IV: dummy variable which takes value 1 during the fourth quarter of 1985.
D1: dummy variable which takes value 1 each year during the first quarter.
D2: dummy variable which takes value 1 each year during the second quarter.

4.3.2 Short-Run Analysis

$$\begin{aligned} DLPP = & -\underset{(0.018)}{0.031}\,(DLPRFPt-2+\ DLPRFPt-3)+\underset{(0.071)}{0.175}\ DLPPt-4 \\ & -\underset{(0.069)}{0.248}\ D85IV-\underset{(0.019)}{0.044}\ D1+\underset{(0.018)}{0.072}\ D2-\underset{(0.104)}{1.088}\ MCEt-1 \end{aligned} \tag{14}$$

R^2 adjusted = 0.78; F = 43.06
S.E. = 0.06; DW = 1.91
Ljung-Box Q(4) = 2.68
Ljung-Box Q(6) = 3.08
Ljung-Box Q(8) = 4.68
Test Jarque-Bera = 1.35
Test ARCH(4) = 4.16
Estimation period: 1975.II; 1990.IV
Estimation method: Ordinary linear square.
Joint non-linear estimation of (13) and (14) is:

$$\text{DLPP}_t = -\underset{(0.020)}{0.37}\,(\text{DLPRFP}_{t-2} + \text{DLPRFP}_{t-3}) + \underset{(0.073)}{0.160}\,\text{DLPP}_{t-4}$$
$$- \underset{(0.071)}{0.257}\,\text{D85IV} - \underset{(0.023)}{0.054}\,\text{D1} + \underset{(0.023)}{0.067}\,\text{D4} - \underset{(0.105)}{1.095}$$
$$- \left(\text{LPP}_{t-1} - \underset{(0.826)}{5.761} - \underset{(0.095)}{0.302}\,\text{LPIB}_{t-1} + \underset{(0.015)}{0.031}\,\text{LPRFP}_{t-1}\right) \quad (15)$$

R^2 adjusted = 0.88; F = 26.96
S.E. = 0.06; DW = 1.91
Ljung-Box Q(4) = 3.03
Ljung-Box Q(6) = 3.43
Ljung-Box Q(8) = 6.74
Test Jarque-Bera = 1.44
Test ARCH(4) = 2.56
Estimation period: 1976.I; 1990.IV

Since the parameter that fits balance is virtually the unit, the former equation can be estimated as follows (see Fernández and Sebastián 1989):

$$\text{LPP}_t = \underset{(0.56)}{5.793} + \underset{(0.087)}{0.297}\,\text{LPIB}_{t-1} - \underset{(0.015)}{0.031}\,\text{LPRFP}_{t-1} - \underset{(0.070)}{0.261}\,\text{D85IV}$$
$$- \underset{(0.023)}{0.057}\,\text{D1} + \underset{(0.023)}{0.068}\,\text{D4} - \underset{(0.020)}{0.037}\,(\text{DLPRFP}_{t-2} + \text{DLPRFP}_{t-1}) \quad (16)$$

R^2 adjusted = 0.62; F = 10.35
S.E. = 0.06; DW = 2.09
Ljung-Box Q(4) = 2.54
Ljung-Box Q(6) = 2.79
Ljung-Box Q(8) = 5.14
Test Jarque-Bera = 1.79
Test ARCH(4) = 3.00
Estimation period: 1976.I; 1990.IV

The results of the elasticities obtained from the short and long-run estimations by the different methods used, are similar. Long-run elasticities are: product elasticity

Table 1 Product, price and cross elasticities of the different sea transport demands

Variables	Long-run			Short-run	
	Production	Price	Price	Price	Price
One stage					
General merchandise	1.100 (0.182)	−0.206 (0.054)	0.590 (0.154)	−0.173 (0.096)	0.754 (0.274)
Solid bulk	0.375 (0.093)	−0.285 (0.045)	0.929 (0.132)	−0.095 (0.036)	0.708 (0.149)
Oil products	0.302 (0.095)	−0.031 (0.015)	–	−0.037 (0.020)	–
Two stages					
General merchandise	1.074 (0.133)	−0.193 (0.053)	0.607 (0.149)	−0.166 (0.0889	0.628 (0.259)
Solid bulk	0.323 (0.099)	−0.297 (0.040)	1.049 (0.112)	−0.091 (0.034)	0.645 (0.155)
Oil Products	0.403 (0.103)	−0.043 (0.018)	–	−0.031 (0.018)	–

Standard deviations in parenthesis

lower than the unit of −0.303; price elasticity of −0.031, which shows that this transport is the least elastic with respect to regulated freight changes. The fact that the regulated freight is not very liable to change, can be useful for regulating bodies and, along with product elasticity, can allow to estimate traffic predictability. However, we must also consider the high aggregation of the results. If it were possible to disaggregate, at least the sea transport services of oil and its derivatives, many of the answers which remain unclear in this study, would provide more information.

In Table 1 we make a summary of the elasticities estimated for short and long-run periods in one and two stages of the different sea transport models analysed in this work.

5 Concluding Remarks

In this work we have estimated a theoretical sea transport model based on former research made on similar transport models. We have used Spanish monthly data on sea transport which have been added to quarterly data in the period 1975.I–1990.IV.

As regards product and price elasticities, we have come to the following conclusions:

- Product long-run elasticities are unitary for "General merchandise", and lower than the unit for "Solid bulk" and "Oil products". Short-run elasticities have not become significant in all the three transport models.
- All short and long-run demand price elasticities are inelastic with values from −0.297 to −0.031. The incidence of international freight in Spanish sea transport is relevant, but this transport is not very much liable to freight changes. Spanish sea transport is not very liable to be replaced by road and railway transport, at least in such aggregate data dealt with in this work. It is worth highlighting the very low value of the price elasticity of "Oil products" regulated freight, which is mainly explained by the fact that freight regulation goes along with the obligation to carry out this service in national ships. For this reason, this transport model is highly inelastic.
- Cross elasticities with respect to substituting sea services, range from 0.607 to 1.049; values which show the possibility of substitution between different sea transport models. However, such substitution is no longer clear with respect to prices relative to other transport services. Again, aggregation is so high that it determines the results. If it were possible to disaggregate each group of services into national and international sea transport, it would be possible to capture the incidence of substitution by other transport modes such as road and railway transport. This would be mainly the case of national sea transport and it would also be possible to captured in continental sea transport from 1986, especially by substituting road for sea transport between Spain and the rest of the Community. If this is possible highly interesting results will be obtained in the future.

These estimations can be useful to carry out short-run transport and revenue predictability.

References

Andrés J, Escribano A, Molinas C, Taguas D (1989) La inversión en España. Econometría con restricciones de equilibrio. Antoni Bosch and Institute of the Institute of Fiscal Studies, Madrid
Coto-Millán P (1986) El transporte marítimo en España: 1974–1983. Doctoral thesis, University of Oviedo. Ed. European Institute of Maritime Studies, Madrid
Coto-Millán P (1988a) Funciones de demanda del transporte marítimo en España. Información Comercial Española 656:125–141
Coto-Millán P (1988b) El transporte marítimo en España (1974–1987): Peculiaridades. Información Comercial Española 658:101–109
Coto-Millán P (1991a) Las funciones de demanda del petróleo: una aproximación empírica. Energía Sept./Oct.: pp 125–132
Coto-Millán P (1991b) El transporte marítimo internacional de Mercancía General: Una aproximación empírica. Revista del Ministerio de Transportes y Comunicaciones 52:74–82
Coto-Millán P, Sarabia JM (1993) Análisis de los servicios de transporte marítimo en España: Demanda, Precios, Renta y Series Temporales, Actas de las IX Jornadas de Economía Industrial. Investigaciones Económicas 52–58
Engle RF, Granger C (1987) Co-integration and error correction: representation, estimation and testing. Econométrica 55:251–276
Fernández I, Sebastián M (1989) El sector exterior y la incorporación de España en la CEE: Análisis a partir de funciones de exportaciones e importaciones. Moneda y Crédito 189:31–73
Friedlaender AF, Spady R (1980) A derived demand function for freight transportation. Rev Econ Stat 62:432–441
Johansen S (1988) Statistical analysis of cointegration vectors. J Econ Dyn Control 12:231–259
Levin R (1978) Allocation in surface freight transportation: does rate regulation matter? Bell J Econ 9(1):18–45
Oum TH (1979) A cross sectional study of freight transport demand and rail-truck competition in Canada. Bell J Econ 10(2):463–482
Pérez T (1993) Un estudio econométrico de la demanda de tráfico telefónico particular en España, 1980–1990. Investigaciones Económicas XVII(2):363–378
Pérez T, Alvarez F, Moreno B (1993) Business telephone traffic demand in Spain: 1980–1991, an econometric approach. Working paper 9318. Instituto Complutense de Análisis Económico, Universidad Complutense, Madrid
Varian H (1978) Microeconomic analysis. Norton Press, NewYork
Winston C (1981a) A disaggregate model of the demand for intercity freight transportation. Econométrica 49:981–1006
Winston C (1981b) A multinomial probit prediction of the demand for domestic ocean containers service. J Transport Econ Pol 15:243–252
Winston C (1983) The demand for freight transportation: models and applications. Transport Res 17a:419–427
Winston C (1985) Conceptual developments in the economics of transportation: an interpretive survey. J Econ Lit XXIII:57–94

Determinants of the Demand of International Maritime Transport

Pablo Coto-Millán, José Baños-Pino, Miguel Ángel Pesquera, Juan Castanedo Galán, and Lucía Inglada-Pérez

Abstract This paper analyses the behaviour of the functions of importing and exporting goods by maritime transport mode in Spain during the period 1994. I–1998. IV Cointegration techniques are used in the estimations to obtain long-term price and income elasticities. These estimations can be used to assess the effects of price modifications in the volume of imports and exports and, hence, to carry out forecasts on international trade and maritime transport.

1 Introduction

This investigation is a continuation of that of Coto-Millán et al. (2005), but with some important innovations. Firstly, maritime transport is considered in terms of its most characteristic types of goods, based on the main port statistics. In other words, we present here models of demand for maritime imports and exports of general cargo, solid bulk, liquid bulk and containers. In Coto-Millán et al. (2005), the analysis referred only to general cargo. Moreover, the time period of the Coto-Millán et al. (2005) study was from 1975.I to 1993.IV, the reason for this being that this period marked the start of the oil crisis, which ushered in a structural change in international trade. The end of the period of study saw the entry into force of the Treaty of the European Union, on 1 November 1993. The reason for analysing this new period, 1994.I–1998.IV, is that the Treaty on European Union came into force in 1994 while the end of the period offers the most updated series of the

P. Coto-Millán (✉), M. Ángel Pesquera, and J. Castanedo Galán
University of Cantabria, Santander, Spain
e-mail: cotop@unican.es

J. Baños-Pino
University of Oviedo, Oviedo, Spain

L. Inglada-Pérez
UNED, Madrid, Spain

P. Coto-Millán et al. (eds.), *Essays on Port Economics*, Contributions to Economics,
DOI 10.1007/978-3-7908-2425-4_5,

Official Customs Agency in pesetas, since operations after 1 January 1999 were in euros.

This paper uses a conventional model to present the results obtained for estimating the income and price elasticities of imports and exports of goods using maritime transport in the Spanish economy. Cointegration techniques are used to perform the estimations. These type of studies are important for determining tariffs for specific products as well as for predicting potential changes caused by economic activity and its cycles. Therefore, price and income elasticities in exports and imports have a broad range of uses, which is the reason why there are many empirical works on total import and export functions in economic literature.

Another advantage of this research is that it can be replicated – with minor modifications – in other countries in Europe and the rest of the world.

2 The Model: Theory and Basic Functions

This model has already been described in Coto-Millán et al. (2005). However, we will now revise some basic concepts because, besides the maritime transport of general cargo – which was estimated in Coto-Millán et al. (2005) – we will also look at the transportation of containers, solid bulk and liquid bulk here. The theoretical model is based on the economic international trade literature. In general terms, the import functions are:

$$\mathrm{M} = \mathrm{M}(\mathrm{Y}, \mathrm{Pm}, \mathrm{P}) \tag{1}$$

where the volume of imports of a given country (M) depend on its monetary income (Y), import prices (Pm) and the prices of national goods and services (P). Assuming that there is no monetary illusion, if we divide the explanatory variables by the price of national goods and services (1), we can describe it thus:

$$\mathrm{M} = \mathrm{M}\left(\frac{\mathrm{Y}}{\mathrm{P}}, \frac{\mathrm{Pm}}{\mathrm{P}}\right) \tag{2}$$

or

$$\mathrm{M} = \mathrm{M}\left(\mathrm{y}, \mathrm{e}^1\right), \text{ where } \frac{\mathrm{Y}}{\mathrm{P}} = \mathrm{y}; \ \frac{\mathrm{Pm}}{\mathrm{P}} = \mathrm{e}^1; \tag{3}$$

where e^1 is the relative price of imports and where the expected signs are:

$$\frac{\partial M}{\partial y} > 0; \ \frac{\partial M}{\partial e^1} < 0$$

The volume of imports (MT_i; where i = general cargo, containers, solid bulk and liquid bulk) will depend on the volume of imports (M) and the prices of transport services (MP). Thus:

$$MT_i = MT(M, MP_i), \tag{4}$$

where these signs are expected:

$$\frac{\partial MT_i}{\partial M} > 0, \frac{\partial MT_i}{\partial MP_i} < 0.$$

Using expressions (3) and (4), it is now possible to state

$$MT_i = MT\left[M\left(y, e^1\right); MP_i\right]$$

or

$$MT_i = \phi\left(y, e^1, MP_i\right), \tag{5}$$

The following signs are expected for the first derivatives:

$$\frac{\partial MT_i}{\partial y} > 0; \quad \frac{\partial MT_i}{\partial e^1} < 0; \quad \frac{\partial MT_i}{\partial MP_i} < 0.$$

For exports, there are three ways of obtaining the relevant function: the demand approach, the supply approach and the combined approach. In the demand approach, the volume of exports of a country (X) is a function of world income (or foreign income), expressed in real terms (y*), and the relative prices of exports, e^2 (e^2 = Px/P*, where Px represents prices of exported goods and services and P* represents world prices). Thus, the exports function can be expressed as follows:

$$X = X\left(y^*, e^2\right), \tag{6}$$

with the necessary condition that the first derivatives must have the correct signs:

$$\frac{\partial X}{\partial y^*} > 0; \quad \frac{\partial X}{\partial e^2} < 0.$$

Based on the above, we have chosen functions (5) and (6) to conduct our estimations, taking into account that the maritime export and import functions in (3) and (6) are just more precise versions of (5) and (6). Moreover, in order to obtain estimations of elasticities, we have adopted the logarithmic-linear functional form of (5) and (6). Thus, the functional specifications to be determined are:

$$LMT_i = \beta_0 + \beta_1 Ly + \beta_2 Le^1 + \beta_3 LMP_i + u_1 \tag{7}$$

for the imports function, and

$$LXT_i = \alpha_0 + \alpha_1 Ly^* + \alpha_2 Le^2 + \alpha_3 LMP_i + u_2 \tag{8}$$

for the exports function.

3 Statistical Sources: The Foreign Trade Series

Statistical information on imports and exports is available from the Spanish Economy foreign sector, supplied by the General Customs Directorate. This information is aggregated and uses different classifications for different economic groups. These series are the main source of statistics for estimating the import and export functions of the Spanish economy. These functions are incorporated into the quarterly econometric model produced by the Studies Service of the Bank of Spain since 1977. Nevertheless, there is an important restriction to our use of these series: since they are in pesetas, unit value indexes have to be calculated in order to obtain series of goods prices. In addition, the use of unit value indexes causes problems with the lack of homogeneity of foreign trade series because of their classifications, regardless of the level of disaggregation. The most common disaggregated series of foreign trade cover the following groups: Food products, Energy, Petrol and lubricant, Intermediate products, Capital goods and Consumer goods.

There are other annual series such as the *Means of Transportation Foreign Trade Statistics,* also published by the General Customs Directorate, which offer information on foreign trade in euros and in tons.

We also have the *Monthly Foreign Trade Series,* publication of which began in 1988. These include information on monthly customs series in tons and pesetas and on monthly means of transportation series with no disaggregation by items (Food products, Energy, Capital goods, etc.) but with data on total imports and exports.

In its *Monthly Foreign Trade Series*, the General Customs Directorate indicates imports and exports of goods by maritime transport in pesetas and tons from 1994.I to 1998.IV. It has therefore been possible to obtain the series of exports and imports of effective demand using the above maritime transport divisions (General Cargo, Bulk Solids, Bulk Liquid and Containers).

4 Variable Definition

The following variables have been considered:

1. LXMGT: Logarithm of volume of exports in tons of 'General Cargo'.
2. LMMGT: Logarithm of volume of imports in tons of 'General Cargo'.
3. LXCONT: Logarithm of volume of exports in tons of 'Containers'.

4. LMCONT: Logarithm of volume of imports in tons of 'Containers'.
5. LXGST: Logarithm of volume of exports in tons of 'Solid Bulk'.
6. LMGST: Logarithm of volume of imports in tons of 'Solid Bulk'.
7. LXGLT: Logarithm of volume of exports in tons of 'Liquid Bulk'.
8. LMGLT: Logarithm of volume of imports in tons of 'Liquid Bulk'.
9. LFRSA: Logarithm of the relative price of scheduled services cargo products (FBL).
10. LPRELSA: Logarithm of the relative price of intermediate goods exports.
11. LPRMM1SA: Logarithm of the relative world prices of intermediate goods.
12. LPRIMEN: Price logarithm of energy imports.
13. LPREXEN: Price logarithm of energy exports.
14. Ly: Logarithm of Spanish Gross Domestic Product in real terms.
15. LFPT: Logarithm of cargo indexes by 'tramp' time of dry cargo (FPT).
16. LFPV: Logarithm of cargo indexes by 'tramp' travel of dry cargo (FPV).
17. LFBL: Index logarithm of the price of scheduled services cargos.
18. LMP: Logarithm of the index of the sum of FPT and FPV.
19. Le^1: Logarithm of the relative price of final goods exports in Spain.
20. Ly*: Logarithm of imports in industrial countries, as a proxy of the logarithm of world income in real terms.
21. Le^2: Logarithm of the relative prices of imports of final goods.

5 Functions of Maritime Imports and Exports: Main Empirical Results in Spain (1994.I–1998.IV)

Equations (7) and (8) have been estimated with cointegration methods.

6 General Cargo Function

The results now follow of the estimations of maritime imports and exports of general cargo in the long term, together with those of containers.

6.1 Import Function of General Cargo

The results for the long-term equilibrium equation are:

$$LMMGT_t = \underset{(-6.28)}{-17.26} + \underset{(10.19)}{2.82\ Ly_t} \underset{(-2.98)}{-0.68\ Le^1{}_t} - \underset{(-2.25)}{0.04\ LMP_t}$$

R^2 adjusted = 0.99; S.E. = 0.61; DW* = 1.87;
ADF (3) = −5.61; DW** = 2.41.

The estimation indicates excellent long-term results. The interpretation of elasticities in the long-term equation is as follows: the value 2.82 indicates that income or product elasticity is positive and greater than 1. Moreover, maritime transport demand for general cargo is income- or product-elastic. A 100% increase in income would increase the demand for the maritime transport of general cargo by 282%. This is reasonable given that Spain's production depends on its imports of products with added value, either for its own consumption or to produce goods with greater added value. The value of −0.68 for cargo price elasticity (for the relationship between the general cargo prices of the countries from which they are imported and the prices of Spanish general cargo) indicates inelastic demand. A 100% increase in price would decrease the demand for maritime transport of general cargo by 68%. This indicates that Spain is heavily dependent on imports to an extent. Lastly, the value of −0.04 for the parameter estimated for the prices of transport services tells us that maritime transport demand is very inelastic. A 100% increase in the price of maritime transport services would decrease demand by just 4%. Demand, then, is scarcely sensitive to changes in transport price, mainly because of the fact that participation of transport price in the total price of transported cargo is around 4%.

6.2 Export Function of General Cargo

The results of the long-term export function are:

$$LXMGT_t = \underset{(8.60)}{19.48} + \underset{(5.60)}{1.39\,Ly_t^*} \underset{(-2.46)}{-0.08\,Le_t^2} - \underset{(-1.92)}{0.04\,LMP_t}$$

R^2 adjusted = 0.92; S.E. = 0.81; DW* = 0.97;
ADF (3) = −5.71; DW** = 2.12.

Estimation of the long-term equation generates excellent new results by observing the econometric tests. Spanish exports depend positively and to an interesting degree on world income. The income or product elasticity value of 1.39 can be interpreted in the sense that a 100% increase in world income would increase Spain's general cargo exports by 139%. Nonetheless, this proportion is lower than that of imports, where a 100% increase in Spanish income would increase Spanish imports by 282%. The above relationships indicate that Spain has a negative balance in its balance of payments for goods. The value of −0.08 for price elasticity values of general cargo (Spain compared to the world) is interpreted economically to indicate that exported goods have an inelastic demand, where major variations in goods prices generate scarce variation in demand. A 100% price decrease only generates a demand increase of 8%. Interpretation of the prices of maritime services is similar to that of exports.

6.3 Import Function of Containers

The results of the long-term equilibrium equation are as follows:

$$\mathrm{LMCONT_t} = \underset{(-6.28)}{-21.54} + \underset{(10.19)}{4.41\,\mathrm{Ly_t}} \quad \underset{(-2.98)}{0.45\,\mathrm{Le_t^1}} - \underset{(-2.25)}{0.28\ \mathrm{LMP_t}}$$

R^2 adjusted = 0.93; S.E. = 0.74; DW* = 1.72;
ADF (3) = −5.38; DW** = 2.01.

Imports of goods by container also offer some interesting results. Firstly, income-product elasticity is high at 4.41, higher than maritime imports of general cargo, as was to be expected, given that goods transported in containers have a greater added value than general cargo. The price elasticity of −0.68 indicates an inelastic demand compared to price, similar to the case of general cargo imports. Lastly, the prices of maritime transport services have a greater influence in this case. The estimated value for the price elasticity of maritime transport services is −0.28, less inelastic than the value estimated for general cargo, −0.04. The result is reasonable because the transportation of goods by container is six times greater than that of goods not transported in containers.

6.4 Export Function of Containers

The results of the long-term export function are as follows:

$$\mathrm{LXCONT_t} = \underset{(8.60)}{5.36} + \underset{(5.60)}{1.77\,\mathrm{Ly_t^*}} \ \underset{(-2.46)}{-0.64\,\mathrm{Le_t^2}} - \underset{(-1.92)}{0.12\ \mathrm{LMP_t}}$$

R^2 adjusted = 0.94; S.E. = 0.81; DW* = 0.96;
ADF (3) = −5.78; DW** = 2.46.

The estimation of income-product elasticity is higher at 1.77 than that of general cargo (1.39), as was to be expected. Again, the interpretation is that goods transported in containers have a greater added value. The value −0.64 indicates that demand is inelastic in relation to the price of the goods, although it is more elastic than that estimated for general cargo (−0.08). Lastly, the value −0.12 tells us that transport price has a higher participation in the total price of goods.

7 Solid Bulk Function

The results now follow of the estimations of maritime imports and exports of solid bulk in the long term.

7.1 Import Function of Solid Bulk

The results for the long-term equilibrium equation are:

$$\mathrm{LMGST_t} = \underset{(-0.80)}{-1.82} + \underset{(7.09)}{1.46}\,\mathrm{Ly_t} \underset{(-4.70)}{-0.51}\,\mathrm{LPRMM1SA_t} - \underset{(-2.36)}{0.16}\,\mathrm{LFRSA_t}$$

R^2 adjusted = 0.91; S.E. = 0.91; DW* = 0.87;
ADF (3) = −5.31; DW** = 2.02.

The estimation of income elasticity, with a value of 1.46, shows that increases in maritime imports of solid bulk are elastic to growth in income, measured by Gross Domestic Product. Price elasticity, estimated at a value of −0.51, indicates that demand is inelastic in relation to the price of goods. For the estimated elasticity of transport services, we obtain the value −0.16 (where LFRSA: logarithm of the relative price of scheduled services cargo products).

7.2 Export Function of Solid Bulk

The results of the long-term export function are:

$$\mathrm{LXGST_t} = \underset{(2.49)}{1.53} + \underset{(9.66)}{1.94}\,\mathrm{Ly_t^*} \underset{(-6.13)}{-0.84}\,\mathrm{LPRELSA_t}$$

R^2 adjusted = 0.72; S.E. = 0.22; DW = 0.50;
ADF (3) = −4.27; DW = 2.02.

In maritime exports of solid bulk, income elasticity and price elasticity obtain the expected results (1.94 and −0.84, respectively) and suggest a similar interpretation to that offered for exports of general cargo. The price variable of maritime transport services was not significant here. It is likely that the low incidence of the percentage of this transport cost (around 1% or 0.5%) generates an estimated coefficient that is not significantly different to zero, with likely values of between −0.01 and −0.005, i.e. transport service price variations increasing by 100% would generate small decreases such as 1% or 0.5% of the required quantity of maritime transport exports of solid bulk.

8 Liquid Bulk Function

The results now follow of the estimations of maritime imports and exports of liquid bulk in the long term.

8.1 *Import Function of Liquid Bulk*

The results for the long-term equilibrium equation are:

$$\mathrm{LMGLT_t} = \underset{(3.20)}{3.06} + \underset{(7.70)}{0.80}\,\mathrm{Ly_t} \; \underset{(-4.20)}{-0.52}\,\mathrm{LPRIMEN}$$

R^2 adjusted = 0.76; S.E. = 0.07; DW = 1.36;
DF = −6.8; DW = 2.06.

The functions of maritime imports of liquid bulk can be explained by the variables of income and price of goods. The value of 0.80 indicates the inelasticity of demand of liquid bulk compared to income. The explanation for this is imports of crude, a commodity not produced in Spain and on which it is clearly dependent. In contrast, the prices of maritime transport services have not been shown to be significant, probably because these prices have very little impact (1–0.5%) on the total cost of importing crude oil.

8.2 *Export Function of Liquid Bulk*

The results for the long-term export function are:

$$\mathrm{LXGLT_t} = \underset{(8.60)}{12.24} + \underset{(5.60)}{0.73}\,\mathrm{Ly_t^*} \; \underset{(-2.46)}{-0.16}\,\mathrm{LPREXEN}$$

R^2 adjusted = 0.97; S.E. = 0.98; DW* = 1.37;
DF = −6.8; DW = 2.06

The functions of maritime exports of liquid bulk can be explained by the variables of world income and the price of energy exports. The value of 0.73 indicates the inelasticity of demand for liquid bulk in comparison to income, the explanation for which lies with exports of products deriving from crude oil, since, as we have explained, crude oil is a commodity not produced in Spain but which is refined and used to produce petroleum products that, besides being consumed in Spain, are also exported. The prices of maritime transport services have not been shown to be significant, probably for similar reasons to those for solid bulk imports and liquid bulk exports. The explanation is that these prices have a very low impact (1–0.5%) on the total cost of imports of crude oil.

Tables 1 and 2 summarise the values of the elasticities estimated for imports and exports in this paper.

9 Summary and Conclusions

Based on the elasticities estimated for the mean imports indicated in Table 1, we can conclude that:

Table 1 Maritime import elasticities

	Income elasticity	Price elasticity of goods	Transport price elasticity
Maritime T of GC	2.82	−0.68	−0.04
Maritime T of C	4.41	−0.45	−0.28
Maritime T of SB	1.46	−0.51	−0.16
Maritime T of LB	0.80	−0.52	–

GC Maritime transport of General Cargo; *C* Maritime transport of Containers; *SB* Maritime transport of Solid Bulk; *LB* Maritime transport of Liquid Bulk
Note: All estimated elasticities are representative up to 95%

Table 2 Maritime export elasticities

	Income elasticity	Good's price elasticity	Transport price elasticity
Maritime T of GC	1.39	−0.08	−0.04
Maritime T of C	1.77	−0.64	−0.12
Maritime T of SB	1.94	−0.84	–
Maritime T of LB	0.73	−0.16	–

GC Maritime transport of General Cargo; *C* Maritime transport of Containers; *SB* Maritime transport of Solid Bulk; *LB* Maritime transport of Liquid Bulk
Note: All estimated elasticities are representative up to 95%

1. Income elasticities of imports had higher values in the empirical literature than that of the interval (1.22 and 1.73). The only exception appeared in one study (Mauleón and Sastre (1995), where the value was 0.66). The income elasticities obtained here are higher than that of the interval (1.46 and 4.41), the only exceptions to this being liquid bulk in maritime transport (0.80). The reason for the income inelasticity of liquid bulk could be that the Spanish economy depends heavily on foreign oil imports, which behaves like a normal product, unlike luxury values, which will provide elasticities of more than one.
2. In the empirical literature we reviewed, price elasticities for imported goods always had negative values lower than that of the interval (−0.39 and −0.75). In our analysis, the results of import estimations are the same in the interval (−0.45 and −0.68) and therefore adapt reasonably well to the results obtained by other research.
3. The price elasticities of maritime services obtained in previous works were significant, negative and less than one if the values used for general cargo were between (−0.03 and −0.19), in the case of solid bulk between (−0.19 and −0.29) and (−0.01 and 0.005) for liquid bulk. In our study, the results were: (−0.04) for general cargo, (−0.28) for goods transported in container ships and (−0.16) for solid bulk. The data obtained for liquid bulk was not significant. The values adjust appropriately to previous results except in the case of liquid bulk, where changes in the regulation of oil maritime transportation required us to use international freights for the approximation from 1994 onwards. This approximation may not be accurate and hence, this elasticity may in fact be significant and around values of (−0.01 and −0.005), as in previous periods. In any case, the interpretation of previous elasticities, all less than one, is that this is an

indication of the low level of substitution of maritime transportation for others means of transportation.

The elasticity values for exports are indicated in Table 2, and the main conclusions are:

1. Income elasticities of goods exports in the empirical literature reviewed for this work showed positive results of more than one in the interval (1.65 and 3.07), with no exceptions. Maritime export income elasticities are always positive in the interval (1.39 and 1.94), with the only exception being exports of liquid bulk, whose value is positive but less than one (0.73).
2. The price elasticities of exported goods all indicate inelastic demands with regards price, which means that these exported products are not very sensitive to price variations.
3. The price elasticities of maritime transport services of exported goods are only significant for exports of general cargo and containers. The most likely explanation for this is that the participation of the cost of transport in solid bulk and liquid bulk is so slight in relation to the cost of the goods (around 1–0.5%), that elasticity or the coefficient would be −0.01 or −0.005, values not significantly different to zero.

In the light of all of the conclusions on imports and exports, the most relevant results are:

Long-term income or product elasticities of maritime exports and imports are very high and significant, except in the case of liquid bulk, for which they are less than one but nevertheless significant.

Maritime transport services demands are inelastic with regard to the price of the goods in maritime imports and exports.

Maritime transport demands are inelastic in terms of the price of the transportation service, with low long-term values because they correspond to a service that cannot easily be replaced in the case of imports and exports.

According to the empirical evidence obtained in this study, the determinants of maritime imports in Spain are (by order of importance): national income, import prices and prices of maritime transport services.

Likewise, the empirical evidence obtained here shows that the determinants of maritime exports in Spain (by order of importance) are: world income, export prices and prices of maritime transport services, as well as the degree of use of Spanish productive capacity.

References

Coto-Millán P, Baños-Pino J, Villaverde J (2005) Determinants of the demand for maritime imports and exports. Transport Res Part E 41:357–372

Mauleón I, Sastre E (1995) El Saldo Comercial en el bienio 1993–94. Información Comercial Española 752:99–103

The Demand for Maritime Transport: A Nonlinearity and Chaos Study

Lucía Inglada-Pérez

Abstract This paper studies the existence of non-linear dynamics and chaos in the Spanish maritime transport services for the period 1992–2007. Using monthly time series data and the Box-Jenkins approach for time series analysis as a preparatory step in order to obtain linear model and applying the BDS test to residuals obtained, we find that a number of sea traffic series – total cargo, solid bulk, liquid bulk, containered and non containered general cargo – do not show significant nonlinear dependence and hence chaos cannot be inferred.

1 Introduction

Linear modeling has been traditionally applied to explain economic theories but in the two last decades, it has been witnessed important changes in the econometric modelling of time series. Interest in nonlinear processes, particularly chaotic processes, nonlinear deterministic processes that look random, has increased dramatically. Many researchers have successfully used nonlinear analysis to model complex series in various fields of natural sciences and engineering as well as economics.[1]

One of the main challenges of econometric models is to forecast a seemingly unpredictable economic series. The traditional linear structural models have not been successful when used for forecasting, particularly in the case of complex series, mainly due to the weaknesses of the economic theory when dealing with complex models of simultaneous equations. If data generating process is nonlinear

[1]For example, see Scheinkman and LeBaron (1989), Frank and Stengos (1989), and Serletis and Gogas (1997).

L. Inglada-Pérez
UNED, Madrid, Spain
e-mail: Linglada@cnio.es

P. Coto-Millán et al. (eds.), *Essays on Port Economics*, Contributions to Economics,
DOI 10.1007/978-3-7908-2425-4_6, © Springer-Verlag Berlin Heidelberg 2010

and chaos exists, it can be shown numerous implications for example on forecasting could exist due to chaos represents a radical change of perspective in the explanation of fluctuations observed in economic and financial time series. In that case, prediction over long periods is all but impossible, due to the property of chaos consistent in sensitive dependence on initial conditions.

The aim of this article is basically to model empirically the time series structure of Spanish sea transport service demand, analyzing the existence of nonlinearity and chaos. Measures of maritime transport service include: total cargo, solid and liquid bulk, and containered and non containered general cargo.

In the literature on maritime transport, there are a lot of papers[2] which deal with the aspects of time series modeling and prediction, but our paper is the only one, to the best of our knowledge, that studies chaos and nonlinearity in the demand for maritime transport at Spanish ports.

In order to explore if linear models explain the maritime traffic reasonably or on the contrary exist a nonlinear stochastic or deterministic chaotic behavior in the traffic series, we firstly model monthly maritime traffic series, applying ARIMA models, for the period January 1992 to December 2007. We then test for non linear and chaos using BDS test. If there is evidence that sea traffic shows nonlinearity, ARIMA models may not be appropriate for forecasting and demand modeling.

The outline of this article is as follows: In Sect. 2 we describe the database used and investigate the univariate time series properties of maritime traffic applying ARIMA models, using monthly data on total, solid and liquid bulk, containerized general and non-containerized cargo traffic for the period January 1992 to December 2007. Sect. 3 provides a description of the key features of the BDS statistic, focusing explicit attention on the test's ability for testing non-linear dependence. In Sect. 4 we describe the results obtained for the five maritime transport service series, discussing the evidence in favor of and against the hypotheses of nonstationarity and nonlinearity in mean and variance. Sect. 5 summarizes some conclusions.

2 Data Analysis and ARIMA Models

2.1 Data

We use monthly Spanish sea transport service series, from 1992:1 to 2007:12 (192 observations), on solid and liquid bulk, containered and non-containered general cargo traffic. Maritime traffic data series have been taken from Spain's Public

[2]For example, Coto-Millan (1995), Coto-Millan and Baños-Pino (1996), Li and Parsons (1997), Cullinane et al. (1999), Kavussanos and Nomikos (2000), Veenstra and Haralambides (2001), Mostafa (2004), Coto-Millan et al. (2004), Batchelor et al. (2007) and Castillo-Manzano et al. (2008).

Works Ministry (http://www.fomento.es/) and are displayed in Figs. 1–5. Table 1 also reports some descriptive statistics of the five series studied.

Before conducting nonlinear dynamical analysis the data must be rendered stationary, delinearized by replacing the stationary data with residuals from the best possible linear model and transformed if necessary.

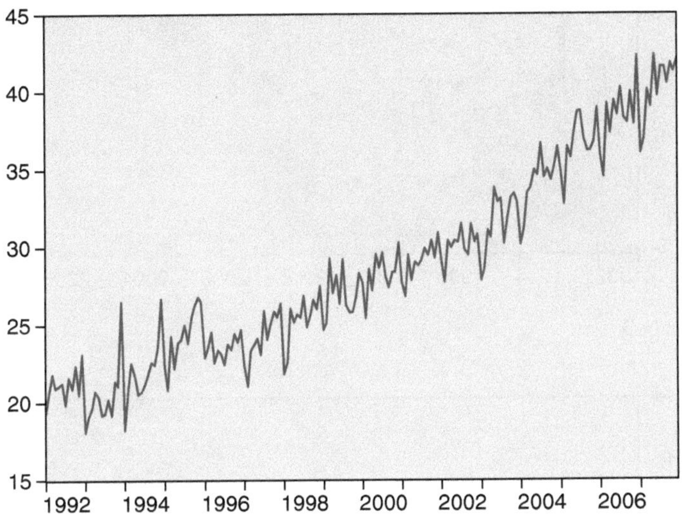

Fig. 1 Total cargo

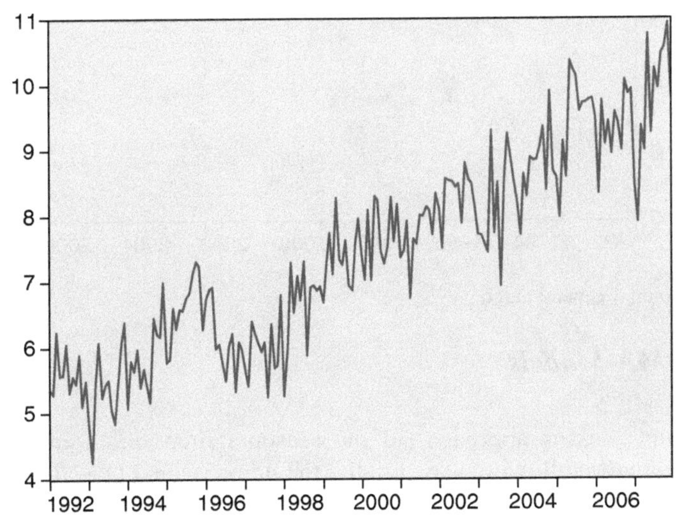

Fig. 2 Solid bulk

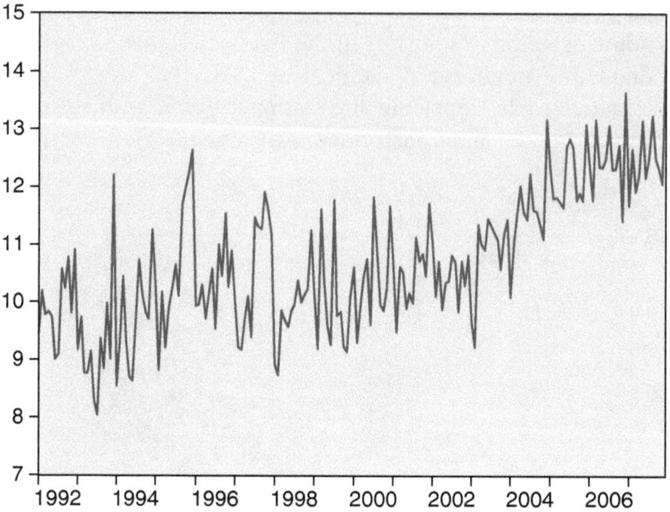

Fig. 3 Liquid bulk

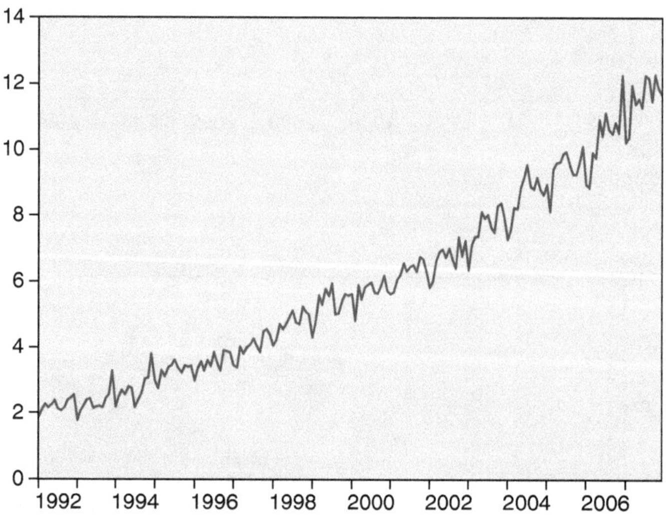

Fig. 4 Containered general cargo

2.2 ARIMA Models

The Box and Jenkins approach[3] to the seasonal time series analysis can be summarized in the following way. Firstly, the non-seasonal $(1 - B)$ and seasonal

[3]See Box and Jenkins (1970) and Box et al. (1994).

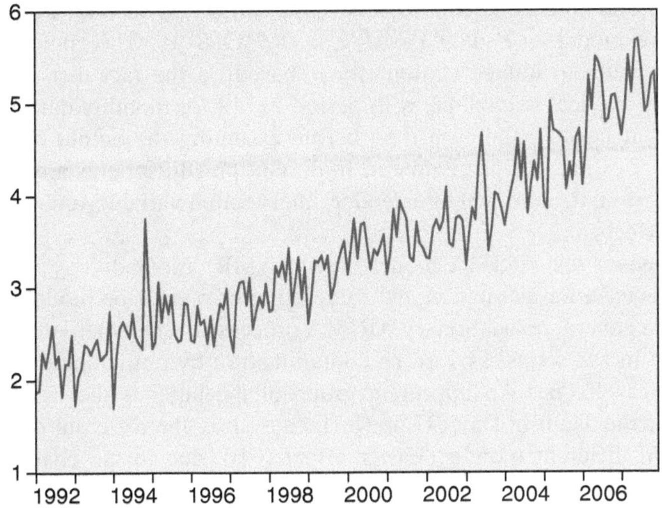

Fig. 5 Non-containered general cargo

Table 1 Descriptive statistics of traffic variables

	Total cargo	Solid bulk	Liquid bulk	Containered	Non contained
Mean	28,511,070	7,426,152	10,711,330	5,902,288	3,421,172
Median	27,584,521	7,350,705	10,578,877	5,611,239	3,323,159
Maximum	42,342,039	10,935,003	14,482,359	12,285,865	5,693,541
Minimum	18,112,191	4,257,251	8,036,282	1,777,358	1,700,501
Standard deviation	6,281,496	1,523,239	1,216,766	2,895,442	941,231,0
Observations	192	192	192	192	192

$(1-B)^s$ differencing operators are used to convert a non-stationary series z_t into a stationary series w_t. It is usually necessary to use d-order non-seasonal and D-order seasonal differencing, that is $w_t = (1-B)^d(1-B^s)^D$(generally d is ≤ 2).

Then, the stationary series w_t is expressed, according the Wold decomposition theorem, as a weighted sum of current and past values of a white noise process

$$w_t = a_t + \sum_{j=1}^{\infty} \psi_j a_{t-j} = \psi(B)a_t \quad \text{where } \psi(B) = 1 + \sum_{j=1}^{\infty} \psi_j B^j.$$

Finally, to achieve parsimonious models the polynomial $\psi(B)$ is approximated by the rational polynomial $\psi(B) = \frac{\theta(B)\Theta(B^s)}{\varphi(B)\Phi(B^s)}$ where $\phi(B) = 1 - \phi_1 B - \cdots - \phi_p B^p$ and $\theta(B) = 1 - \theta_1 B - \cdots - \theta_q B^q$ are the non-seasonal autoregressive and moving-average polynomials which describe the dependence between consecutive data, and $\Phi(B) = 1 - \Phi_1 B - \cdots - \Phi_P B^P$ and $\Theta(B) = 1 - \Theta_1 B - \cdots - \Theta_Q B^Q$ are the seasonal autoregressive and moving-average polynomials describing the dependence between data which are s periods apart.

Therefore, the nonstatinonary seasonal time series z_t is described by the general multiplicative model, $\varphi(B)\Phi(B^s)\nabla^d\nabla_s^D z_t = \theta(B)\Theta(B^s)a_t$. The choice of the seasonal differencing to induce stationarity is based on the fact that seasonal time series show a cyclical behaviour, with period $s = 12$ for monthly data.

However, it is often the case that, before assuming the output of an ARIMA model, the series need prior treatment. Important preadjustments are for example, outlier correction, the removal of calendar, intervention variable, and other possible regression effects.

In this sense, the ARIMA-model-based (AMB) methodology[4] for seasonal adjustment is used for estimation and forecasting of regression models with errors that follow in general nonstationary ARIMA processes, when there may be missing observations in the series, as well as contamination by outliers and other special (deterministic) effects.[5] An important group of the latter is the Calendar effect, composed of the Trading Day (TD) effect, caused by the different distribution of week-days in different months, Easter effect (EE), due to the changing date of Easter, leap year (LY) effect, and holidays effect.

The AMB approach to the seasonal time series analysis can be sketched as follows. If B denotes the lag operator, such that $\mathrm{B\ x(t) = x(t-1)}$ and s the number of observations per year, given the observations $\mathrm{y = (y(t_1), y(t_2), \ldots, y(t_m))}$ where $0 < \mathrm{t_1} < \cdots < \mathrm{t_m}$, we fit the general model:

$$\mathrm{y(t)} = \sum_{\mathrm{i=1}}^{\mathrm{n_{out}}} \omega_\mathrm{i}\,\lambda_\mathrm{i}(\mathrm{B})\mathrm{d_i(t)} + \sum_{\mathrm{i=1}}^{\mathrm{n_c}} \alpha_\mathrm{i}\,\mathrm{cal_i(t)} + \sum_{\mathrm{i=1}}^{\mathrm{n_{reg}}} \beta_\mathrm{i}\,\mathrm{reg_i(t)} + \mathrm{x(t)} \tag{1}$$

where $\mathrm{d_i(t)}$ is a dummy variable that indicates the position of the i-th outlier, $\lambda_\mathrm{i}(\mathrm{B})$ is a polynomial in B reflecting the outlier dynamic pattern, $\mathrm{cal_i}$ denotes a calendar-type variable, $\mathrm{reg_i}$ a regression or intervention variable, and x is the ARIMA error. The parameter ω_i is the instant i-th outlier effect, α_i and β_i are the coefficients of the calendar and regression-intervention variables, respectively, and $\mathrm{n_{out}}, \mathrm{n_c}$ and $\mathrm{n_{reg}}$ denote the total number of variables entering each summation term in (1).

In matricial notation, (1) can be rewritten as

$$\mathrm{y(t) = z'(t)\ b + x(t)} \tag{2}$$

where b is the vector with the ω, α and β coefficients and $\mathrm{z'(t)}$ denotes a matrix with columns the variables:

$$\left(\mathrm{cal_1(t)}, \ldots, \mathrm{cal_{n_c}(t)}, \lambda_1(\mathrm{B})\mathrm{d_1(t)}, \ldots, \lambda_{\mathrm{n_{out}}}(\mathrm{B})\mathrm{d_{n_{out}}(t)}, \mathrm{reg_1(t)}, \ldots, \mathrm{reg_{n_{reg}}(t)}\right)$$

[4]See, for example Hillmer and Tiao (1982), Bell and Hillmer (1983 and 1984), Maravall and Pierce (1987), and Gómez and Maravall (1996, 1998, 2000a, b).

[5]See, for example Box and Tiao (1975), Chang et al. (1988), Chen and Liu (1993), Gómez and Maravall (1994), and Gómez et al. (1999).

The first term of the addition in (2) represents the effects that should be removed in order to transform the observed serie into a serie that can be assumed to follow an ARIMA model; In matricial form, the ARIMA model for x(t) can be written as $\varphi(B)\,\delta(B)\,x(t) = \theta(B)\,a(t)$ where a(t) denotes the $N(0, V_a)$ white-noise innovation, and $\varphi(B)$, $\delta(B)$ and $\theta(B)$ are finite polynomials in *B*.

The first one $\varphi(B)$ contains the stationary autoregressive (AR) roots, $\delta(B)$ contains the nonstationary AR roots, and $\theta(B)$ is an invertible moving average (MA) polynomial. Often they assume the multiplicative form:

$$\delta(B) = \nabla^d \nabla_f^{d_s}$$
$$\varphi(B) = \left(1 + \varphi_1 B + \cdots + \varphi_p B^p\right)\left(1 + \Phi_1 B^f + \cdots + \Phi_{p_s} B^{p_s f}\right)$$
$$\theta(B) = \left(1 + \theta_1 B + \cdots + \theta_p B^q\right)\left(1 + \Theta_1 B^f + \cdots + \Theta_{q_s} B^{q_s f}\right)$$

where $\nabla = 1 - B$ and $\nabla_f = 1 - B^f$ are the regular and seasonal difference operators and s denotes the number of observations per year.

The model may contain a constant μ, equal to the mean of the differenced series. In practice, this parameter is estimated as one of the regression parameters.

We use program TRAMO[6] in order to test for the log/level transformation, for the possible presence of calendar-type and Easter effects, and for three types of outliers (namely, additive outliers (AO), transitory changes (TC), and level shifts (LS)). Hence we identify and estimate by maximum likelihood the reg-ARIMA model.

3 Application to Maritime Traffic Series

3.1 ARIMA Models

Since we are interested in nonlinear dependence, we remove any linear dependence in the data by fitting the best possible linear model, applying ARIMA-model-based (AMB) methodology as described in Sect. 2. The results obtained for the five maritime traffic series during the period September 1992 to December 2007 (192 observations) are as follows.

3.1.1 Total Cargo (TC)

The model obtained for TC is:

$$\underset{(t-\text{values}):}{TC_t} = 3.36\,\underset{(3.84)}{I_t^{1993:12}} - \frac{3.35}{1 - B}\,\underset{(-3.52)}{I_t^{1996:12}} + N_t \tag{3}$$

[6]TRAMO, SEATS, and program TSW, a Windows version that integrates both programs, are available at http://www.bde.es, together with documentation.

$$\underset{\text{t-values:}}{\nabla\nabla_{12}N_t} = (1 - \underset{(-5.33)}{0.706}\ B)\ (1 - \underset{(-10.78)}{0.717}B^{12})a_t \qquad (4)$$

Where B is the backward operator, such that $B_j(z_t) = z_{t-j}$; ∇ and ∇_{12} represent the operator's regular difference = (1 – B) and seasonal difference = 1 – B^{12}, respectively.

With $\sigma_a = 0.14$ and $I_t^{1993:12} = 1$ (1993:12), $I_t^{1996:12} = 1$ (1996:12) and zero otherwise. The first is additive outlier (AO) and the second is a level shift (LS).

Equation (3) specifies the regression variables, this is, the deterministic part of the series, while (4) specifies the ARIMA model – the stochastic part. Accordingly, the model estimated corresponds to the so-called airline model (ARIMA (0, 1, 1) (0, 1, 1)$_{12}$), popularized by Box and Jenkins (1970) in levels and with no mean, where several adjustments are made to isolate 'outliers effect'. Figure 6 displays the residuals obtained.

3.1.2 Solid Bulk (SB)

The model obtained is:

$$\underset{\text{(t-values):}}{SB_t} = \underset{(10.65)}{22.59}\ TD_t + N_t$$

$$\underset{\text{(t-values):}}{\nabla\nabla_{12}N_t} = (1 - \underset{(-11.54)}{0.663}\ B)\ (1 - \underset{(-10.97)}{0.789}B^{12})a_t$$

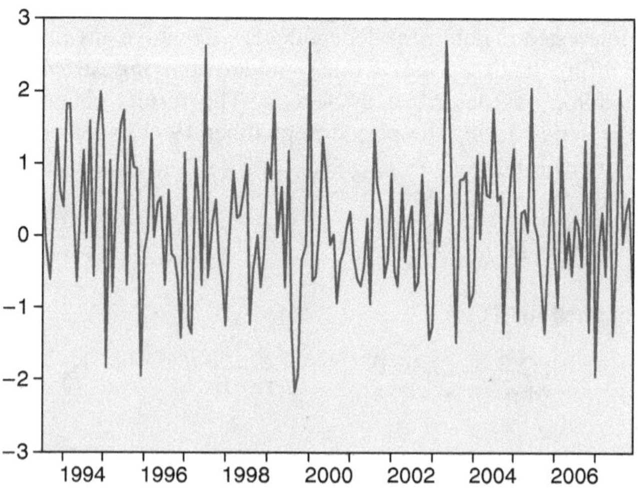

Fig. 6 Total cargo model residuals

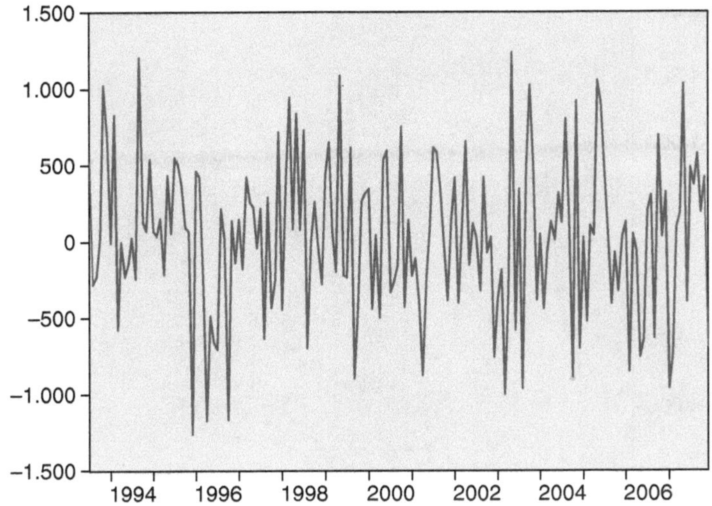

Fig. 7 Solid bulk model residuals

With $\sigma_a = 35.85$ and significant trading day (Working days) effect:

$$\begin{aligned} \mathrm{TD_t} &= \text{(Number of working days (M, T, W, Th, F)} \\ &\quad - \text{Number of weekend days (Sat, Sun))} \times 5/2 \end{aligned}$$

Accordingly, the model fitted corresponds to the ARIMA $(0,1,1)(0,1,1)_{12}$ model in levels and with no mean, where several adjustments are made to isolate 'calendar effects'. Figure 7 displays the residuals obtained.

3.1.3 Liquid Bulk (LB)

The model obtained is:

$$\underset{\text{(t-values):}}{\mathrm{LB_t}} = \underset{(-4.75)}{\frac{-1,783}{1-\mathrm{B}}}\, \mathrm{I_t^{1997:12}} + \underset{(4.15)}{\frac{1,972}{1-0.70\mathrm{B}}}\, \mathrm{I_t^{1995:9}} + \underset{(3.61)}{\frac{1,359}{1-\mathrm{B}}}\, \mathrm{I_t^{1997:7}} + \underset{(3.28)}{1,783\, \mathrm{I_t^{1999:7}}}$$

$$+ \underset{(3.07)}{1,665\, \mathrm{I_t^{1999:3}}} + \mathrm{N_t}$$

$$\underset{\text{(t-values):}}{\nabla\nabla_{12}\mathrm{N_t}} = (1 - \underset{(-16.04)}{0.782}\ \mathrm{B})\ (1 - \underset{(-11.78)}{0.751}\mathrm{B}^{12})\mathrm{a_t}$$

With $\sigma_a = 63.10$ and $\mathrm{I_t^{y:m}} = 1(y\text{:m}) = 1$ and zero otherwise.

Therefore the model estimated corresponds to ARIMA $(0, 1, 1)(0, 1, 1)_{12}$ in levels and with no mean, where several adjustments are made to isolate 'outliers effect'. Figure 8 displays the residuals obtained.

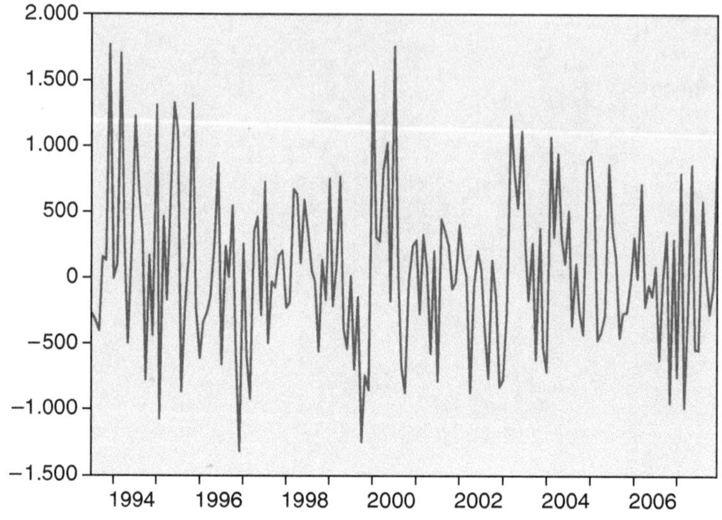

Fig. 8 Liquid bulk model residuals

3.1.4 Containered General Cargo (CC)

The model obtained is:

$$\underset{\text{(t-values):}}{CC_t} = \underset{(2.51)}{5.05} + \underset{(5.85)}{1{,}477}\ I_t^{2006:12} + N_t$$

$$\underset{\text{(t-values):}}{\nabla\nabla_{12}N_t} = (1 - \underset{(-15.02)}{0.756}\ B)\ (1 - \underset{(-9.96)}{0.676}B^{12})a_t$$

With $\sigma_a = 2.87$ and $I_t^{2006:12} = 1(2006{:}12) = 1$ and zero otherwise.

Accordingly, the model fitted corresponds to ARIMA $(0, 1, 1)(0, 1, 1)_{12}$ in levels and with mean, where several adjustments are made to isolate 'outliers effect'. Figure 9 displays the residuals obtained.

3.1.5 Non-Containered General Cargo (NC)

The model obtained is:

$$\underset{\text{(t-values):}}{NC_t} = \underset{(3.58)}{16.348}TD_t + \underset{(7.60)}{1119}\ I_t^{1994:11} + \underset{(5.26)}{765}\ I_t^{2003:6}$$

$$+ \underset{\text{(t-values):}}{N_t} \left(1 + \underset{(10.33)}{0.728}B + \underset{(6.24)}{0.439}B^2\right)\nabla\nabla_{12}N_t = (1 - \underset{(-8.73)}{0.751}B^{12})a_t$$

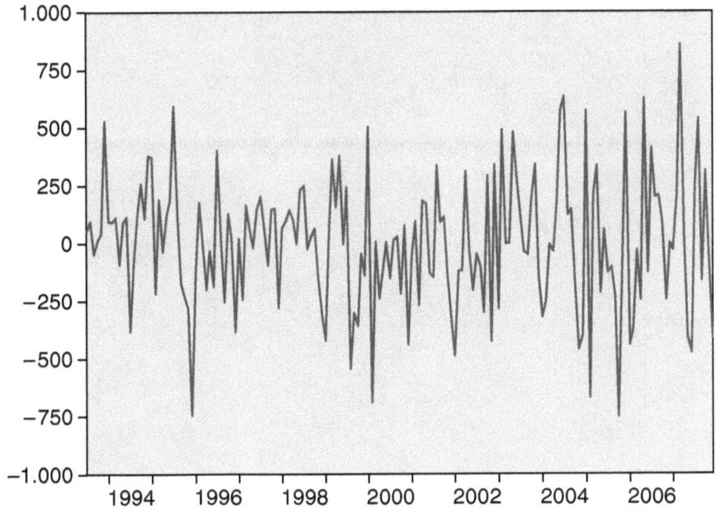

Fig. 9 Containered general cargo model residuals

With $\sigma_a = 4.48$ and $I_t^{y:m} = 1(y{:}m) = 1$ and zero otherwise and with significant trading day (Working days) effect:

$$\begin{aligned} TD_t = {} & (\text{Number of working days (M, T, W, Th, F)} \\ & - \text{Number of weekend days(Sat, Sun)}) \times 5/2. \end{aligned}$$

Therefore the model fitted corresponds to ARIMA $(2, 1, 0)(0, 1, 1)_{12}$ in levels and with mean, where several adjustments are made to isolate 'outliers effect'. Figure 10 displays the residuals obtained.

3.2 *Residual Diagnostics*

One important feature of what we are doing is to present the results of a diagnostic test in order to verify the appropriateness and accuracy of these models obtained for five data series – total cargo, solid bulk, liquid bulk, containered and non containered general cargo.

Summary diagnostics for residuals of the five sea traffic series models are presented in Table 1 and all diagnostics are acceptable. The residuals can be comfortably accepted as zeromean, uncorrelated, normally distributed, with zero skewness and kurtosis equal to 3; they do not contain residual seasonality, nor nonlinearity of the ARCH-type, and their signs are randomly distributed. Therefore, the residuals are free of linear structure and all five models for maritime transport are very parsimonious and provide a good fit.

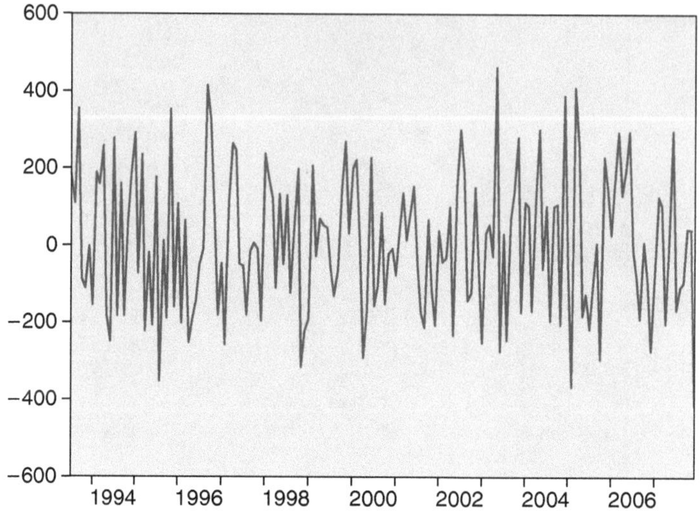

Fig. 10 Non-containered general cargo model residuals

4 Testing for Nonlinearity and Chaos

In this section, we test for nonlinearity and deterministic chaos in the maritime traffic. We will apply the test to the residuals of the ARIMA model estimated in Sect. 3 to analyze whether any deterministic nonlinearity remains in the series. The residuals of the model should be in principle linear independent, and therefore any dependency found in the residuals must be due to ignored nonlinearity.

The test chosen for testing nonlinear dependence is the BDS test. Nonlinear dependence may be chaotic (i.e., nonlinear deterministic) or stochastic, and hence is a necessary condition but not sufficient for chaos.

Chaos is able to generate complex behavior, which appears random, thereby representing a change of perspective in the explanation of fluctuations in economic time series. Tests for chaos will determine whether there is a pattern, a complex and deterministic one, in the residuals. If this is the case, there will be room for improvement in forecasting by applying flexible nonlinear models.

The absence of chaos will be implied if it is demonstrated that the nonlinear structure in the data arises from a well known non-deterministic system.[7] In our case, the BDS test is run on the residuals from ARIMA model, and if the null hypothesis has not been accepted, it can be concluded that the ARIMA process is able to explain any non-linear structure in the data.

[7]See Brock and Sayers (1988) and Brock et al. (1993).

4.1 BDS Test

The BDS test[8] is a statistical version of the correlation dimension test for randomness or "whiteness" against the alternative general dependence in a series. Brock et al. (1987) employed the correlation integral to obtain a statistical test that has been shown to have strong power in detecting various types of nonlinearity as well as deterministic chaos. The BDS test can be used for testing against a variety of possible deviations from independence and has been shown to have strong power in detecting various types of nonlinearity as well as deterministic in data.

The test can be applied to a series of estimated residuals to check whether the residuals are independent and identically distributed (iid). For example, the residuals from an ARMA model (as in our case) can be tested to see if there is any nonlinear dependence in the series after the linear ARMA model has been fitted. If the null of iid is rejected, then a general dependence in the residuals exists which may be due to the neglected non-linearity in the estimation process. In this case, further investigation is needed to narrow down the alternative and determine the causes of the failure of the linear process.

The basic concept behind this test is simple. To perform the test,[9] we firstly choose a distance, ε. We then consider a pair of points and if the observations of the series truly are iid, then for any pair of points, the probability of the distance between these points being less than or equal to epsilon will be constant. We denote this probability by $c_1(\varepsilon)$.

We can also consider sets consisting of multiple pairs of points. One way of choosing sets of pairs is to move through the consecutive observations of the sample in order. That is, given an observation s, and an observation t of a series X, we can construct a set of pairs of the following form:

$$[(X_s, X_t), (X_{s+1}, X_{t+1}), (X_{s+2}, X_{t+2}), \ldots, (X_{s+m-1}, X_{t+m-1})]$$

where m is the number of consecutive points used in the set, or "embedding dimension". We denote the joint probability of every pair of points in the set satisfying the epsilon condition by the probability $c_m(\varepsilon)$.

The BDS test proceeds by noting that under the assumption of independence, this probability will simply be the product of the individual probabilities for each pair. That is, if the observations are independent, $c_m(\varepsilon) = c_1^m(\varepsilon)$.

When working with sample data, we do not directly observe $c_1(\varepsilon)$ or $c_m(\varepsilon)$ and we can only estimate them from the sample. As a result, we do not expect this relationship to hold exactly, but only with some error. The larger the error, the less likely it is that the error is caused by random sample variation. Thus the BDS test provides a formal basis for judging the size of this error.

[8]See Brock et al. (1996).

[9]See EViews 6 User's Guide.

To estimate the probability for a particular dimension, we simply go through all the possible sets of that length that can be drawn from the sample and count the number of sets which satisfy the condition. The ratio of the number of sets satisfying the condition divided by the total number of sets provides the estimate of the probability. Given a sample of n observations of a series X, we can state this condition in mathematical notation,

$$c_{m,n}(\varepsilon) = \frac{2}{(n-m+1)(n-m)} \sum_{s=1}^{n-m+1} \sum_{t=s+1}^{n-m+1} \prod_{j=0}^{m-1} I_\varepsilon(X_{s+j}, X_{t+j})$$

where I_ε is the indicator function:

$$\mathrm{I}_\varepsilon(\mathrm{x}, \mathrm{y}) = \begin{cases} 1 & if |x-y| \leqslant \varepsilon \\ 0 & otherwise \end{cases}$$

In summary, this test is based on the concept of correlation integral[10] used in tests for chaos and non-linearity. The statistics $c_{m,n}$ are often referred to as correlation integrals.

We can then use these sample estimates of the probabilities to construct a test statistic for independence: $b_{m,n}(\varepsilon) = c_{m,n}(\varepsilon) - c_{1,n-m+1}(\varepsilon)^m$ where the second term discards the last observations from the sample so that it is based on the same number of terms as the first statistic.

Under the assumption of independence, we would expect this statistic to be close to zero. In fact, Brock et al. (1987) showed that is asymptotically normally distributed. That is, $\left(\sqrt{n-m+1}\right)\frac{b_{m,n}(\varepsilon)}{\sigma_{m,n}(\varepsilon)} \to N(0,1)$ where $\sigma^2_{m,n}(\varepsilon) = 4\left[k^m + 2\sum_{i=1}^{m-1} k^{m-j}c_1^{2j} + (m-1)^2 c_1^{2m} - m^2 k c_1^{2m-2}\right]$ and where c_1 can be estimated using $c_{1,n}$; k is the probability of any triplet of points lying within of each other, and is estimated by counting the number of sets satisfying the sample condition:

$$k_n(\varepsilon) = \frac{2}{n(n-1)(n-2)} \times \sum_{t=1}^{n} \sum_{s=t+1}^{n} \sum_{r=s+1}^{n} [I_\varepsilon(X_t,X_s)I_\varepsilon(X_s,X_r) + I_\varepsilon(X_t,X_r)I_\varepsilon(X_r,X_s) + I_\varepsilon(X_s,X_t)I_\varepsilon(X_t,X_r)]$$

To carry out the test, we must choose ε, the distance used for testing proximity of the data points, and the dimension m, the number of consecutive data points to include in the set. ε is frequently specified as a multiple of the standard deviation of the series. Brock et al. (1993) examine the finite sample distribution of the BDS statistic and find the asymptotic distribution will well approximate the distribution

[10]This concept is used for example by Grassberger and Procaccia (1983).

of the statistic when the embedding dimension is selected to be 5 or lower and is selected to be between 0.5 and 2 standard deviations of the data.

Nonlinearity and therefore chaos will be established if the BDS statistic is significant for a stationary series void of linear dependence. The absence of chaos will be implied if it is demonstrated that the nonlinear structure arises from a known nondeterministic system.

4.2 Analysis of Results

Since we are interested in nonlinear dependence, in Sect. 3 we have removed any linear dependence in the stationary data by fitting the best possible linear model. Although the residuals of the ARIMA are uncorrelated, their variance may not be constant over time. Therefore in Sect. 3, as first step for checking if some nonlinearity not captured by the linear model exists, we have carried out on the residuals of the ARIMA model, the McLeod-Li test (McLeod and Li 1983) on linearity of the process versus bilinear or ARCH-type structures.

To shed more lights on the underlying data generating process of the maritime traffic, we carried out BDS test for nonlinearity and deterministic chaos in the residual series. Table 2 reports the BDS test results on the residuals of the ARIMA models for the five maritime traffic series (1992–2007).[11]

As the results of the BDS test applied to the residuals of the five ARIMA models presented in Table 3 show, we do not reject the null of iid for the residuals of ARIMA models. This reveals that nonlinear structures do not remain in the residuals and that all linearity from the data has been eliminated appropriately applying ARMA filter. Hence the results of the BDS test also indicate that it does not follow a chaotic process.

5 Summary and Conclusions

In this paper, we have analyzed the statistical features of the monthly maritime traffic series in 1992–2007. Moreover, we have examined the time series structure of sea transport service demand for the existence of chaos and nonlinearity. If the data generating process of the maritime traffic was nonlinear, the traditional modeling for estimation and forecasting would be misleading.

Before conducting such a nonlinear analysis, the data were rendered stationary and appropriately filtered, in order to remove any linear as well as nonlinear stochastic dependence. We firstly estimate the most robust ARIMA model for the five monthly maritime traffic series – total cargo, solid bulk, liquid bulk,

[11]We used EViews 6 for fitting of BDS test statistics.

Table 2 BDS statistics

ARIMA residuals	€/σ	M							
		2		3		4		5	
		BDS stati.	z-stat. prob.	BDS stati.	z-stat. prob.	BDS stati.	z-stat. prob.	BDS stati.	z-stat. prob.
Total cargo	0.5	0.00055	0.340	−0.00049	−0.495	0.00001	0.016	0.00016	0.839
			0.734		0.621		0.988		0.401
	1.0	0.00292	0.746	0.00020	0.044	−0.00007	−0.017	0.00050	0.162
			0.455		0.965		0.986		0.872
	1.5	0.00289	0.632	0.00098	0.135	0.00283	0.327	0.00727	0.803
			0.527		0.893		0.744		0.422
	2.0	0.00331	1.028	0.00226	0.371	0.00376	0.437	0.00834	0.784
			0.304		0.711		0.662		0.433
Solid bulk	0.5	−0.00255	−1.278	−0.00257	−1.150	−0.00064	−1.101	0.00014	0.587
			0.201		0.251		0.271		0.557
	1.0	−0.00458	−0.958	−0.00813	−1.433	−0.00550	−1.090	−0.00338	−0.859
			0.338		0.152		0.276		0.390
	1.5	−0.00682	−1.363	−0.01458	−1.830	−0.01273	−1.338	−0.00985	−0.991
			0.173		0.067		0.181		0.322
	2.0	−0.00515	−1.568	−0.00986	−1.588	−0.00917	−1.044	−0.00698	−0.642
			0.117		0.112		0.296		0.521
Liquid bulk	0.5	−0.00153	−0.744	−0.00133	−0.992	0.00053	0.809	0.00042	1.520
			0.457		0.321		0.418		0.129
	1.0	−0.00381	−0.786	−0.00199	−0.341	0.00300	0.573	0.00328	0.795
			0.432		0.733		0.567		0.427
	1.5	−0.00151	−0.295	0.00036	0.043	0.00875	0.873	0.01584	1.493
			0.768		0.966		0.383		0.135
	2.0	−0.00403	−1.122	−0.00461	−0.675	0.00292	0.301	0.01166	0.965
			0.262		0.500		0.763		0.335
Containerized general cargo	0.5	0.00139	0.569	0.00262	1.624	0.00210	2.619	0.00094	2.697
			0.569		0.104		0.009		0.007

	1.0	0.00245	0.453	0.00445	0.680	0.00204	0.344	0.00164	0.347
			0.651		0.497		0.731		0.729
	1.5	0.00542	0.972	0.01390	1.539	0.01357	1.239	0.01325	1.140
			0.331		0.124		0.215		0.254
	2.0	0.00117	0.319	0.00591	0.848	0.00759	0.766	0.00884	0.718
			0.750		0.396		0.443		0.473
Non-container.	0.5	−0.00052	−0.546	−0.00072	−1.231	−0.00018	−0.678	−0.00012	−1.151
general cargo			0.585		0.218		0.498		0.250
	1.0	−0.00002	−0.006	0.00056	0.170	0.00278	0.970	0.00242	1.116
			0.995		0.865		0.332		0.265
	1.5	−0.00653	−1.607	−0.00413	−0.645	0.00164	0.217	0.00367	0.469
			0.108		0.519		0.829		0.639
	2.0	−0.00717	−2.242	−0.00506	−0.838	0.00203	0.238	0.00675	0.637
			0.025		0.402		0.812		0.524

Note: m is the embedding dimension; ε the distance used for testing proximity of the data points and is equal to 0.5, 1, 1.5 and 2 times the standard deviation of the residual series

Table 3 ARIMA fit: summary diagnostics for different ARIMA fit models

	Total cargo	Solid bulk	Liquid bulk	Containers	Non containers	CV (95%)
$t(\mu_a)$	1.94	0.95	1.37	0.14	0.32	$\|t\| <2$
$Q_a(24)$	25.74	19.88	12.09	21.36	29.91	<34
N_a	1.2	0.27	5.09	0.67	0.49	<6
Skewness	0.15	−0.05	0.42	0.02	0.03	$\|Sk\| <2 \times SE$
	(SE = 0,19)	(SE = 0,19)	(SE = 0,19)	(SE = 0,19)	(SE = 0,19)	(<0.38)
Kurtosis	2.73	2.84	3.09	3.30	2.74	$\|Ku\| <3+2 \times SE$
	(SE = 0.37)	(SE = 0.37)	(SE = 0.37)	(SE = 0.37)	(SE = 0.37)	(<3.74)
$Q_{as}(2)$	3.24	0.10	0.10	0.10	0.62	<6
$Q_{as}(24)$	17.55	23.32	25.15	21.61	20.20	<34
t_a(runs)	0.45	0.15	0.01	−0.76	0.60	$\|t\| <2$

Note:

$t(\mu a)$ is the *t*-value associated with H_0: the mean of the residuals is zero

$Q_a(24)$ is the Ljung-Box (Ljung and Box 1978) test for residual autocorrelation, computed with 24 autocorrelations (in all cases, asymptotically distributed (a.d.) as $\chi^2(22\ d.f.)$)

N_a is the Bowman-Shenton test (Bowman and Shenton 1975) for Normality of the residuals (a.d. as $\chi^2(2\ d.f.)$)

Sk is the value of skewness statistic (residuals): ($|Sk| <2 \times SE$)

Ku is the value of kurtosis statistic (residuals) ($|Ku| <3+2 \times SE$)

Q_{as} (2) is the Pierce test (Pierce 1978) for the presence of seasonality in the residual autocorrelation, (a.d. as $\chi^2(2\ d.f.)$)

Q_{a2} (24) is the McLeod-Li, (McLeod and Li 1983) test on linearity of the process versus bilinear or ARCH-type structures (a.d. as $\chi^2(22\ d.f.)$)

ta (runs) is the *t*-value associated to H_0: signs of the residuals are random.The 95% critical value for each test is given in the last row

containered and non containered general cargo – at Spanish ports during the 1992–2007 period in which we have carried out a number of interventions to isolate effects such as calendar, Easter, leap year, and outliers.

To shed more light on the underlying data generating process of the maritime traffic, we carried out BDS test for nonlinearity and deterministic chaos. From the results of the BDS test applied to the residuals of the five ARIMA models we do not reject the null hypothesis of independent and identically distributed for the residuals of ARIMA models. This reveals that nonlinear structures do not remain in the residuals and that all linearity from the data has been eliminated appropriately applying ARMA filter. Hence the results of the BDS test also indicate that it does not follow a chaotic process and we can conclude that the hypothesis of the nonlinearity and chaotic processes can be strongly rejected for a number of maritime traffic series studied.

References

Batchelor R, Alizadeh A, Visvikis I (2007) Forecasting spot and forward prices in the international freight market. Int J Forecast 23(1):101–114

Bell WR, Hillmer SC (1983) Modelling time series with calendar variation. J Am Stat Assoc 78(383):526–534

Box GEP, Jenkins GM (1970) Time series analysis: forecasting and control. Holden-Day, San Francisco

Box GEP, Tiao GC (1975) Intervention analysis with applications to economic and environmental problems. J Am Stat Assoc 70:177–193

Box GEP, Jenkins GM, Reinsel GC (1994) Time series analysis, forecasting and control, 3rd edn. Prentice Hall, Engle Woods, New Jersey

Bowman K, Shenton L (1975) Omnibus test contours for departures from normality based on b1 and b2. Biometrika 62:243–250

Brock WA, Dechert W, Scheinkman J (1987) A test of independence based on the correlation dimension. Unpublished Manuscript (Technical Report 8702), University of Wisconsin, Madison, University of Houston, and University of Chicago

Brock WA, Sayers CL (1988) Is the business cycle characterized by deterministic chaos? J Monet Econ 22:71–90

Brock WA, Hsieh DA, LeBaron B (1993) Nonlinear dynamics, chaos, and instability: statistical theory and economic evidence. MIT Press, Cambridge, MA

Brock WA, Dechert WD, Scheinkman J, LeBaron B (1996) A test for independence based on the correlation dimension. Econom Rev 15:197–235

Castillo-Manzano JI, Lopez-Valpuesta L, Perez JJ (2008) Economic analysis of the Spanish port sector reform during the 1990s. Transp Res Part A 42:1056–1063

Chang I, Tiao GC, Chen C (1988) Estimation of time series parameters in the presence of outliers. Technometrics 30(2):193–204

Chen C, Liu LM (1993) Joint estimation of model parameters and outlier effects in time series. J Am Stat Assoc 88:284–297

Coto-Millan P (1995) The conditioned demands of Spanish sea transport 1975–1990. Int J Transp Econ XXII(3):325–346

Coto-Millan P, Baños-Pino J (1996) Derived demands for general cargo shipping in Spain: 1975–1992, an economic approach. Appl Econ Lett 3:175–178

Coto-Millan P, Baños-Pino PJ, Villaverde J (2004) Determinants of the demand for maritime imports and exports. Transp Res-E 41:357–372

Cullinane K, Mason KJ, Cape M (1999) A comparison of models for forecasting the Baltic freight index: Box-Jenkins revisited. Int J Marit Econ 1:73–86

Frank M, Stengos T (1989) Measuring the strangeness of gold and silver rates of return. Rev Econ Stud 56(4):553–567

Gómez V, Maravall A (1994) Program TRAMO: time series regression with ARIMA noise, missing observations, and outliers, instructions for the user, EUI Working Paper Eco No. 94/31, Department of Economics, European University Institute

Gómez V, Maravall A (1996) Programs TRAMO and SEATS. Instructions for the user. Working Paper 9628, Research Department, Bank of Spain

Gómez V, Maravall A (1998) Guide for using the programs TRAMO and SEATS. Working Paper 9805, Research Department, Bank of Spain

Gómez V, Maravall A (2000a) Automatic modelling methods for univariate series. Working Paper 9808, Research Department, Bank of Spain

Gómez V, Maravall A (2000b) Seasonal adjustment and signal extraction in economic time series. Working Paper 9808, Research Department, Bank of Spain

Gómez V, MaravalL A, Peña D (1999) Missing observations in ARIMA models: skipping approach versus additive outlier approach. J Econom 88:341–364

Grassberger P, Procaccia I (1983) Measuring the strangeness of strange attractors. Physica 9:189–208

Hillmer SC, Tiao GC (1982) An Arima-model based approach to seasonal adjustment. J Am Stat Assoc 77:63–70

Kavussanos MJ, Nomikos NK (2000) Constant vs. time varying hedge ratios and hedging efficiency in the BIFFEX market. Transp Res Part E 36:229–248

Li J, Parsons MJ (1997) Forecasting tanker freight rates using neural networks. Marit Pol Manag 24:9–30

Ljung GM, Box GEP (1978) On a Measure of lack of fit in time series models. Biometrika 65:297–303

Maravall A, Pierce DA (1987) A prototypical seasonal adjustment model. J Time Ser Anal 8:177–193

McLeod AI, Li WK (1983) Diagnostic checking ARMA time series models using squared residual autocorrelations. J Time Ser Anal 4:269–273

Mostafa MM (2004) Forecasting the Suez canal traffic: a neural network analysis. Marit Pol Manag 31(2):139–156

Pierce DA (1978) Seasonal adjustment when both deterministic and stochastic seasonality are present. In: Zellner A (ed) Seasonal analysis of economic time series. U.S. Department of Commerce, Washington DC, pp 242–269

Scheinkman JA, LeBaron B (1989) Nonlinear dynamics and stock returns. J Bus 62:311–337

Serletis A, Gogas P (1997) Chaos in East European black-market exchange rates. Res Econ 51:359–385

Veenstra AW, Haralambides HE (2001) Multivariate autoregressive models for forecasting seaborne trade flows. Transp Res Part E 37:311–319

Part II
Supply

Productivity in Maritime Transportation

María Jesús Freire

Abstract This study aims to analyze basic maritime traffic in the area of the economy and its relationship with growth. The content analysis includes an introduction to fundamental concepts of economic theory and its application to maritime business especially with regard to productivity, the costs involved and the difference in the supply of tonnage by type of vessel in response to the cycle time global economic.

1 Introduction

The widely accepted definition of production, from an economic point of view, relates to the idea that production is any process that increases the suitability of goods to meet human needs. This concept covers not only the strict sense corresponding to the technical point of view, but also, more broadly, everything that facilitates the use of these goods in terms of time or space.

Production is defined as any process that converts or transforms goods into other different goods. Normally, the goods and services used in the production process are called factors of production. These include both the factors of production and the raw materials and intermediate products. The goods and services produced are called products.

In economics, the word 'production' covers a field of activity that is much broader than the usual concept. For an economist, production is any process that converts or transforms goods or services into other different goods or services. It follows from this approach that all of the processes listed below are production:

M. Jesús Freire
University of La Coruña, A Coruña, Spain
e-mail: maje@udc.es

P. Coto-Millán et al. (eds.), *Essays on Port Economics*, Contributions to Economics,
DOI 10.1007/978-3-7908-2425-4_7, © Springer-Verlag Berlin Heidelberg 2010

1. Transportation of goods between two places.
2. Storage of goods.
3. Disembarkation, packing and retail.

Equally, in economics goods can be both physical objects and services or intangible objects.

2 The Production Function in the Company

Production is not simply the manufacture of physical objects, but also transportation, storage and sale, as well as intangible services. The concept of production or production activity requires the intervention of humans who know how the factors of production operate, and also some kind of equipment, which must be prepared before the process begins and which, in general, is described as capital by economists.

Production is analysed on the basis of the complete recipe, or in other words by determining the factors, including the work, capital and land, and the ingredients or raw materials needed to obtain a given amount of the product. This complete recipe defines a production process or activity and its characteristics can be specified in terms of: (1) whether by doubling the amount of factors used, it is possible to double the product; (2) whether there is more than one way of obtaining the product; (3) whether it is possible to obtain different amounts of products with different processes.

The list of all known processes is the available technology. Evidently, this number is now very high and nobody can know all of the technology of an advanced society, but we can presume that, in the interests of production, the starting hypothesis is that technological information transmission is perfect, so that the available technology is the same for all companies within the economy. The direction of technological evolution and its past and future expansion in particular is a historic question. This technological variation is an important factor and must be taken into account in the theories of long-term economic growth, but the typical microeconomic context tends to refer to the period in which the available technology is considered fixed. This means that, in the current state of affairs, a technically inefficient process will never be used. By the same token, processes that are technically inefficient are completely supplanted by the more efficient ones and can be forgotten without the scope of the available technology being reduced.

The acquisition of the factors and their introduction into the production process is carried out by the economic production unit, which generally are firms. These production units do not produce the products for their own consumption; they sell them on the market. Consequently, their aim is not to meet their own needs, but to obtain a profit, hence they can be described as profit-making economic units, since their objective is to make a return.

The production function can be defined as the relationship between physical quantities of factors of production in the production process and the quantity of product obtained, for a given technology. It can be expressed by means of the following function:

$$X = F(f_1, f_2, \ldots . f_n) \tag{1}$$

where: F is the technology used. X is the product obtained. f_1, f_2,. . .f_n are the quantities of the various factors of production used.

This relationship between the product obtained and the factors of production used is defined by technical laws. It is a technological relationship. If the technology advances, a large quantity of product can be obtained with the same quantity of factors of production, although if this is the case the production function changes.

Firms, when it comes to deciding their production quantity, face a number of alternatives, which forces them to make a number of unalterable decisions. Certain choices must be made within a short space of time. For instance, in response to an increase in demand, the firm can increase shipped goods and improve their utilisation of premises. Another series of choices have a broader timeframe for decision-making, such as changing technology and expanding or building new plants.

This range of decisions can be grouped into three timeframes:

- Short term, in which the plant is not modified, just used in a better way.
- Long term, in which it is possible to modify plants and premises with the existing technology.
- Very long term, in which everything changes, including the technology.

The short term is the period of time in which some factors of production cannot be altered. In fact, factors can be classed as fixed or variable. The fixed factors are those that the company must have, regardless of the quantity produced. Normally they are part of the firm's capital. The variable factors are those whose use can vary in the short term and they depend on the quantity produced.

Within a firm, as long as there is at least one fixed factor, the company is operating in the short term. However, the short term is not necessarily a defined as a short period of time, weeks or months; it will depend on the type of company. The long term is the period in which all of the factors of production can vary, without changing the existing technology. There are no fixed factors, they are all variable. Likewise, it does not correspond to a specific period of time. In the long term, the firm, in its planning decisions, can choose any proportion of factors of production, although once the choice is made the firm is once again operating in the short term.

In the very long term, it is assumed that the basic technology is also variable. One of the most important characteristics of society in recent years has been the rapid technological changes, with the emergence of new raw materials and new and more efficient manufacturing methods.

In the analysis, the concept of economic production efficiency must be taken into consideration, which is a relative concept. It is obtained through comparison with other available alternatives, taking into account the resources used to obtain the results. It consists, therefore, of an economic concept that is justified by the traditional shortage of resources that have alternative uses. It is not of an absolute nature, since it is determined by the existing alternatives. When it comes to evaluating the normal economic activity of firms, the economy ensures efficiency in the production process, since it has proven to be effective.

Productive efficiency is the combination of factors chosen that lead to an output of goods and services. The science of economics is concerned with studying the efficiency with which firms achieve this production. However, technical efficiency is a multi-faceted process, given that there are not one but several types, depending on the objective proposed by the company. Thus, we can talk about cost efficiency, income efficiency and profit efficiency. To evaluate these kinds of efficiency, information on the market prices is needed, since these determine the optimum action to take in each case. However, whatever the criteria guiding the action taken by firms, and whatever the current prices, there is a basic kind of efficiency, unrelated to any kind of economic objective, which consists of the adequate use of the resources utilised; this kind of efficiency is called technical efficiency.

Technical efficiency was defined as the feasible combination of inputs and outputs which is technologically impossible to improve, without reducing the output or increasing inputs. Other authors propose both the construction of a resource utilisation coefficient and assignative efficiency. Both of these authors have contributed significantly to economic theory, making them the most influential scholars in the field of productive efficiency. This methodology is supported by recent literature and has undergone significant growth in recent years. Despite this development, efficiency methodology is still a long way from being a closed field of study.

Furthermore, it is important to note that, in production, the factors and products are flow variables, or in other words they are measured in quantities per unit of time. However, the choice of a given period of time to represent the process imposes a restriction on firms with regard to the various combinations of factors that can be used to produce, unless the period under consideration is very long. Therefore, the production function expresses the functional relationship that exists between the quantities of factors used by the firms and the quantity of product obtained.

Moreover, since the firm's profit is the difference between its income and its expenses, these terms should be analysed in order to know what variables affect the profit. The income obtained by companies for the sale of their goods or services constitutes the gross takings that are known on occasions as the volume of sales or turnover. This amount is equal to the price of the product multiplied by the quantity of product produced and sold. The total cost of producing this quantity also varies for each volume, but there is a part of this cost that the company has to meet whether it produces or not: the fixed costs, also known as the unavoidable costs.

In any country or region, a capitalist company, by producing goods or services, incurs a series of costs, which it will have to assess properly in order to ensure that the difference between the income generated by the sale of the product and the total costs involved in its production is as great as possible. The impact of production costs on company profits occurs through various channels, the two most significant of which are: when the costs have repercussions for the final price of the products, in which case the rise in prices could reduce demand and consequently the total income and when the rise in costs does not have an impact on the price, in which case the profits will be reduced by the same amount in which the costs increase.

The basic decision that all companies must make is how much to produce, and this depends on the selling price and the cost of production. The firm's desire to maximise profits leads it to determine the quantity of product that will be placed on the market.

The basic function of the company is to transform the factors of production and convert them into goods and services suitable for consumption or investment. The main objectives that companies pursue are:

1. To obtain maximum profits.
2. To expand activity.
3. To satisfy customers and suppliers.

The firm normally takes advantage of the benefits of mass or large-scale production. To do so, it combines the necessary resources and organises all of its production activities. These tasks are the responsibility of the company administrators and, at the heart of the company, the administrators organise the production and make the business decisions, for the purposes of which they acquire the necessary market information.

A concept of great significance, from a business point of view, relates to the efficiency of a production unit. This can be defined as the quality of the business initiative that consists of minimising the use of resources to achieve a given objective.

Inefficiencies can be defined as costs that exceed the minimum cost achievable, and they can be explained by one or more of the following causes. Firstly, there is technical or productive inefficiency. This is reduced whenever possible by increasing the production of goods without increasing consumption of resources, or by reducing the production of other products. Secondly, price inefficiency occurs when, given the price of the products and factors, as well as the marginal productivities of the factors, the choice of produced quantities is not optimal.

Neoclassical economic theory shows that, under certain circumstances, the economy will position itself at a Pareto Optimum. The necessary conditions for this include the maximisation of profits as the objective of producers, which in turn implies production efficiency. The assumptions of perfect competition are a working hypothesis and do not fully materialise in the real world; however, a failure to fulfil a certain initial hypothesis does not completely invalidate the conclusions, and studying the reality according to this competition model can be of value from a normative point of view.

3 The Concept of Total, Average and Marginal Productivity

In this introductory analysis, it is assumed that the product increases whenever all of the factors increase, thereby obtaining different quantities of the total product. The total product can be defined as the total amount produced by using all of the factors in a certain period of time. If the quantities of all the factors used are constant with the exception of one, then the result is defined as total productivity and, in this case, it will vary in accordance with the amount of the variable factor that is used.

$$P_t = P\ (f_1) \tag{2}$$

where: P_t is the total productivity; f_1 is the variable factor.

The concept of productivity in economics explains the quantity of goods and services produced by a factor in a given period of time. Productivity is positively correlated with economic growth and variations in productivity generate important economic differences.

One interesting interpretation is to examine how the quantity of the product obtained varies when the quantities of one factor vary and the others are kept constant. In this case, the productivity of the factor, in a generic sense, is defined as the capacity or aptitude that it possesses to produce goods. However, logically, the productivity provided by a factor varies according to the quantities of the other factors used, which are assumed to be constant in each case, even if they can vary at times.

In countries where the workers produce a large amount of goods per working hour, the situation leads to the majority of citizens enjoying a good quality of life. However, in countries where the production system does not develop or is conducive to low productivity among workers, the majority of citizens will have a low capacity to acquire goods and services.

This important relationship leads us to the pivotal question of economic growth, in which the aim is to know which factors determine the capacity to produce more or less goods and services. In this regard, the key variables are: access to better technology, having ideal machines and tools, having intellectually prepared and well-trained workers, having good administrators and using efficiently designed production systems. The economic reality experienced by the various countries and regions of the world has been extremely varied. The factors that determine productivity and growth rates are, as mentioned above, a combination of universal factors and ones that are specific to each country.

The average productivity is the total product divided by the quantity of the variable factor used, which can be expressed as:

$$P_t^* = \frac{P(f_1)}{f_1} \tag{3}$$

where: P_t^* is the average productivity of the variable factor and f_1 is the quantity of the variable factor used.

The marginal product is the variation of the total product resulting from using an additional unit of the variable factor, while the quantities of the other factors remain constant. The most common expression is:

$$P_t' = \frac{\Delta P(f_1)}{\Delta f_1} \tag{4}$$

where: P_t' is the marginal productivity and $\frac{\Delta P(f_1)}{\Delta f_1}$ is the amount by which the total productivity increases every time the variable factor is increased by one unit.

Evidently, the total, average and marginal productivity of a factor are of different values depending on the quantities applied. Thus, the relationship that exists between the quantity of the product obtained and the amount of resources used is defined as the average or unitary productivity of a factor of production. This concept of productivity is generally what is being referred to in the modern campaigns that advocate an increase in productivity. The marginal productivity of a factor is the limit of the relationship between the product increase and the increase in this factor, when this increase is close to zero and the other factors remain constant or, roughly, the increase in product due to the last unit of the respective factor used.

Among the factors used to obtain a product, there are some that possess the property that with each dose of the factor no more than a certain quantity of product can be obtained, no matter how much the other factors' use is increased.

Between the applied quantities of these factors and the quantities of product obtained there is, therefore, a biunique correspondence; or in other words, each quantity of product can only be obtained with a given quantity of each factor and that with each dose of these used factors only a given quantity of the product can be obtained. The factors that possess this property are called limiting factors because, without increasing the quantity used, the quantity of product obtained cannot be increased; the factors that do not possess this quality are called replaceable factors, because they are not indispensable, in given doses, for obtaining a quantity of product, but are replaced by others.

In addition to these extreme cases, to distinguish whether the factors are complementary or substitutive, Pareto and Edgeworth's norm on the second mixed derivative can be applied to verify that it is positive if the factors are complementary, zero if they are independent and negative if they are substitutive.

The function of total productivity can adopt various forms in its initial phases:

(a) Growing and concave downwards
(b) Linear
(c) Concave upwards

However, whatever the initial form, after the point of inflexion it always becomes concave downwards. Thus, in all cases, starting from a given value, the marginal productivity decreases, at least for quantities of factors used that exceed certain quantities (Table 1, Graphs 1 and 2). This is the important economic law of diminishing marginal productivity, confirmed through experience, which is logically justifiable because there is always a combination of factors that brings about the most favourable output. Consequently, if one of the factors is used in a greater proportion, the increase obtained with each additional unit, or in other words the marginal productivity, gradually diminishes.

If this law was not fulfilled in nature, the composition of the world would be radically different to its current one. All factors of production would be replaceable without restriction, in which case none would possess their own value, since they

Table 1 Average productivity, marginal productivity

Units of variable factor L	Total product P_t	Average productivity p_t^*	Marginal productivity P_t'
1	12	12.00	
2	28	14.00	16.00
3	52	17.33	24.00
4	74	18.50	22.00
5	92	18.40	18.00
6	104	17.33	12.00
7	114	16.29	10.00
8	120	15.00	6.00
9	122	13.56	2.00
10	116	11.60	−6.00

Source: Compiled by author

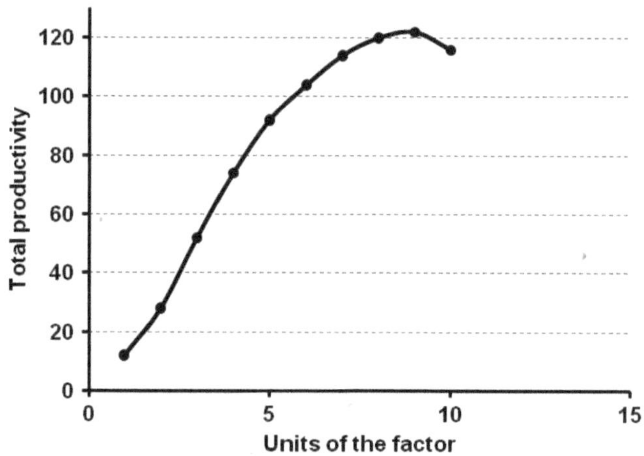

Graph 1 Total productivity

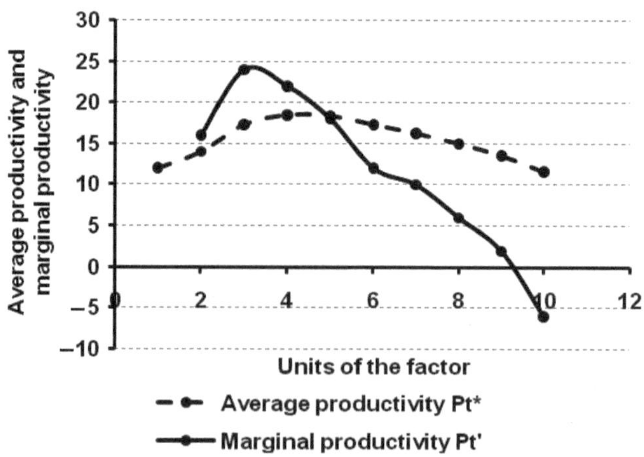

Graph 2 Average productivity and marginal productivity

could perfectly well be replaced by others. The opposite situation can also be analysed, in which, to obtain any quantity of product, a precise quantity of each factor must be used; in this case, if all the factors are limiting, the problem of production is resolved entirely within the technical sphere.

The technical scope is much smaller that it would be in all other cases of replaceable factors. To obtained a product, the laws of chemistry require the application of these components in precise proportions, but the manufacturing process can be carried out according to different procedures or, even when following the same procedure, using different equipment or processes, which broadens the scope of the technical indeterminacy, which in turn is precisely what is resolved by applying the principle of economics. This principle is applied wherever there is technical indeterminacy and wherever there is a variety of combinations of factors with which a given volume of production can be obtained, which, in this case, is resolved by choosing the least expensive combination.

If the target is the desired quantity of product obtained, the chosen combination will be the cheapest from among all those that can produce the desired quantity of the product. Production theory puts a premium on the hypothesis that the most favourable combination is denoted the breakeven point as the result of the contrast between the tendency to obtain the largest quantity of product and lowest cost, which act as opposing forces.

In reality, this situation expresses the law of equally weighted marginal productivity, which is fundamental in production theory. The most intuitive, albeit approximate, approach is that the marginal productivity of a factor of production is defined as the increase in output due to the last unit of the respective factor used, when the quantities of the other factors remain invariable. This is the physical marginal productivity, and by dividing this by the price of the factor in question, which is the number of monetary units expressing the value of the unit of the factor, we obtain what we call the weighted marginal productivity, or in other words the increase in product corresponding to the increase in the factor by a quantity whose value is the monetary unit.

The law of equally weighted marginal productivity expresses that the last monetary unit invested in each of the factors gives rise by itself to the same increase in the product. The businessperson in charge of production looks for the most favourable combination of factors, which is the one in which the last monetary unit invested in each factor gives rise to the same increase in product.

4 The Concept of Production in Maritime Transportation

An analysis of the production function in maritime economics reveals significant differences compared to production in other economic sectors. In this case, an important element is the diversity of the sectors that are engaged in trading goods and the one-to-one relationship that exists between cargo, transportation and distribution. Furthermore, each of these production systems has its own unique features and must be broken down into the type of goods loaded and unloaded, the type of ship, country of origin and destination of the maritime trade.

To study closely the production of world maritime transportation from an economic point of view, the decision was made to analyse, initially in a disaggregated manner, the type of goods loaded and unloaded. Moreover, a detailed analysis was carried out specifying the percentage of the various goods within the total traffic. Imports and exports by country group are also broken down according to the various classifications and the percentage of each group within the total is analysed.

Table 2, Graph 3 presents a breakdown of the data on loaded goods by product type, according to whether the goods are oil, bulk goods or other goods, from 1980 to 2007. An analysis of the figures reveals that oil's share in world trade has changed, both in absolute terms and in proportion to the total. In 1980 it represented 37.27% of the loaded goods and its share diminished to 33.42% in 2007, while dry cargo increased from 62 to 66.58% during the same period. Maritime traffic, for all goods, increased year on year with some minor exceptions: petroleum cargo in 2000 and 2001 fell year on year by between 0.5 and 0.75%. In the case of bulk goods, the reductions took place in 1992 with a fall of 1.53% and in 1995 with a less

Table 2 World maritime traffic. Goods loaded (millions of tonnes transported)

Year	Oil	Variation %	Bulk cargo	Variation %	Other goods	Variation %	Total	Variation %
1980	1,227		745		1,320		3,292	
1985	1,263		752		1,370		3,385	
1987	1,283		788		1,390		3,461	
1988	1,367	6.55	848	7.61	1,460	5.04	3,675	6.18
1989	1,460	6.80	875	3.18	1,525	4.45	3,860	5.03
1990	1,526	4.52	881	0.69	1,570	2.95	3,977	3.03
1991	1,573	3.08	927	5.22	1,610	2.55	4,110	3.34
1992	1,624	3.24	913	−1.51	1,660	3.11	4,221	2.70
1993	1,714	5.54	915	0.22	1,710	3.01	4,339	2.80
1994	1,796	4.78	1,021	11.58	1,870	9.36	4,687	8.02
1995	1,870	4.12	1,019	−0.20	1,970	5.35	4,859	3.67
1996	1,929	3.16	1,093	7.26	2,070	5.08	5,092	4.80
1997	1,937	0.41	1,170	7.04	1,966	−5.02	5,073	−0.37
1998	1,965	1.45	1,196	2.22	2,008	2.14	5,169	1.89
1999	2,027	3.16	1,288	7.69	2,119	5.53	5,474	5.90
2000	2,017	−0.49	1,331	3.34	2,165	2.17	5,513	0.71
2001	2,002	−0.74	1,383	3.91	2,210	2.08	5,595	1.49
2002	2,085	4.15	1,475	6.65	2,280	3.17	5,840	4.38
2003	2,113	1.34	1,475	0.00	2,545	11.62	6,133	5.40
2004	2,265	7.19	1,587	7.59	2,690	5.70	6,542	6.70
2005	2,308	1.90	1,686	6.24	2,790	3.72	6,784	3.70
2006	2,595	12.44	1,876	11.27	3,181	14.01	7,652	12.80
2007	2,681	3.31	1,997	6.45	3,344	5.12	8,022	4.83

Source: Lloyd's Statistical Tables
Compiled by author

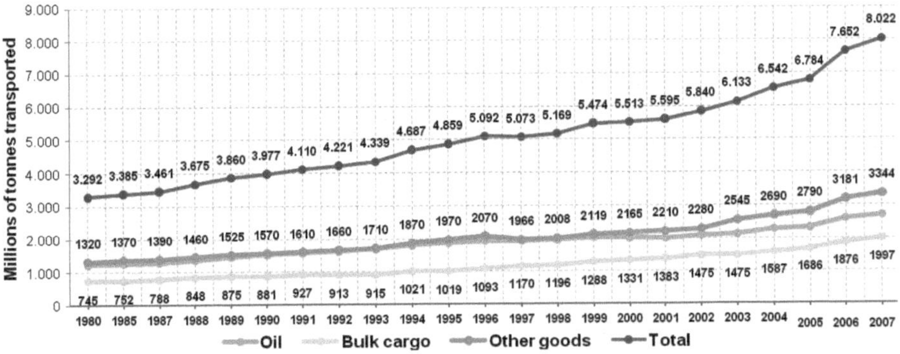

Graph 3 World maritime traffic. Goods loaded

Table 3 World maritime traffic by cargo type (world total in millions of tonnes)

Year	Goods loaded				Goods unloaded			
	Oil		Dry cargo	Total	Oil		Dry cargo	Total
	Crude oil	Derivatives		loaded	Crude oil	Derivatives		unloaded
1970	1,109	232	1,162	2,504	1,101	298	1,131	2,529
1980	1,527	344	1,833	3,704	1,530	326	1,823	3,679
1990	1,287	468	2,253	4,008	1,315	466	2,365	4,126
1998	1,548	534	3,549	5,631	1,515	548	3,855	5,918
1999	1,553	532	3,598	5,683	1,543	510	3,955	6,007
2000	1,605	544	3,736	5,885	1,633	513	4,097	6,242
2001	1,678	497	3,717	5,891	1,702	552	3,913	6,167
2002	1,629	500	3,819	5,948	1,713	550	4,014	6,276
2003	1,690	533	4,257	6,480	1,743	536	4,324	6,603
2004	1,783	534	4,527	6,845	1,807	557	4,528	6,893
2005	1,856	565	4,686	7,108	1,853	573	4,696	7,122
2006	1,802	792	5,057	7,651	1,929	839	4,993	7,761
2007	1,866	815	5,341	8,022	1,963	839	5,230	8,032

Sources: United Nations Conference on Trade and Development (UNCTAD). Compiled by author

abrupt fall of 0.20%. As far as other goods are concerned, the rates were positive throughout the period analysed with the exception of 1997, when the drop reached a substantial 5.29%; this affected all loaded goods to produce a negative variation. Every other year the quantities increased and 2006 in particular saw exceptional two-digit increases for all goods types.

Table 3, Graph 4 presents the total loading and unloading figures for oil (crude oil and derivatives) and dry cargo worldwide. The worldwide figures are presented and corroborate the fact that the loaded and unloaded petroleum has reduced proportionately within maritime traffic as a whole, going from 55.3% in 1970 to 33.42% in 2007. The figures for unloaded petroleum behaved in a similar manner and the claim can be made that maritime traffic has moved away from petroleum and towards other goods.

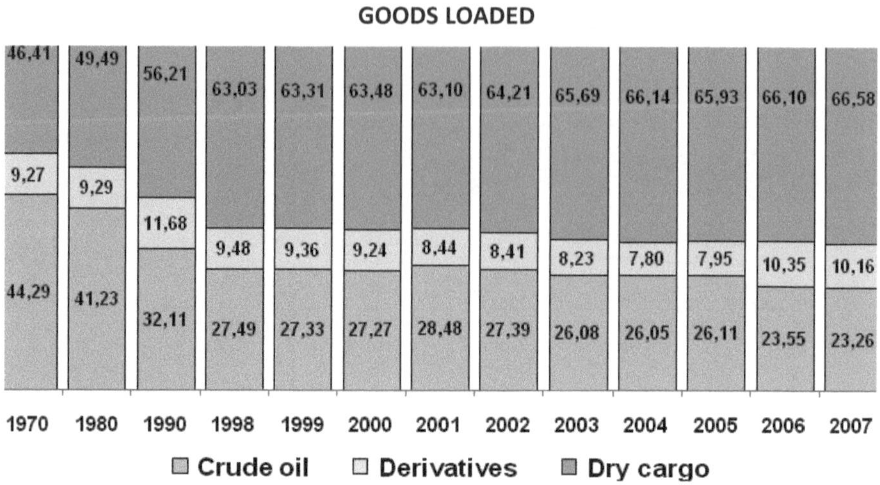

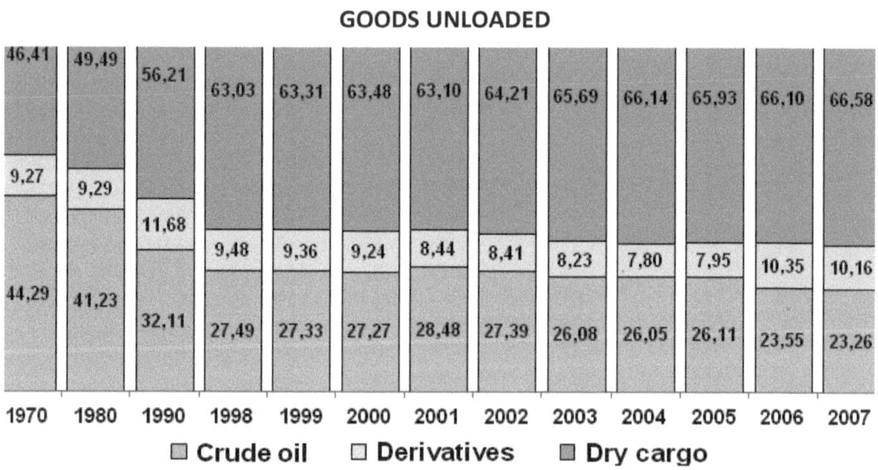

Graph 4 World maritime traffic by cargo type. Percentage distribution

5 Factors of Production in the Maritime Transportation Industry

The maritime transportation industry, like industry in general, uses a wide range of different factors of production, and in many production processes the products of an industry are used as factors for other industries. However, in essence, the economic expression can be generalised by saying that there are three factors of production in the maritime transportation industry: the work factor, the capital factor (the ship and its components) and a broader concept which incorporates all of the finished products of other industries used as factors of production in this sector.

This means that, if the aim is to increase production quickly in the industry, there will be some of these factors that cannot be increased in the short term and the increase in production will only be possible using greater quantities of variable factors.

These factors that can be increased in the short term are called variable factors and they are labour, raw materials and also some material resources whose utilisation in better conditions can be feasible in a short period of time. However, by increasing the quantity of a factor used, while all the other factors remain constant, the 'law of diminishing marginal returns' comes into play. According to this law, starting from a given rate of increase of the variable factor, diminishing quantities of product are obtained.

In the decision-making process and when it comes to demanding a specific factor like capital, companies compare the benefits and costs associated with the decision to undertake the project. The elements included in the investment's cost function are the price of the capital goods to be acquired and the quantities used, as well as the financing costs that would be incurred if external financing were used.

The benefits, in theory, materialise with the flow of future net returns that the investment is expected to generate. The difficulty that all of these firms have in calculating the flow of returns is owed to the fact that many risk factors must be taken into account. Another drawback is that all sectors are interdependent, or in other words the development of a sector is closely related to what happens in the other sectors. Thus, an increase in the use of a factor in a given sector, while all the other factors remain constant, can lead to diminishing returns in the industry under certain circumstances. From this perspective, a company of a certain sector that requires capital will invest if the net present value of the expected returns exceeds the cost of acquiring the capital goods.

When there are a limited number of firms in each sector in the market, they become aware of their strategic interdependence. Any decision adopted by a company affects the situation of the others and it is very likely that they will react to the initial decision of the leading firm, altering the situation initially seen as the starting point. This is why, logically, in worldwide maritime transportation, oligopolistic firms, when it comes to making decisions, take into account the foreseeable reaction of their competitors.

6 The Capital Factor: The World Fleet by Ship Type

To analyse the impact of the economic factors that affect production and growth in maritime trade, firstly one must analyse what the theory has to say about the subject, before incorporating the available information in order to cross-reference it. In particular, this study analyses the world fleet as the most important factor of production in the long term.

Before presenting the capital factor in maritime transport, the units of measurement and the most important ship types, as well as the goods they can transport, must be defined. As far as the units of measurement are concerned, the most commonly used are deadweight tonnage (DWT), metric tonnes of cargo (t) and the volume of a ship in gross tonnage (GT).

The most widely used unit in maritime transportation is the deadweight tonnage (DWT), defined as the sum of all the weights transported by a ship, including the cargo, the ballast (if there is any), fuel and lubricants, water for boilers and drinking water and a quantity known as K (which comprises the sum of all the weights generally considered to be constant, such as crew, supplies and replacement parts). Shipyards usually also use a unit of measurement known as lightship displacement, which is the bare weight of the ship as it comes off the slipway. This concept includes the basic machine fluids, such as boiler water and the oil inside the hydraulic systems.

The unit of cargo weight most commonly used in maritime trade is the metric tonne. This unit of measurement equivalent to 1,000 kg (approximately 2,500 pounds) denotes a unit of mass in the decimal metric system and the current International System of Units. Its official abbreviation is *t*.

In addition, for many years shipyard output was measured in gross tonnage (GT). This unit of measurement is used to represent the load capacity, or in other words the capacity allocated to the ship after discounting weight from the tonnage, which until a few years ago varied from country to country, but are now the same regardless of the ship's nationality. The standard international measurement for the size of a ship, under the Universal Tonnage Measurement System, defined in the Tonnage Regulations of 1969, is the gross tonne. Gross tonnage (GT) is not a measurement of weight but of volume (2.78 m^3).

The various stages of the construction of a ship as a means of transport can help to clarify the various units of measurement that the complementary information can refer to. When the ship has just been launched and is at the shipyard's wharf, the unit used would be the lightship displacement. Then the owner provisions the ship with the elements needed for it to sail, in safe stability conditions, such as the necessary ballast, fuel, replacement water, drinking water and the K constant (namely crew, supplies and replacement parts), whereupon the unit used would be the ship's deadweight.

Ship design also now plays an important role in the economics of maritime transportation, since it can be very specific, although in general terms designs can be defined as liquid and solid cargo. The type of naval construction included in the liquid cargo group is characterised by the fact that the goods must be transported in sealed containers so that they cannot leak out accidentally. The need for specialisation in accordance with the product to be transported is much more common for these ships. In these cases the priority, for both economic and technical reasons, is to know what the vessel will initially be used for. The shipowner that requires a specific type of transport and the shipyards that specialise in building ships of certain specifications can benefit enormously if it is stipulated in the construction contract what the vessel will transport in the future – a great deal of interior space and materials can be saved.

In the case of ships for solid cargo, they can transport a wide variety of cargo types and are limited only by the demands of the insurance companies and by the requirements imposed both to enter ports and to pass through certain canals. This situation gives rise to various transportation conditions particular to each case, taking into account the characteristics of the product to be transported and its possible packaging. From a physical point of view, in general we can distinguish between two types of shipload: the bulk carriers and the multipurpose cargo ships, which will be classed as different types of ship here.

Table 4 and Graph 5 present the figures for the distribution of the world fleet by ship type from 1980 to 2007. The ships are categorised as: oil tankers, bulk carriers

Table 4 Distribution of the fleet by ship type (millions of DWT)

Year	Oil tankers	% of total	Bulk carriers	% of total	Other	% of total	Total	Annual variation rate
1980	320	46.50	200	29.00	170	24.50	690	
1990	240	36.50	236	35.80	182	27.60	658	
1998	280	35.50	275	35.00	233	29.50	788	
1999	282	35.00	276	34.50	241	30.50	799	1.40
2000	282	35.20	276	34.80	242	30.00	800	0.13
2001	285	35.27	281	34.78	242	29.95	808	1.00
2002	286	34.62	295	35.71	245	29.66	826	2.23
2003	304	36.02	300	35.55	240	28.44	844	2.18
2004	317	36.99	307	35.82	233	27.19	857	1.54
2005	336	37.50	321	35.83	239	26.70	896	4.55
2006	354	36.90	346	36.00	260	27.10	960	7.14
2007	382	36.73	367	35.29	291	27.98	1,040	8.33
2008	408	36.53	391	35.00	318	28.47	1,117	7.40

Sources: United Nations Conference on Trade and Development (UNCTAD). Compiled by author

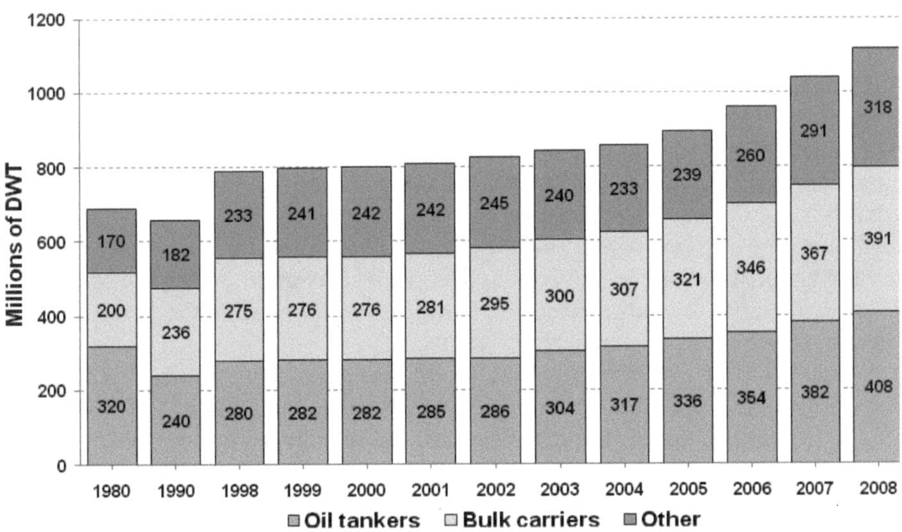

Graph 5 Fleet distribution by ship type

and others (cargo ships, container ships etc). The available figures on the world fleet of merchant ships show that they transported 1,117 million deadweight tonnes (DWT) in the year 2008. This figure amounts to a variation rate of 6.89% compared to the previous year and a growth trend has been observed since 2001 with increasing variation rates. The tonnage of the world fleet increased from 1980 despite the slump that the fleet underwent in the late 1980s. In 1990 the DWT (658 million DWT) of the world fleet was considerably lower than 10 years earlier (690 million DWT). In general terms, it can be said that this increase in the size of the merchant fleet also increased the average capacity of the transportation. In particular, the biggest increase has taken place among container ships, both in terms of the number of ships and their capacity. These modifications have reaffirmed the growing proportion of products that are generally transported in this kind of ship.

The figures presented in the above table show that in the period under consideration there has been a substantial redistribution of the fleet between the various ship types. The fleet of oil tankers in 1980 was of 320 million DWT and currently it is of 408 million DWT, or in other words it has increased by just 88 million DWT in the 29 years that have gone by. These ships have reduced their share in the world total from 46.5% in 1980 to 36.53% in 2008. The bulk carrier fleet has increased from 200 million DWT in 1980 to 391 million DWT in 2008 and its share in relative terms also increased, from 29% in 1980 to 35% in 2004. The fleet made up of other types of ship behaved in a similar manner to the bulk carriers: their capacity increased from 170 million DWT in 1980 to 318 DWT in 2008 and their share in the total increased from 24.5 to 28.47%.

Table 5 and Graphs 6 and 7 present the figures for the evolution of the world merchant fleet from 1980 to 2007. The unit of measurement used is the volume of cargo, or the tonnage. This series of figures clearly shows the reduction in the oil tanker fleet throughout the 1980s. It shrank from 175,004 million tonnes in 1980 to 123,726 million tonnes in 1989, or in other words an average annual fall of 3.25%. For bulk carriers the figures were good in the early 1980s, but from 1987 the trend changed and they underwent a gradual reduction, going from 131,000 million tonnes to 129,482 in 1989. However, since 1990 the growth has been almost continuous, with just two exceptions in 1998 and 1999. With regard to other types of ship, the trend has been one of growth throughout the period, with a very high annual growth average of 4.65%. The general state of affairs, however, has been of a more moderate growth of the fleet, with an average annual rate of 2.11%. This situation has to a large extend been the result of the fall in the number of oil tankers registered, in particular in the 1980s.

The 1990s were a decade of growth for all ship types and logically this increase is reflected in the total: in a period of 15 years the fleet grew from 423,627 million tonnes in 1990 to 633,321 million tonnes. By ship type, the oil tankers have undergone annual growth of 2.03%, the bulk carriers a little over 2.59% and, lastly, the other types of ship grew by 5.5%.

Graph 7 shows the figures for the age of the world fleet and demonstrates that the ships have aged over the 28 years under analysis. The fleet went from an age of

Table 5 World merchant fleet and average age of ships (millions of gross tonnes)

Year	Oil tankers	Bulk carriers	Other	Total	Average age
1980	175,004	109,600	135,307	419,911	15
1985	134,861	134,000	147,408	416,269	15
1986	124,140	132,900	147,870	404,910	16
1987	122,718	131,000	149780	403,498	17
1988	122,388	129,635	151,383	403,406	17
1989	123,726	129,482	157,273	410,418	17
1990	128,678	133,190	161,759	423,627	17
1991	132,438	135,885	167,703	436,026	18
1992	138,149	139,042	167,977	445,169	17
1993	143,077	140,915	173,922	457,914	18
1994	144,595	144,914	186,350	475,859	18
1995	143,521	151,694	195,447	490,662	18
1996	146,366	157,382	204,125	507,873	19
1997	147,108	162,169	212,920	522,197	19
1998	151,036	158,565	222,292	531,893	19
1999	154,092	158,958	230,560	543,610	20
2000	155,429	161,186	241,439	558,054	20
2001	156,068	168,000	250,483	574,551	20
2002	154,559	169,954	261,070	585,583	22
2003	159,273	173,071	272,874	605,218	21
2004	165,345	181,444	286,532	633,321	22
2005	174,467	193,213	307,436	675,116	22
2006	182,380	204,827	334,648	721,855	22
2007	193,059	206,981	355,261	755,301	22

Source: Lloyd's Statistical Tables, compiled by author

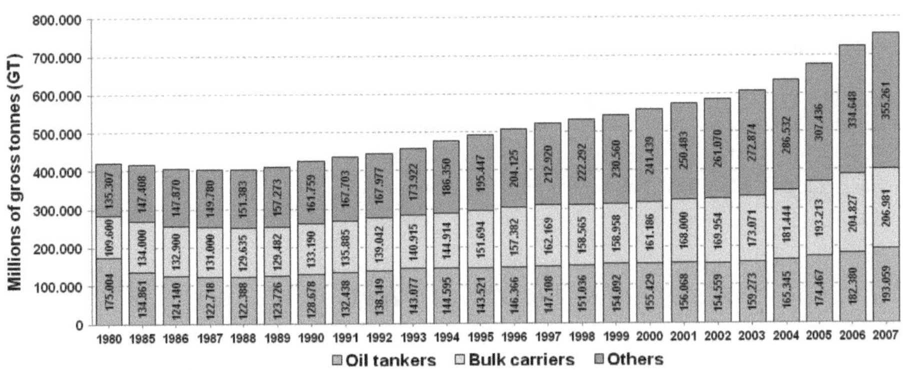

Graph 6 Distribution of the fleet by ship type

15 years in 1980 to 20 years in 1999 and by 2007 they had reached 22 years. The current state of affairs has been caused by variables that affect maritime trade and that are interrelated, including the rise in the cost of shipbuilding, the reduction in freight prices, the uncertainty of the oil market and the increase in Asian trade.

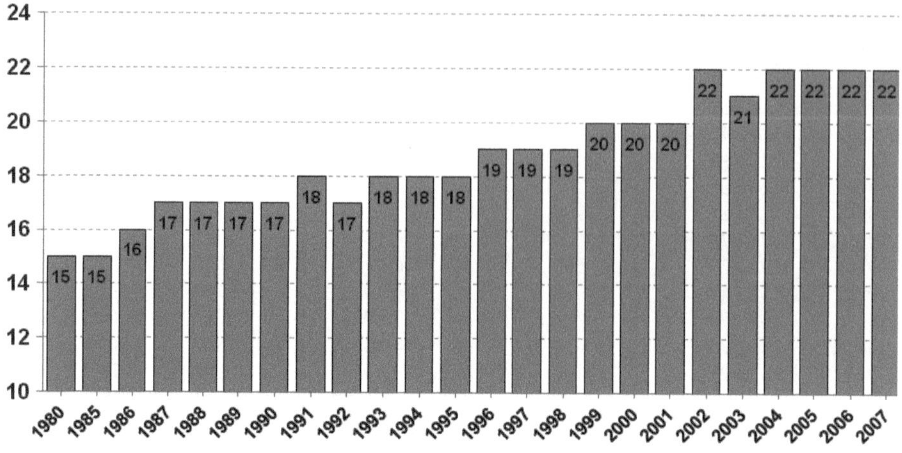

Graph 7 Average age of ships

Table 6 World maritime traffic in tonnes per mile (billions of tonnes transported per mile)

Year	Oil	% variation	Bulk cargo	% variation	Other goods	% variation	Total	% variation
1970	6,487		2,049		2,118		10,654	
1975	9,727		2,826		2,810		15,363	
1980	9,405		3,652		3,720		16,777	
1985	5,157		4,480		3,428		13,065	
1990	7,821		5,259		4,041		17,121	
1995	9,170		5,953		5,065		20,188	
2000	10,265		6,638		6,113		23,016	
2001	10,179	−0.84	6,782	2.17	6,280	2.73	23,241	0.98
2002	9,898	−2.76	6,879	1.43	7,395	17.75	24,172	4.01
2003	10,580	6.89	7,454	8.36	7,810	5.61	25,844	6.92
2004	11,235	6.19	8,065	8.20	8,335	6.72	27,635	6.93
2005	11,749	4.57	9,119	13.07	8,730	4.74	29,598	7.10
2006	12,130	3.24	9,976	9.40	9,341	7.00	31,447	6.25
2007	12,440	2.56	10,827	8.53	9,665	3.47	32,932	4.72

Source: Lloyd's Statistical Tables, compiled by author

6.1 The Productivity of the World Fleet

The concept of the productivity of the world fleet denotes the quantity of transportation services occurring in a given period of time. Productivity is positively correlated with economic growth and the variations in productivity generate important economic differences. To analyse productivity, the main productivity indicators for the world fleet are needed, and these are: tonnes of cargo, tonnes per mile and the deadweight tonnes of the fleet.

Table 6, Graph 8 presents the data on the tonnes of cargo per mile travelled in relation to the world fleet of oil tankers, bulk carriers and other merchant ships, as

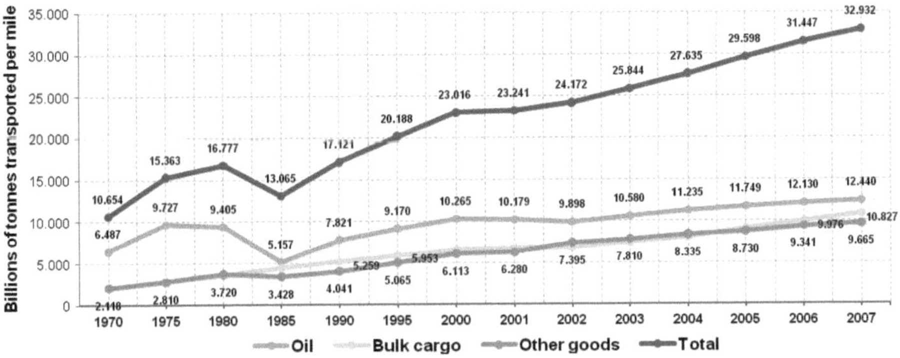

Graph 8 World traffic in tonnes/miles

Table 7 Average productivity of the fleet and the transported cargo

Year	World fleet (in millions of DWT)	Total transported cargo (millions of tonnes)	Tonnes per mile (billions)	Tonnes of transported cargo/ DWT of fleet	Tonnes per mile /DWT of fleet
1970	640.8	2,566	10,654	4.00	16.63
1980	690.3	3,704	16,777	5.37	24.30
1990	658.4	4,008	17,121	6.09	26.00
1993	710.6	4,330	18,235	6.09	25.66
1995	734.9	4,651	20,188	6.33	27.47
1997	775.9	4,953	21,825	6.38	28.13
1999	799.01	5,668	24,114	7.09	30.18
2000	808.4	5,871	23,016	7.26	28.47
2001	825.7	5,840	23,241	7.07	28.15
2002	844.2	5,888	23,251	6.97	27.54
2003	857.15	6,480	25,844	7.56	30.15
2004	895.8	6,758	27,635	7.54	30.85
2005	960.5	7,109	29,045	7.40	30.24
2006	960	9,652	31,447	8.00	32.80
2007	1,040	8,022	32,932	7.70	31.60

Sources: United Nations Conference on Trade and Development (UNCTAD). Compiled by author

well as the variation rates from 1970 to 2007. During this time, there was a significant crisis in the 5-year period from 1980 to 1985, in particular for oil tankers (with average annual variation rates of 16.47%) and other merchant ships (with a rate of 1.70%), and as a result the total value of world maritime traffic reduced in average terms by 5.70%. In spite of the significance of this contraction, in the 38 years under analysis the index that reflects the cargo transported per mile travelled has grown by an average of 5.50%.

Table 7, Graph 9 reflects the information obtained on the tonnes of cargo, the miles travelled and the world fleet in DWT; these variables, in general terms, have grown over the period under analysis with slight decreases in certain years. The

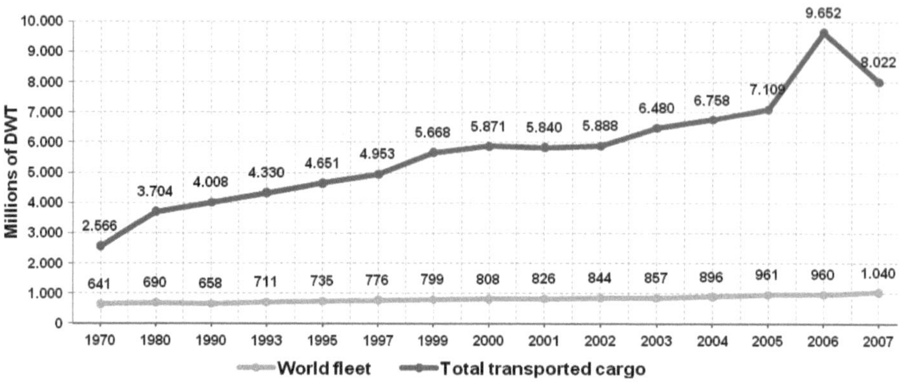

Graph 9 Average productivity of the fleet and transported cargo

average productivity measured by the tonnes of total transported cargo divided by the deadweight tonnes of the world fleet has grown from a rate of 4 in 1970 to 7.7 in 2007. The figure for the productivity of cargo has fluctuated significantly in recent years; in 2003, for instance, the highest level, with a productivity of 8, was reached, making it the best year in the history of maritime transportation. The falls in productivity that have taken place in recent years must be seen in relative terms and are related to the reduction in cargo transported in comparison to the enlargement of the fleet.

There is also an emphasis on analysing the performance of the merchant fleet in terms of the concept of average productivity as the quotient of the tonnes of cargo transported per mile divided by the deadweight tonnes of the world fleet. This indicator grew from 1970 to 1999 when it reached a maximum value of 30.18. In the 38 years that have gone by, the value of the index went from 16.63 to 31.60, but this indicator is highly volatile, much more so than the previously analysed indicator, with consecutive maximums and minimums until the year 2007. These changes can be explained in part by the slumps in maritime traffic.

Table 8, Graphs 10 and 11 show the productivity figures broken down into the type of cargo transported and the type of ship (i.e. oil tankers, bulk carriers and the rest of the fleet) for the 1970 to 2007 period. In the case of oil tankers and bulk carriers, productivity fell drastically between 1970 and 1980 due to the major transport crisis, in particular because of the situation arising in the Persian Gulf. The productivity index for oil tankers reduced from 9.74 to 5.51, a situation explained by the loss of cargo suffered in this period by the supertankers and an existing fleet that could not be phased out in the short term. The situation also intensified for bulk carriers, with a considerable reduction from 6.21 in 1970 to 4.29 in 1980. The rest of the fleet maintained an upward trend in this period. Average productivity grew from 6.38 to 6.57. In the following decades, the oil tankers and bulk carriers maintained a very similar trend with continual growth in productivity, but without reaching 1970 levels. In these two cases the indicators reached values of 7.66 and 4.67 in the year 2000, while for the rest of the fleet the pattern remained

Table 8 Estimated productivity of the world fleet by ship type (tonnes transported per DWT)

Year	Tonnes of oil transported in ships of over 50,000 DWT (millions)	Tonnes of oil transported/ DWT of tanker	Tonnes of bulk cargo transported/ DWT (millions)	Tonnes of bulk cargo transported/ DWT of bulk carrier	Tones transported by rest of fleet (millions)	Tonnes transported by rest of fleet/ DWT of rest of fleet
1970	1,442	9.74	448	6.21	676	6.38
1980	1,871	5.51	796	4.29	1,037	6.57
1990	1,755	7.14	963	4.13	1,285	7.23
1998	1,985	7.10	1,137	4.40	2,379	10.20
1999	1,995	7.10	1,167	4.50	2,375	9.90
2000	2,163	7.66	1,288	4.67	2,532	10.53
2006	2,595	7.33	1,876	5.42	3,181	12.24
2007	2,681	7.00	1,997	5.43	3,344	11.76

Sources: United Nations Conference on Trade and Development (UNCTAD). Compiled by author

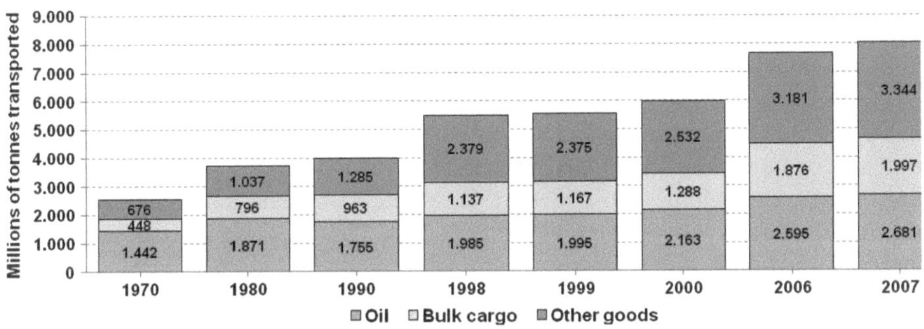

Graph 10 Estimated productivity of the world fleet by ship type

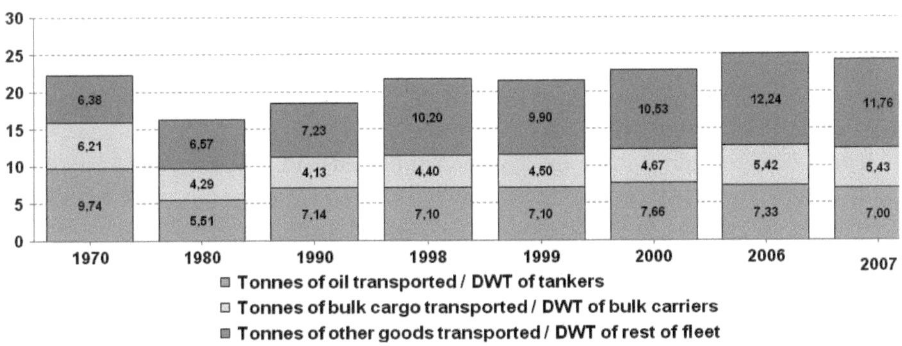

Graph 11 Tonnes transported/DWT of the ships

one of growth throughout the entire period, reaching a productivity of 10.53 in 2000. In the remaining years until 2007, the three indices became very volatile with rises and falls each year.

7 The Concept of Production Costs

The production cost concept refers to the sum of a firm's spending to supply itself with the factors of production it needs. The price that the firm must pay per unit factor must be good enough to attract the additional (marginal) unit of factor.

Within this framework of equally weighted marginal returns, the quantity of each factor used is a function of the quantity of product that the firm aims to obtain. The cost of each quantity of product is expressed in terms of the factors, or in other words the lowest cost with which this quantity can be obtained when the prices of the factors are given. However, when a company produces a product, some of the means of production used tend to be determined in fixed quantities. This gives rise to so-called fixed installations, wherein the firm cannot alter the applied quantities of the factors of production involved in the processes, which leads to fixed costs.

The variable replaceable factors can also give rise to fixed costs, when to obtain any quantity of product, no matter how small, certain minimum quantities of these factors must be used. These fixed costs, caused by the replaceable factors, are always reflected, so those that derive from fixed factors must be added, together with those corresponding to the limiting factors, to obtain the production costs function.

Production costs vary according to various factors, the most significant of which are: the volume of production, the variations in the price of the factors of production and the technical variation. Of all of the factors, the most influential as far as cost is concerned is the volume of production. In these circumstances the variation of the cost in accordance with production is distinguished by two types of cost: short-term and long-term costs.

It is common, due to the nature of facilities, for firms' adaptation to new quantities of product to be only partial; or in other words, when there are variations in the volume of production, the quantities of some of the factors used cannot be altered, or at least not immediately. In general, when these variations occur, the firm considers the quantities that it must use of certain factors as fixed, and partially adapts, varying the quantities used of the other factors. Only when the variation in volume becomes permanent and the firm can consider it to be definitive will it vary the quantities of the factors classed as fixed. This is done in the long term rather than immediately.

The total short-term costs are divided into fixed and variable; the former are characterised by their independence from the volume of production and the latter by their close correlation to production. To analyse the short-term costs, the starting hypothesis is that the price of the factors of production is given, or in other words the supply of factors is completely elastic.

The total, average and marginal costs are defined below. The total cost is the cost of producing a given quantity of product (x) and is a function of the product obtained.

$$Ct = f(x) \tag{5}$$

where: Ct is the total costf(x) is the quantity of product obtained.

The form in which this function is adopted derives from the law of diminishing physical marginal returns. The total cost is made up of the fixed cost and variable cost:

$$Ct = Cf + Cv \tag{6}$$

The division of the costs into fixed and variable can only be established in the short term, in which the partial adaptation of the company to the volume of production takes place. In the long term there are no fixed costs; rather, they can all be deemed variable, since the company ensures that all of the factors are applied in the appropriate quantity to fulfil the law of equally weighted marginal productivity, which is essential for ensuring the cost is as low as possible. This is why long-term adaptation is of a complete nature.

The development of total costs is easily analysed bearing in mind that they are obtained by adding together the variable and fixed costs. The resulting S-shaped curve reflects the diminishing returns. In the first section, from zero to xa, the total product grows by a higher proportion than the costs. In the second section, from xa, the costs grow by a higher proportion than the product, due to the law of diminishing returns.

The average costs are obtained by dividing the total cost by the respective volume of production. The average variable cost will also be obtained, bearing in mind that only the variable costs will feature in the numerator. Firms are interested in determining the volume of production which corresponds to the lowest possible average total costs. This minimum coincides when the cost of the last unit and the average cost per unit are the same. The volume of production which entails the minimum total average cost is referred to as the optimal point of production, because it represents the best utilisation that the company can achieve with the factors of production and techniques available.

This concept, which initially seems somewhat abstract, is easily justified by geometrical considerations, bearing in mind that, for this volume of production, the radius vector that measures the total average cost is lower and, in turn, this radius vector coincides with the tangent on the curve that measures the marginal cost or the cost of the last unit produced.

The respective average cost functions are the average total costs, the average variable costs and the average fixed costs.

$$Ct^* = Ct/x \tag{7}$$

where: Ct* is the average total cost. Ct/x is the total cost divided by the number of units produced.

$$Cv^* = Cv/x \tag{8}$$

where: Cv* is the average variable cost. Cv/x is the variable cost divided by the number of units produced.

$$Cf^* = Cf/x \tag{9}$$

where: Cf* is the average fixed cost. Cf/x is the fixed cost divided by the number of units produced.

There is an important correspondence between analyses of production from an economic perspective on the one hand and a technical point of view on the other. A technical analysis easily reveals the line on which technical optimums are located, corresponding to the various output functions, and all one has to do is plot the line corresponding to the relationship between the prices of the factors, assumed to be the data, to select among the points of the line the one that gives rise to the lowest average total cost, or in other words the one that resolves the technical indeterminacy.

In the analysis carried out thus far, we have assumed that the factors used in production could vary continuously. The divisibility of the factors that are involved in the production process would allow the desired quantity of the product to be obtained through the corresponding adaptation of the factors. But within the sphere of the economics of maritime transportation, it is common for some of the factors to be variable only in a discontinuous or staggered manner, and they must necessarily vary only by whole numbers. In this case, for each portion of the discontinuous factor, there is a technical product maximum.

The curve representing the costs in production that uses discontinuous factors is sawtooth-shaped, expressing a relative minimum of the average costs and descending as the combination of factors becomes more harmonious, until the absolute minimum of these costs is reached, corresponding to the harmonious minimum, where all of the factors are used in the right dose.

In general, the curve representing the costs is not reversible, which means that for each volume of production the cost is different, given that the volume is reached by growing from other smaller volumes or shrinking from larger volumes. This relationship is almost always present, especially if discontinuous factors of production are involved, because a degree of delay in the adaptation of certain factors to the variations in the volume of production is commonplace. The most evident example would be when production increases, in which case the delay in appointing new staff or in introducing new elements is compensated by the intensification or effort of the existing factors, which can only be maintained temporarily and scarcely translates into an increase in costs. In contrast, for decreasing outputs, the delay in laying off staff or in reducing other factors causes higher costs. This disparity is referred to as cost hysterisis due to its similarity with the well-known phenomena of magnetism and elasticity.

7.1 Estimation of the Total Costs of the Fleets

The importance attached to business transactions is nothing new and various theories support this consideration in a very positive manner. Under this perspective, open economies grow at a faster pace than closed economies. For some theories, growth is provided by commercial liberalisation, which increases firms' investments by a larger extent so that they can be more competitive. For others, growth is a result of the fact that openness increases the quality and productivity of investments in the most expansive sectors and this attracts a higher volume of foreign investment once the overseas sector has been liberalised.

However, the real-life situation has some gaps that are difficult to fill with the information available, given that the integration of the production and financial markets due to trade should increase the simultaneity of the economic cycles of countries for various reasons. Firstly, better commercial integration brings with it a transmission of the national disturbances of certain countries to others, and consequently the economic cycle of each country becomes much more volatile. Secondly, international trade tends to increase specialisation and each country specialises in the goods and services in which it has a greater comparative advantages. This leads to a situation that augments both individual and common economic disturbances. This change confirms the fact that economic differences between the developed and the developing countries are becoming greater.

In the current context, international trade involves various types of service such as supply, production, marketing, transactions and transportation. Naturally, the comparative advantage is determined by technology more than by the intensity of the capital and manifests itself in the costs that affect imports of all high-quality services. The same can be argued in relation to the cost of maritime transport, in which the more efficient and productive countries, or in other words the most developed nations, incur lower costs than the less developed countries. The advantage of low wages in poor countries can be outweighed by their lower productivity, making convergence unviable.

Table 9, Graph 12 shows the estimated cost of import transportation valued in millions of current dollars, both worldwide and by groups of developed and developing countries, from 1980 to 2007. In the period from 1980 to 2000 the cost of transportation increased by an annual average of 8.40%, from 123,264 million dollars in 1980 to 330,400 million in 2000. However, in the seven remaining years until 2007 the annual average increased to 16.02%, from 384,013 million dollars in 2001 to 836,349 million dollars in 2007.

In 1980 the estimated transportation costs of developed countries accounted for 63.5% of the worldwide total and, 10 years later, in 1990, they increased to 67.6%. But in 2000 this percentage went down to 60.72% and this decreasing trend continued in succeeding years to reach a floor in 2003 of 51.45% of the total worldwide cost. In 2004 there was a very significant upturn among the developed countries with a percentage similar to that of 2000 (61.49%), however, in the remaining years it returned to the value of 53.93% of the worldwide total. In the developing countries, the average annual variation in transportation costs increased by 26.14% over the 28 years under analysis. In the period between 1980 and 2000, the annual average increased to 9.01%, but from 2004 to 2007 the percentage shot up to 28.93%.

Table 9 Estimation of the total cost of import transportation in world trade by country group (millions of US dollars)

Country group	1980	1990	2000	2001	2002	2003	2004	2005	2006	2007
World total	123,264	173,102	330,400	384,013	364,008	379,200	481,800	632,400	727,260	836,349
Developed countries	78,286	117,004	200,800	233,784	221,248	195,100	296,300	341,100	392,265	451,105
Developing countries	44,978	56,098	126,100	150,229	142,760	184,100	173,500	259,900	311,880	374,256
Distributed between:										
Africa	10,432	9,048	12,354	14,447	13,806	17,900	21,600	24,600	29,520	35,424
America	10,929	9,626	26,658	34,624	33,895	39,200	18,100	19,400	23,280	27,936
Asia	21,979	35,054	76,925	98,364	92,023	122,700	146,600	153,000	183,600	220,320
Europe	1,320	1,909	2,103	2,182	2,428	3,500	20,000	24,000	28,800	34,560
Oceania	318	461	638	612	608	800	2,000	800	960	1,152

Sources: United Nations Conference on Trade and Development (UNCTAD). Compiled by author

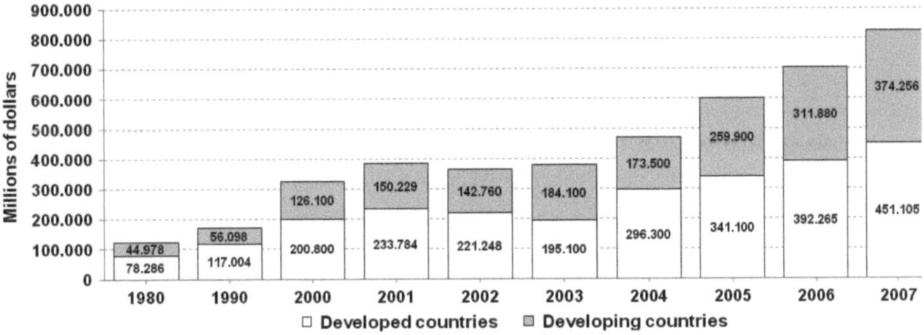

Graph 12 Total cost of import transportation in world trade

8 The Supply and Demand of the World Fleet

In the maritime transportation market, all of the operations of suppliers and clients determine the services that will be provided and the freight charges. Supply and demand are two concepts that economists use with great frequency since they are the forces that drive the market economy. Their interaction in the markets determines the quantities of the various goods and services that will be produced and the price they must be sold at. The factors that affect the quantities supplied and demanded per unit of time are various and determined independently on the market taking into account various parameters.

Buyers and sellers agree on the price of goods or services so that given quantities of these are exchanged for a quantity of money that is also pre-established. The decisions of producers and consumers are coordinated in the exchange. The prices that are determined in the market can, under certain circumstances, lead to a surplus in the capacity of the unused fleet.

The quantities demanded of the goods that the consumers want and can buy are referred to as the demand for the goods in question. There are a number of factors that determine the quantities that consumers wish to acquire of each type of goods per unit of time. It is possible to hypothesise that all of these factors remain constant with the exception of the price of the goods themselves, in which case we are referring to the demand schedule, or in other words the relationship between the quantity in demand and the price.

The demand for maritime transportation is calculated on the basis of the cargo, and logically this varies over time and depends on various factors: the freight charges per mile, the behaviour of world trade, the exchange rates and the behaviour of the world economy.

The quantity of goods offered is what the producers/sellers want to sell and are able to sell. The economic information on the quantity of goods offered and the price is shown in the supply schedule.

The supply schedule, as indicated, is the relationship that exists between the price of goods and the quantities that the firms want to offer of these goods per unit of time, while all other factors remain constant. Logically, global or market supply figures are obtained on the basis of the individual suppliers, adding together the quantities that the firms in the market want to offer for each price. At very low prices production costs are not covered and producers will not produce anything; as the prices increase production is stimulated and at very high prices production will be higher. However, growth in supply has short-term limits due to the law of diminishing returns and the relationship between production factors and the goods produced in the production process. It follows that there are diminishing returns in the production of goods if there is a gradual decline in the quantity of additional product obtained when we add successive equal quantities of certain factors in relation to other factors that remain fixed.

Supply in the maritime transportation industry is represented by the capacity of the existing merchant fleet and newly built ships, minus the ships that have been scrapped in the period under analysis. To analyse the supply, the existing capacity of the fleet according to vessel type must be examined. This function can vary considerably over time.

However, when the quantity in demand is equal to the supply, the equilibrium price and quantity is obtained. This is located at the intersection of the supply and demand curves and has no shortage or surplus. This situation takes place at a given time and any modification in the quantities of supply or demand at a given price will lead to a surplus in supply or demand, normally known as a shortage.

Given an equilibrium price, when the market establishes a lower price there is a surplus in demand, which sets in motion a number of forces that tend to lift the price. Surplus in demand is the situation in which the quantity in demand is higher than the quantity offered. When the price is higher than the equilibrium price there is a surplus in supply and it tends to lower the price. Surplus in supply is the situation in which the quantity offered is higher than the quantity in demand. Thus, in a free market, prices tend to move towards the level of equilibrium.

In economics, market equilibrium is seen as the situation in which there are no inherent forces that induce change. Once this has been reached, only external factors can alter the status quo. Nevertheless, when the factors that underlie the supply or demand curves change, the curves move and changes in the equilibrium price and quantity take place.

In maritime economics, supply is represented by the merchant fleet and demand by the activity of the fleet. If the difference between them is positive, there is a surplus in the fleet capacity. In this market, at different times, there can be a surplus in supply, a surplus in demand or market equilibrium. It all depends on whether the external factors that affect the two functions are modified.

This market, in recent years, has experienced an exceptional climate in which the supply has been close to the demand with a gradual decline in the surplus capacity, so that equilibrium has almost been reached.

Table 10, Graphs 13 and 14 show the figures broken down by vessel type, the evolution of the quantities offered, the surpluses and the surplus ratio with regard to the fleet from 1995 to 2007. In aggregate terms, in this period the merchant fleet

Table 10 Analysis of the supply of tonnage by ship type (millions of DWT)

	Oil tankers			Bulk carriers			General cargo			Unitised ships[a]		
	Fleet	Surplus	Fleet surplus %	Fleet	Surplus	Fleet surplus %	Fleet	Surplus	Fleet surplus %	Fleet	Surplus	Fleet surplus %
1995	277.0	28.8	10.4	252.9	17.9	7.1	62.0	2.0	3.2	143.0	0	0
1996	285.1	28.8	10.1	257.2	17.2	6.7	62.7	1.4	2.2	153.2	0	0
1997	290.6	17.0	5.8	260.9	10.3	3.9	62.0	1.7	2.7	162.4	0	0
1998	291.0	17.3	5.9	257.1	5.8	2.3	60.5	1.6	2.6	180.1	0	0
1999	281.8	14.0	5.0	245.7	7.9	3.2	59.9	1.8	3.0	211.6	0	0
2000	279.4	13.5	4.8	247.7	3.8	1.5	59.3	1.1	1.8	222.0	0	0
2001	280.2	17.9	6.4	255.3	2.9	1.1	57.8	0.7	1.2	232.3	0	0
2002	267.7	19.1	7.1	258.8	2.2	0.9	57.3	0.4	0.7	260.4	0	0
2003	286.0	6.0	2.1	297.5	3.6	1.2	43.4	0.7	1.6	230.1	0	0
2004	298.0	3.4	1.1	307.0	2.5	0.8	43.0	0.7	1.6	247.8	0	0
2005	313.0	4.5	1.4	340.0	2.0	0.6	45.0	0.7	1.6	262.0	0	0
2006	367.4	6.1	1.7	361.8	3.4	0.9	44.7	0.6	1.4	268.4	0	0
2007	393.5	7.8	2.0	393.4	3.6	0.9	43.8	0.7	1.6	287.1	0	0

Sources: United Nations Conference on Trade and Development (UNCTAD). Compiled by author

[a]The fleet of unitised ships comprises: fully cellular container ships, mixed container ships, ro-ro ships and lighter aboard ships

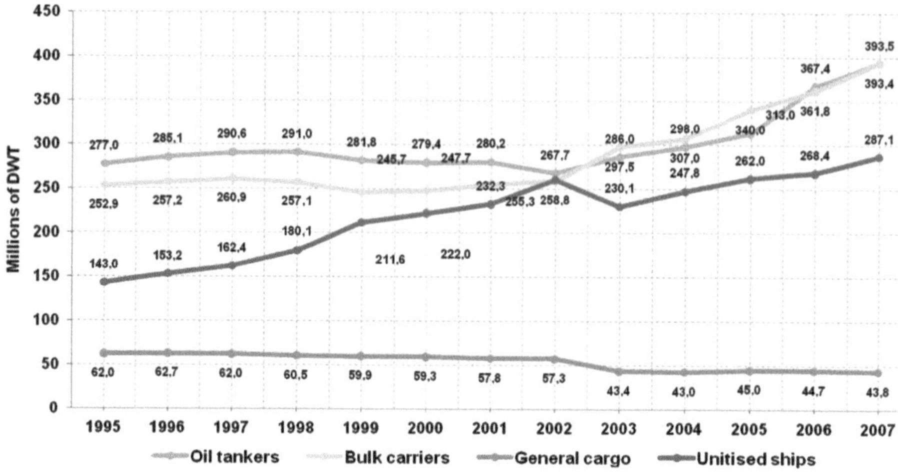

Graph 13 Analysis of tonnage supply by ship type

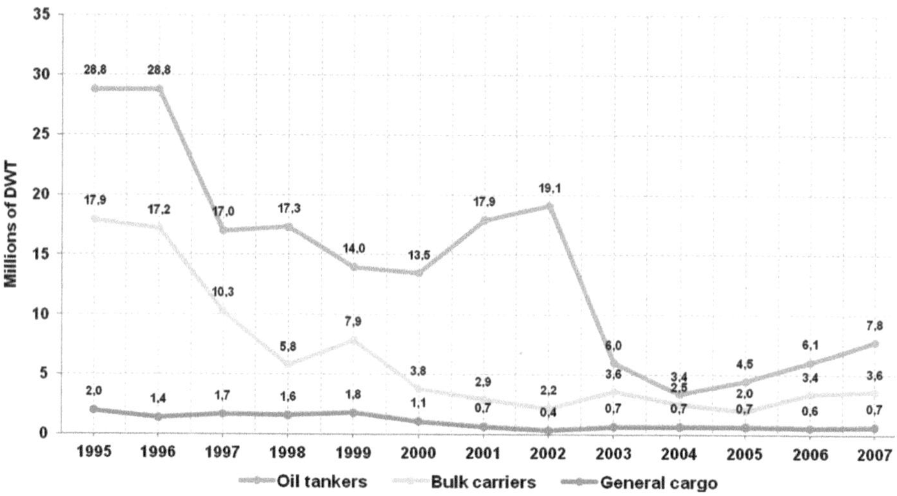

Graph 14 Analysis of surplus by ship type

supply grew by an average annual rate of 3.88% due to the rise in freight charges and the reduction in fleet surplus.

The fleet supply is calculated by taking into account the ships in existence at the beginning of the period, plus the new ships produced by the shipyards, minus the ships scrapped or lost in that year, in millions of DWT. The world fleet went from 658.4 million DWT in 1990 to 1,117.8 million DWT in 2007. This situation is representative of the improvements in maritime transportation activity generated by a widespread growth in all of the sectors that depend directly or indirectly upon it.

The ships have been categorised on the basis of their specialisation: oil tankers, bulk carriers, general cargo ships and unitised ships. The unitised ships comprise: fully cellular container ships, mixed container ships, ro-ro ships and lighter aboard ships. In this context, each ship type's surplus and the relationship between this ship type and the fleet is also analysed.

The situation of oil tankers from 1995 to 2007 is analysed. With this kind of ship, the capacity supply increased from 277 DWT in 1995 to 291 DWT in 1998, with an average annual growth rate of 1.68%, but, from 1999, a decreasing trend began and lasted until 2003, when the cycle changes. During this period the fleet shrank from 281.8 million DWT in 1999 to a floor of 267.7 million DWT in 2002. In this period, it is possible to verify that oil tanker construction went down, while the change in the maritime cycle increased the active fleet. This situation improves the fleet surplus ratio, which reduces, and in the years 2003 and 2004 it reaches figures close to the demand with a fleet surplus of 1.1%. As far as the current situation is concerned, a new crisis is taking place, with increases in surpluses and considerable availability of ships.

The supply of bulk carriers is also analysed. This type of ship behaves similarly to oil tankers, but with some minor differences. In 1998 and 1999, the capacity of the fleet shrank, and then it recouped slightly in 2000, following which it grew with average rates over the 5 years reaching 6%; or in other words, in this period there was a boom in the maritime business with a sufficiently long-lasting shockwave. The surplus capacity shrank over the period under analysis with some volatility, reflected in the figures of 17.9 million inactive DWT in 1995 and 2 million DWT in 2005. This is due to the recovery of world trade and the considerable increase in dry cargo shipping. The surplus capacity indicator for bulk carriers in relation to the fleet has gradually shrunk, with a slight upturn in 1999 (3.2%), but it fell once again to 0.6% in 2005. In 2006 and 2007 there was surplus tonnage and the bulk transportation cycle started to take a negative course with increases in surplus, or in other words part of the fleet was not being used.

Next, the disaggregated behaviour of the general cargo fleet will be analysed. The capacity supply of the cargo ships was stable from 1995 to 1997, following which a decreasing trend began going from 62 million DWT in 1997 to 43 million DWT in 2004, with an average annual rate of decline of 3.06%. During this period, the general cargo ships experienced a fairly broad (3.06%) and long-lasting (up to 7 years) cycle of depression. In the remaining years the fleet underwent a slight increase.

Lastly, the evolution of the unitised fleet was analysed. The capacity of the fleet went from 143 million DWT in 1995 to 287.1 million DWT in 2007. The supply rose by an average annual rate of growth of 7.75%. This rate is certainly eye-catching; only a far-reaching change in the way world transportation is carried out allows us to assess this market and, subsequently, demand a more in-depth analysis of the container ships.

As far as the behaviour of the demand for this kind of ship is concerned, it is possible to confirm, with the figures available, that it coincides with supply throughout the period under analysis. Technically, it can be said that the market is in a state of equilibrium, since the supply and demand in existence coincide. The practical reason for this is the higher level of traffic among these ships in goods transportation and this situation means that all of the tonnage supplied is absorbed; consequently there is no surplus and in this sub-sector the entire fleet is used.

References

Ades A, Glaeser E (1999) Evidence on growth, increasing return, and the extent of the market. Q J Econ 114(3):1025–1046
Aghion P, Howitt P (1992) A model of growth through creative destruction. Econometric 60:325–351
Anderson J, Wincoop E (2001) Border, trade and welfare. NBER Working Paper 8515, Cambridge
Anderson J, Wincoop E (2004) Trade costs. J Econ Lit 42(3):691–751
Arasa C, Andreu J (1999) Desarrollo Económico. Dykinson, Madrid
Banco Mundial (1999) The evolution of ports in a competitive world. Port Reform Tool Kit, Module 2. World Bank, Washington
Castañeda J (1979) Lecciones de Teoría Económica. Aguilar de Ediciones, Madrid
Clark X, Dollar D, Micco A (2002) Maritime transport cost and port efficiency. Policy Research Working Paper no 2781. The World Bank, February
Coto-Millán P (1988) Funciones de demanda del transporte marítimo en España. Información Comercial Española 656:125–141
De La Dehesa G (2000) Comprender la globalización. Alianza, Madrid
De Langen P (2000) Time centrality in transport. Int J Marit Econ 1(2):41–56
De Salvo J (1994) Mesearing the direct impacts of a port. Transp J 33:33–42
De Salvo J, Fuller D (1995) The role of price elasticity of demand in the economic impact of a port. Rev Reg Stud 25:13–35
Commission of the European Communities (1997) Libro verde sobre los puertos y las infraestructuras marítimas. COM (97) 678 final. Brussels 10.12.1997
Commission of the European Communities (2001) Documento de trabajo de los servicios de la Comisión sobre los regímenes de financiación pública y los sistemas de tarificación en el sector portuario de la Comunidad. SEC (2001) 234. Brussels 14.02.2001
Fernandez De Castro J, Tugores Ques J (1987) Fundamentos de Microeconomía. Mc Graw Hill/ Interamericana de España, Madrid
Freire Seoane MJ, Gonzalez Laxe F (2003) Economía del transporte marítimo. Netbiblo, Coruña
Freire Seoane MJ, Gonzalez Laxe F (2007) Fletes y comercio marítimo. Netbiblo, Coruña
Gonzalez Fernandez S (Coord.) (1993) Organización económica internacional. McGraw-Hill, Madrid
Gonzalez Laxe F (dir) (1999) Análisis económico del sistema portuario gallego. Instituto de Estudios Económicos de Galicia. Fundación Barrié, A Coruña
Gonzalez Laxe F (2000) Perspectivas de los tráficos marítimos y competitividad portuaria. Boletín Económico de ICE, Ministerio Economía, Madrid. vol 2666. pp 7–16
Knowles RD (1994) New horizons in transport geography. J Transp Geogr 2(2):83–6
Krugman P (1995a) Economía internacional. McGraw-Hill, Madrid
Krugman P (1995) Growing world trade: causes and consequences. Brooking papers on economic activity, Brookings Institution Press, Whasington D.C. vol 1, pp 327–362
Landes DS (1990) Why are we so rich and they so poor? American Economic Association. Papers and Proceedings, vol 80, may, pp 1–13
Lloyd' Register (2008) World fleet statistics. United Kingdom. (Several years)
Micco A, Perez N (2001) Maritime transport costs and port efficiency. Inter-American Development Bank, Washington

Mochón Morcillo F (2002) Principios de Economía (Segunda edición). Mc Graw Hill/Interamericana de España, Madrid
OCDE (2001) Maritime transport. (Several years)
Sepulveda D (2000) Competitividad, eficiencia y productividad portuaria. Comisión Interamericana dc Puertos
UNCTAD (2008) Review of maritime transport. United Nations, Geneva

Analysis of the Returns to Scale, Elasticities of Substitution and Behavior of Shipping (General Cargo) Production

Pablo Coto-Millán, José Baños-Pino, Miguel Ángel Pesquera, Pedro Casares, and Juan Castanedo

Abstract The empirical analysis of production functions can be directed at least in two different ways which may have the same results. Firstly, we may directly assume a particular and flexible production function, and then test the different restrictions stage by stage, in order to find the most suitable functional form. An alternative way would be to assume cost, profit or input conditioned demand functions, which, after satisfying the corresponding duality theorems must provide the same information as the production function. Therefore, if the production integrability problem allows us to change from a cost function – for instance – to a production function, the technology can be perfectly detected from such a cost function.

In this study, I will use the former method in which a flexible functional form such as the logarithmic transcendental or translog is directly assumed, which allows us to execute different parametrical tests of the various properties of the production technology. This choice requires information about the amount of output and of productive input. Some interesting results can be obtained from this function, such as the input elasticities of substitution, the returns to scale of the production function, the marginal products of inputs, the output elasticities and different economic hypotheses.

P. Coto-Millán (✉), M. Ángel Pesquera, P. Casares, and J. Castanedo
University of Cantabria, Santander, Spain
e-mail: cotop@unican.es
J. Baños-Pino
University of Oviedo, Oviedo, Spain

P. Coto-Millán et al. (eds.), *Essays on Port Economics*, Contributions to Economics,
DOI 10.1007/978-3-7908-2425-4_8,

1 The Model

As said before, the functional form chosen to estimate the production function will be the translog, which can be written as:

$$\log Q = \alpha_0 + \beta_L \log L + \beta_k \log K + \beta_E \log E + \frac{1}{2}\gamma_{LL}(\log L)^2 + \frac{1}{2}\gamma_{KK}(\log K)^2 + \frac{1}{2}\gamma_{EE}(\log E)^2 + \delta_{LK} \log L \log K + \delta_{LE} \log L \log E + \delta_{KE} \log KE \quad (1)$$

where Q, L, E and K represent the amounts of output, labour, energy and capital respectively. Functional form (1) is, by definition, an approximation made from a second-order Taylor series, to any arbitrary point of the technological space. This functional form does not impose homotheticity, unitary elasticity of substitution or input linear separability. On the other hand, homogeneity of degree one will be assumed for prices, which means that (1) must be estimated along with the following restrictions in (2):

$$\begin{aligned} \beta_{\mathrm{L}} + \beta_{\mathrm{K}} + \beta_{\mathrm{E}} &= 1 \\ \gamma_{\mathrm{LL}} + \delta_{\mathrm{LE}} + \delta_{\mathrm{LK}} &= 0 \\ \gamma_{\mathrm{KK}} + \delta_{\mathrm{LK}} + \delta_{\mathrm{KF}} &= 0 \\ \gamma_{\mathrm{EE}} + \delta_{\mathrm{LE}} + \delta_{\mathrm{KE}} &= 0. \end{aligned} \quad (2)$$

In this kind of production model, (1) and (2) are usually estimated along with two share equations in order to increase the efficiency of the estimation. It is a matter of indifference which equation is to be excluded. Share equations are given by:

$$\begin{aligned} S_L &= \frac{\partial \mathrm{LogQ}}{\partial\ \mathrm{Log\ L}} = \beta_{\mathrm{L}} + \gamma_{\mathrm{LL}}\ \mathrm{Log\ L} + \delta_{\mathrm{LK}}\ \mathrm{Log\ K} + \delta_{\mathrm{LE}}\ \mathrm{Log\ E} \\ S_K &= \frac{\partial \mathrm{LogQ}}{\partial\ \mathrm{Log\ K}} = \beta_K + \gamma_{KK}\ \mathrm{Log\ K} + \delta_{\mathrm{LK}}\ \mathrm{Log\ L} + \delta_{\mathrm{KE}}\ \mathrm{Log\ E} \\ S_E &= \frac{\partial \mathrm{LogQ}}{\partial\ \mathrm{Log\ E}} = \beta_E + \gamma_{\mathrm{EE}}\ \mathrm{Log\ E} + \delta_{LE}\ \mathrm{Log\ L} + \delta_{\mathrm{KE}}\ \mathrm{Log\ K} \end{aligned} \quad (3)$$

As well as homogeneity of degree one in prices – assumed here – positive monotonicity restrictions and strict quasi-concavity guarantee that the production function is well-behaved. Monotonicity does not have to be assumed, and can be tested once the production function has been estimated by verifying the positive sign of the various marginal products of inputs. On the other hand, the strict quasi-concavity requires that the marginal rates of substitution be decreasing (that is to say, that the principal minors of the relevant bordered Hessian must alternate in sign) and can be tested for the functional form estimated.

On the other hand, the output elasticities of the factors labor (ε_L), energy (ε_E) and capital (ε_k), can be estimated from function (1) and (3) as follows:

$$\varepsilon_L = S_L;\ \varepsilon_K = S_K \text{ and } \varepsilon_E = S_E \tag{4}$$

Once we know the different values of the output elasticities of the factors in (4), the following returns to scale of production can be obtained

$$\varepsilon_L + \varepsilon_K + \varepsilon_E = \varepsilon \tag{5}$$

If value ε is higher, equal or lower than the unit the returns to scale obtained will be increasing, constant or decreasing.

The Allen-Uzawa elasticities of substitution of factors can be obtained as follows:

$$\sigma A_{ij} = [\ \delta_{ij} + S_i S_j]/S_i S_j;\quad i \neq j \tag{6}$$

where σA_{ij} is the input elasticity i with respect to input j.

Once we have obtained the values of (6) the demand-price elasticities for the production factors are calculated as follows,

$$E_{ij} = S_i\, \sigma A_{ij};\ \text{for all } i = j \text{ or } i \neq j \text{ and } E_{ij} \neq E_{ji} \tag{7}$$

2 The Data

The data have been obtained from the records of the companies in various Registers of business companies, for 1992. Further information obtained from the companies themselves and other institutions such as the associations of ship owners, shipping agents and shipbuilders, as well as from the association of the Merchant Navy officers, has been added to these data (Coto-Millán 1990; Coto-Millán 1995; Coto-Millán 1996b). The number of litres of fuel has been obtained dividing the total cost of fuel by the average price of fuel-oil paid by each ship in 1992. Analogously, in order to estimate the capital, I have calculated the dead weight tonnage (DWT) in each ship, which actually measures the cargo capacity of each ship, so we have something which is not either the typical capital stock which corresponds with the amount of equipment, or the flow of goods or services which correspond with the depreciation generated in the ship by the cargo of each goods – corrected by its corresponding stowage factor. The amount of labor is estimated through the number of workers of each company. To measure the output, I have applied the number of tons of general cargo goods transported by each company that year.

All the data have been obtained from 41 companies, of which 27 are dry cargo companies with ships of less than 16,000 DWT, 2 are transoceanic regular line companies and the 12 remaining are national coastal trade regular line companies. Thus, the dry cargo sector with less than 16,000 DWT is covered in 86.3% of its tons, the transoceanic regular lines in 76.81% and the national coastal trade regular line sector in 81.42%. The sum of these three sectors provides what we might call the "general cargo" sector of maritime transportation.

3 Empirical Results

The translog production function estimated appears in Table 1. From this estimation I have carried out the tests of hypotheses of linear homogeneity in the prices as shown in Table 2. As can be observed, the assumption of such a hypothesis is not rejected at the 0.01 level. In Table 2 other restrictions such as homotheticity, homogeneity and homogeneity and unitary elasticity of substitution (linear separability of inputs) are tested. In view of the results of these tests all these hypotheses are rejected.

Table 1 Coefficients estimated for translog production function

	Coefficients
α_0	−0.7114 (−1.2321)
β_L	0.0523 (2.3127)
β_K	0.1729 (4.5126)
β_E	0.8758 (12.321)
γ_{LL}	0.1612 (2.341)
γ_{KK}	0.1431 (1.932)
γ_{EE}	0.0617 (2.3421)
δ_{LK}	0.0143 (4.212)
δ_{LE}	0.018 (6.275)
δ_{KE}	0.0025 (0.7891)
Log likelihood	273.512
S.E. dependent variable	1.472
S.E. regression	0.162

Note: Statistics t-student within brackets

Table 2 Tests of likelihood rate

	χ^2 calculated	N° restrictions	Critical values 10%	5%	1%
Homotheticity	21.31	3	4.60	5.99	9.21
Homogeneity	9.86	3	6.25	7.81	11.34
Unitary elasticity of substitution	65.24	3	6.25	7.81	11.34
Homotheticity and unitary elasticity of substitution	88.22	5	9.23	11.07	15.08
Homogeneity and unitary elasticity of substitution (Cobb-Douglas technology)	108.34	6	10.64	12.59	16.81

Table 3 Allen-Uzawa elasticities

	Labor	Capital	Energy
Labor	−0.852145	−0.02983	0.07156
Capital	−0.02983	−1.023145	0.610234
Energy	0.07156	0.610234	−0.92131

Table 4 Own price and cross elasticities for Inputs demands

	Labor	Capital	Energy
Labor	−0.413126	0.28132	0.223254
Capital	0.31452	−0.62781	0.193185
Energy	0.216852	0.210815	−0.281876

From Table 1 we can estimate the corresponding returns to scale taking the average value of 1.32 for all the companies, which provides the sector with increasing returns to scale and, therefore, the companies exert power of the market, and do not make optimal use of the productive capacity.

The Allen-Uzawa elasticities of substitution – in Table 3 – present low average values, there is little scope for substitution between the factors labor, capital and energy (Table 4).

References

Coto-Millán P (1990) Determinants of private demand for sea transport in relation to the international market: an empirical approach. University of Oviedo, Department of Economics. Working paper

Coto-Millán P (1995) The conditioned demands of Spanish sea transport 1975–1990. Department of Economics. Working paper

Coto-Millán P (1996a) Derived demands for 'general cargo' shipping in Spain, 1975–1992, an economic approach. Appl Econ Lett 3:175–178

Coto-Millán P (1996b) Maritime transport policy in Spain (1974–1995). Transport policy. vol 3. Unit of Transport Studies, Oxford University, pp 37–41

Returns to Scale, Elasticities of Substitution and Behavior of Shipping (Dry Bulk) Transport Costs, Some Empirical Evidence

Pablo Coto-Millán, José Baños-Pino, Miguel Ángel Pesquera, Rubén Sainz, and Juan Castanedo

Abstract In the present section, we estimate cost functions for Solid Bulk shipping transport by 34 private firms for 1991. The functional form used is the Translog drawn by Christensen et al. (1973). We also estimate the Allen and Morishima elasticities of substitution, and compare and interpret the results as well.

1 Introduction

The wish to know the business reaction to the relative prices of production factors has led to studies of cost and production functions. The elasticity of substitution, formulated by Hicks in 1932 (Ferguson 1979), is the key concept provided by such a function. In Christensen et al. (1973) extended the concept of elasticity of substitution to multiproduct technologies, generating a new concept which carries his name in the economics literature. Subsequently, according to this, there have been some contributions on constant elasticity of substitution production functions, such as the work by Arrow et al. (1961). However, more recent studies question the accuracy of the Allen elasticity of substitution, since it is a one-factor one-price elasticity and is a derived demand elasticity divided by a share proportion. Thus, other authors state that the Morishima elasticity (Blackorby and Russell 1981) provides some information about economics more relevant than Allen's. Blackcorby and

P. Coto-Millán (✉), M. Ángel Pesquera, R. Sainz, and J. Castanedo
University of Cantabria, Santander, Spain
e-mail: cotop@unican.es
J. Baños-Pino
University of Oviedo, Oviedo, Spain

P. Coto-Millán et al. (eds.), *Essays on Port Economics*, Contributions to Economics, DOI 10.1007/978-3-7908-2425-4_9,

Russell (1989) point out the need to assume both of these elasticities of substitution, and warn seriously against the use of the Allen elasticities of substitution when multifactor technologies are being studied.

Maritime transport is managed in Spain by private companies (although there are also some public firms), which provide passenger and goods transport service. Transport prices are fixed mainly by the market.

2 Model

The present model is based upon Keeler's (1974) research and Caves et al. (1981).

It is assumed that the production function has only one output and three inputs. The shipping of solid bulk is multiproduct as regards goods, services, departures and arrivals, timetables and so on. For this reason, the production function must have a vector of products. There are also many other factors, but the availability of data forces us to consider only three.

It is also assumed that the prices of production input are exogenous.

Due to these two assumptions: product homogeneity and exogeneity of input prices; it is possible to represent a transformation surface of combinations of the three production factors in order to obtain the output implicitly:

$$\phi(Q,\ K,\ L,\ E) \tag{1}$$

Function ϕ in (1) represents variable Q of the output, K of the capital, L of the labor and E of the energy.

If ϕ meets the good behavior conditions (monotonicity, quasi-concavity and homogeneity of degree-one), we can apply Shepard's Lemma and obtain the satisfactory combinations of inputs and products from the cost function derivatives with respect to the factor prices.

In other words, there is a dual cost function:

$$C = \psi(Q,\ m,\ w,\ e) \tag{2}$$

where m, w and e are the prices of the respective inputs and C the output production costs.

In order to estimate cost function (2) we can test several functions, of which the most common ones are: Cobb-Douglas, CES, Diewert and Translog. Following the Okun principle we choose the Translog, the function which postulates the least number of restrictions and which is the most flexible one. Moreover, through it we can show the particular functions mentioned above.

The Translog cost specification used is:

$$
\begin{aligned}
LnC = {} & \alpha_0 + \alpha_Q LnQ + \frac{1}{2}\alpha_{QQ}(LnQ)^2 + \beta_L Lnw + \beta_K Lnm + \beta_E Lne \\
& + \frac{1}{2}\gamma_{LL}(Lnw)^2 + \frac{1}{2}\gamma_{KK}(Lnm)^2 + \frac{1}{2}\gamma_{EE}(Lne)^2 + \delta_{LK}(Lnw)(Lnm) \\
& + \delta_{LE}(Lnw)(Lne) + \delta_{KE}(Lnm)(Lne) + \rho_{LQ}(Lnw)(LnQ) \\
& + \rho_{KQ}(Lnm)(LnQ) + \rho_{EQ}(Lne)(LnQ)
\end{aligned} \tag{3}
$$

In this kind of cost model, (3) is usually estimated along with two share equations in order to increase the efficiency of the estimation. It is a matter of indifference which equation is to be excluded. Share equations are given by:

$$S_L = \frac{\partial \mathrm{LnC}}{\partial\ \mathrm{Ln\ w}} = \beta_{\mathrm{L}} + \gamma_{\mathrm{LL}}\ \mathrm{Ln\ w} + \delta_{\mathrm{LK}}\ \mathrm{Ln\ m} + \delta_{\mathrm{LE}}\ \mathrm{Ln\ e} + \rho_{\mathrm{LQ}}\ \mathrm{Ln\ Q}$$

$$S_K = \frac{\partial \mathrm{LnC}}{\partial\ \mathrm{Ln\ m}} = \beta_K + \gamma_{KK}\ \mathrm{Ln\ m} + \delta_{\mathrm{LK}}\ \mathrm{Ln\ w} + \delta_{\mathrm{KE}}\ \mathrm{Ln\ e} + \rho_{\mathrm{KQ}}\ \mathrm{Ln\ Q}$$

$$S_E = \frac{\partial \mathrm{LnC}}{\partial\ \mathrm{Ln\ e}} = \beta_E + \gamma_{\mathrm{EE}}\ \mathrm{Ln\ e} + \delta_{LE}\ \mathrm{Ln\ w} + \delta_{\mathrm{KE}}\ \mathrm{Ln\ m} + \rho_{\mathrm{EQ}}\ \mathrm{Ln\ Q}$$

The following restrictions guarantee that (3) presents homogeneity of degree one in the input prices,

$$
\begin{aligned}
\beta_{\mathrm{L}} + \beta_{\mathrm{K}} + \beta_{\mathrm{E}} &= 1 \\
\rho_{\mathrm{LQ}} + \rho_{\mathrm{KQ}} + \rho_{\mathrm{FQ}} &= 0 \\
\gamma_{\mathrm{LL}} + \delta_{\mathrm{LE}} + \delta_{\mathrm{LK}} &= 0 \\
\gamma_{\mathrm{KK}} + \delta_{\mathrm{LK}} + \delta_{\mathrm{KF}} &= 0 \\
\gamma_{\mathrm{EE}} + \delta_{\mathrm{LE}} + \delta_{\mathrm{KE}} &= 0.
\end{aligned} \tag{4}
$$

The Allen elasticities of substitution σA_{ij} between factors are defined as follows (Uzawa 1962):

$$\sigma A_{ij} = \frac{C\left(\frac{\partial^2 C}{\partial P_i \partial P_j}\right)}{\left(\frac{\partial C}{\partial P_i}\frac{\partial C}{\partial P_j}\right)}$$

where C are the costs and P_i and P_j, are the input prices (P_i, P_j: m, w, e).

For the Translog cost function, Allen partial elasticities of substitution Blackorby and Russell (1989) can be estimated as shown,

$$\sigma A_{ij} = \frac{(\delta_{ij} + S_i S_j)}{S_i S_j};\ with\ i \neq j$$

where δ_{ij} refers to parameters in (4): δ_{LK}, δ_{LE} and δ_{KE}. However, variables S_i and S_j account for the different shares: S_L, S_K and S_E.

As regards multiple factors, the Morishima elasticities σM_{ij} are defined as follows (Blackorby and Russell 1989):

$$\sigma M_{ij} = \frac{\partial \mathrm{Ln}\left(\frac{C_i}{C_j}\right)}{\partial\ \mathrm{Ln}\left(\frac{P_i}{P_j}\right)}$$

where C_i and C_j account for the production costs with the respective factors i and j; and P_i and P_j account for the prices of factors m, w and e.

For the Translog cost function, the Morishima elasticities of substitution can be obtained from the following expressions (Blackorby and Russell 1989):

$$\sigma M_{ij} = \left(\frac{\delta_{ij}}{S_i}\right) - \left(\frac{\gamma_{jj}}{S_j}\right) + 1$$

$$\sigma M_{ji} = \left(\frac{\delta_{ij}}{S_j}\right) - \left(\frac{\gamma_{ii}}{S_i}\right) + 1$$

where $\delta_{ij} = \delta_{ji}$, and $\sigma M_{ij} \neq \sigma M_{ji}$. This means that, unlike the Allen elasticity of substitution, which verifies that $\sigma A_{ij} = \sigma A_{ji}$, the Morishima elasticity is not symmetrical, which can be interpreted as follows: the various relative prices of factors provide different elasticities of substitution depending on the price (i^{th} or j^{th}) of the modified factor.

3 The Data

The data used comes from the official statistics of the Ministry of Transport as well as Private Companies' Memoranda.

The output variable has been approximated by the indicator tons of goods transported. By adding different outputs (as many as possible departures and arrivals in the port network, and for the different timetables) we can see the great heterogeneities in the amount of services. However, such aggregation depends on the availability of data.

Three inputs have been considered: labor, capital and energy. The price of labor will be represented by (w), that of capital by (m), and the price of input energy will be (e).

Finally, the total costs C, consist of the amount of the factors used, multiplied by their respective prices.

4 The Empirical Results

A Translog model has been used with degree one homogeneity in prices. The results of the estimation are shown in Table 1.

Table 1 Coefficients estimated for translog cost function

	Eq. without restrictions
α_0	12.4957 (86.681)
α_Q	0.3039 (2.233)
α_{QQ}	0.1754 (0.505)
β_L	0.5463 (7.854)
β_K	0.2880 (10.004)
β_E	0.1656 (2.355)
γ_{LL}	0.0373 (2.494)
γ_{KK}	0.0929 (5.226)
γ_{EE}	0.1047 (65.464)
δ_{LK}	−0.0127 (−1.045)
δ_{LE}	−0.0245 (−1.881)
δ_{KE}	−0.0801 (−5.371)
ρ_{LQ}	−0.0028 (−0.123)
ρ_{KQ}	0.0825 (2.876)
ρ_{EQ}	−0.0796 (−3.199)
Log likelihood	66.43
S.E. dependent variable	0.475
S.E. regression	0.406

Table 2 Tests of likelihood rate

	χ^2 calculated	Nº restrictions	Critical values		
			10%	5%	1%
Homotheticity	11.45	2	4.60	5.99	9.21
Homogeneity	11.53	3	6.25	7.81	11.34
USE	41.42	3	6.25	7.81	11.34
H and USE	42.66	5	9.23	11.07	15.08
(Cobb-Douglas)	42.69	6	10.64	12.59	16.81

Moreover, an analysis has been carried out on the optimal behavior of the cost function. The results obtained give evidence of well behaved since the cost function tested satisfies the following conditions:

- Monotonicity in factor prices
- Homogeneity of degree one with respect to factor prices
- Quasi-concavity with respect to input prices

In Table 2 the following conditions have been tested: homotheticity, homogeneity, unitary elasticity of substitution, homotheticity and unitary elasticity of substitution, and homogeneity and unitary elasticity of substitution (Cobb-Douglas technology).

The Allen and Morishima elasticities of substitution have been estimated from the coefficients of the estimation in Table 1. In Tables 3–6, we show the elasticities obtained by this procedure. The estimations of the Allen and Morishima elasticities of substitution give evidence that factors are all substitutive but not complementary between themselves.

Table 3 Allen elasticities

	Labor	Capital	Energy
Labor	−1.6361 (0.1562)	0.8847 (0.2586)	0.7815 (0.8031)
Capital	0.8847 (0.2586)	−1.1806 (0.1130)	0.2701 (0.8720)
Energy	0.7815 (0.8031)	0.2701 (0.8720)	−1.0564 (0.2092)

Table 4 Morishima elasticities

	Labor	Capital	Energy
Labor	0	0.6796 (0.0973)	0.6131 (0.2760)
Capital	0.8503 (0.0852)	0	0.4425 (0.3047)
Energy	0.8154 (0.2743)	0.4773 (0.2901)	0

Table 5 Own price and cross elasticities for inputs demands

	Labor	Capital	Energy
Labor	−0.5518 (0.0527)	0.2911 (0.0851)	0.2607 (0.2679)
Capital	0.2984 (0.0872)	−0.3885 (0.0397)	0.0900 (0.2909)
Energy	0.2636 (0.2709)	0.0888 (0.2869)	−0.3524 (0.0698)

The comparison between elasticities supports the results of previous works such as Blackorby and Russell (1989) and McMillan and Amoako-Tuffour (1991); the Allen elasticities overestimate substitution relationships.

5 Summary and Conclusions

The behavior of the costs of private firms for 1991 has been studied in this section. The Translog function with the restriction of homogeneity of degree one in factor prices has been analyzed and acceptable results have been obtained for the output tons of goods. The cost function estimated has a good behavior since it satisfies the conditions of monotonicity, quasi concavity and homogeneity of degree one in factor prices.

Table 6 Scale economies

1	0.495189
2	0.490410
3	0.565623
4	0.690091
5	0.726141
6	0.706385
7	0.857494
8	0.834654
9	0.839357
10	0.730124
11	0.771997
12	0.786064
13	0.751192
14	0.570201
15	0.538653
16	0.468404
17	0.568659
18	0.583951
19	0.724467
20	0.676536
21	0.664386
22	0.657617
23	0.625997
24	0.679806
25	0.786789
26	0.822612
27	0.731166
28	0.670107
29	0.657031
30	0.837710
31	0.755245
32	0.724776
33	0.759895
34	0.917812

The homotheticity hypothesis can only be rejected in a 10% so that the homothetic cost function is also estimated.

It has been tested in both cases that all the direct price elasticities and the Allen elasticities of substitution are always negative. Therefore, the cost function is monotonic in the factor prices.

Moreover, the matrix of the Allen elasticities of substitution is negative semi-definite in the average value of the data, so that the cost function is quasi-concave.

The comparison between elasticities support the results in previous research since it can be seen that the Allen elasticities of substitution with respect to the Morishima elasticities overestimate both substitution and complementary relationships.

References

Arrow KJ, Chenery HB, Minhas BS, Solow RW (1961) Cápital-labor substitution and economic efficiency. Rev Econ Stat 43:225–254

Blackorby C, Russell R (1981) The Morishima elasticity of substitution: symmetry, constancy, separability, and its relationship to the Hicks and Allen elasticities. Rev Econ Stud 48:147–158

Blackorby C, Russell R (1989) Will the real elasticity of substitution please stand up? (A comparison of the Allen/Uzawa and Morishima elasticity of substitution). Am Econ Rev 79:882–888

Caves D, Christenson L, Swanson J (1981) Productivity growth, scale economies and capacity utilisation in US rail-roads 1955–1974. Am Econ Rev 5:994–1002

Christensen LR, Jorgeson DW, Lau LJ (1973) Transcendental logarithmic production frontiers. Rev Econ Stat 55:28–45

Ferguson CE (1979) The Neoclassical theory of production and distribution. Cambridge University Press, Cambridge

Keeler TE (1974) Railroads costs, returns to scale and excess capacity. Rev Econ Stat 61:201–208

McMillan ML, Amoako-Tuffour J (1991) Demands for local public sector outputs in rural and urban municipalities. Am J Agric Econ 73:313–325

Pindyck RS (1979) Interfuel substitution and the industrial demand for energy: an international comparison. Rev Econ Stat 61:69–179

Uzawa H (1962) Production functions with constant elasticities of substitution. Rev Econ Stud 29:291–299

Cycles in the Ship Building Industry: An Empirical Evidence

Pablo Coto-Millán, José María Sarabia-Alegría, and Lucía Inglada-Pérez

Abstract In the following short paper the possible cycles of the ship industry are studied through univariate autoregressive integrated moving average time series models. The adjusted model with data from the world's fleets (1924–1994) presents an empirical contrast which seems to confirm the coexistence of long and short cycles in the maritime transport of 4 and 12.7 years, respectively. The result seems to confirm the well known Cobwed theorem in the case of short-run cycles and long-run stock cycles.

1 Introduction

In the following study the historic evolution of the world's fleets for the 1924–1994 period is studied. This evolution shows certain cyclic behaviour which can be empirically contrasted. Static classic economics does not explain the existence of cycles. However, their most simple dynamic versions, such as the Cobwed theorem, explain the possibility of short cycles. The new classic economics explanains the possibility of short-run cycles, through the behaviour theory, the formation of expectations and the stock cycles, which are also contrasted by this study of the world's fleets for the same period.

P. Coto-Millán (✉) and J.M. Sarabia-Alegría
University of Cantabria, Santander, Spain

L. Inglada-Pérez
U.N.E.D., Madrid, Spain

P. Coto-Millán et al. (eds.), *Essays on Port Economics*, Contributions to Economics,
DOI 10.1007/978-3-7908-2425-4_10,

2 Arima Models and Fleet Cycles

The use of autoregressive integrated moving average (ARIMA) time series models to detect cycles and pseudo-periodic behaviour is not new within the frame of the economic time series. Geweke (1988) uses a third-order autoregressive model with a pair of complex roots, in order to modelize the logarithm of the gross domestic product per capita and to thus obtain an adequate breakdown of the series into two components, a secular and a cyclic one. Other references are Nelson and Plosser (1982) and Campbell and Mankiw (1987). Harvey (1985) proposes models made up by sums of invisible components for the gross national product series, using the Kalman filter. King et al. (1991) consider cointegration relations among different macroeconomic time series. As an alternative to the stationary linear models, Hamilton (1989) and recently Goodwin (1993), propose an alternative class of non-linear models called Markov-Switching models, which allow us to contrast different hypothesis about the economic cycle from processes with discrete changes in the media between high-growth and low-growth states. Sarabia and Coto-Millán (1993) point out the possibility of pseudo-cyclic behaviour in the historic evolution of the Spanish and world's fleets.

The underlying idea for the detection of cycles consists in identifying when the characteristic polynomial associated to the autoregressive part of an ARIMA model has complex roots. It is well known that these type of roots lead to an oscillatory and pseudo-periodic behaviour, both in the structure of the series itself and in its correlogram. When z is a complex root of the characteristic polynomial associated to the autoregressive part the periodicity of the complex roots,

$$T_z = \frac{2\pi}{\tan^{-1}[\mathrm{Im}(z)/\mathrm{Re}(z)]} \tag{1}$$

indicates the periodicity in the cycle. In this formula, $\tan^{-1}$ is the principal branch between 0 and π, hence $2 \leq Tz < \infty$. The amplitude of the complex roots $Az = |z|^{-1}$ indicates the exponential rate at which the cycle induced by a shock decays, or grows if $Az > 1$.

The fleet data comes from Lloyd's (1994). This publication presents separated statistics for all the countries (pabellones or flags) expressed in Gross Register Tons (GRT) for ships of 100 GRT and more. Annual data corresponding to the total world fleet have been used. Data from the 1940 to 1947 period, non available on publication, have been completed assuming a constant annual growth rate between 1939 and 1948.

The series of the world data clearly corresponds to a non-stationary series, which shows an increasing tendency until 1983 with a maximum GRT of 423 millions, a decrease until 1988 and an apparent recovery after 1989. A first difference in the logarithms series seems sufficient to reach a stationary plane. In this case the model identified through the usual Box-Jenkins methodology, has been an ARIMA (5, 1, 0),

Table 1 Time series models for the World's Fleets (1924–1994)

Model 1

$$\underset{(2.54)}{(1+0.317B}-\underset{(2.44)}{0.293B^3}-\underset{(3.23)}{0.383B^4}-\underset{(1.74)}{0.209B^5})\nabla \log z_t = \underset{2.45}{0.032}+a_t$$

$R^2 = 0.234$

$\sigma_a = 0.045$

SSE = 0.121

Q(11) = 10.2

Comments: Cycles of T = 4 and T = 12.7 years

Model 2

$$\log z_t = \underset{(2.07)}{0.03} - \underset{(45.13)}{0.211I_{31}} + \underset{(24.92)}{0.12I_{65}} - \underset{(2.69)}{0.02IP_{74}} + \frac{a_t}{\nabla(1+\underset{(14.11)}{1.04B}-\underset{(3.10)}{0.45B^4}+\underset{(2.52)}{0.32B^5})}$$

$R^2 = 0.944$

$\sigma_a = 0.012$

SSE = 0.008

Q(11) = 13.44

Notes: B is the backshift operator $B^k x_t = x_{t-k}$ and 1−B is the difference operator; z_t is the observed series; a_t is a white-noise process of uncorrelated variables with zero mean and constant variance. The parenthetical numbers under the estimated parameters are the t-statistics for these parameters. Q is the Ljung-Box statistics (11 degrees of freedom); σ_a is the residual variance of the model and SSE is the sum of the square of the errors. I_k is an indicator variable with $I_k = 1$ and $I_j = 0$ for j_k, where k = 1931, 1965 and 1974. $IP_{74} = 1$ if t_1974 and 0 otherwise.

which leads to an explanation of the first difference in the logarithms series through a linear combination of the difference values in the five previous periods. The model estimated, together with the most important diagnosis statistics, are represented in Table 1. The autoregressive structure of the model allows us to obtain information about the pseudo-periodic behaviour of the series. The fifth-order characteristic polynomial has a real root and two pairs of complex roots,[1] which leads us to both a 4-year and an approx. 12.7-year pseudo-cycles, using formula (1).

In order to complement the previous model an intervention analysis of the series has been carried out. With that aim, three key years in the evolution of the world's fleets have been highlighted: 1931, 1965 and 1974. The two first years have been described as having a singular effect, while 1974 has been considered as having a permanent effect.

The 1931 intervention corresponds to a decrease of the world's fleets as a consequence of the 1929 crisis, which is shown with a 2 years delay. The 1965 intervention has a different explanation. In 1956 the Suez Canal was nationalised, causing the oil tankers to use the Cape of Good Hope route for the journeys from the Persian Gulf bound for European ports. This was the reason for oil freight to be multiplied by almost three during that year. In order to reduce transport costs, shipowners started to order ships of a considerable larger size. Thus, economies of scale provided competitive freights again. As an example, the world's fleets in 1965, in GRT, increased by 18 per cent, against previous growth rates of around

[1]The roots of the characteristic polinomial associated to Model 1 are given by ($I^2 = -1$): $B_1 = 1.11396$. $B_2, B_3 = -1.38592 \pm 0.74749$ I. $B_4, B_5 = -0.08402 \pm 1.31173$ I.

5 per cent. The oil tanker fleet orders almost doubled from 1961 to 1965 (UNCTAD 1968) not only as a consequence of the delayed adjustment of the nationalisation, but of the expected closure of the Suez Canal, as finally happened in 1967. Due to the increase of the oil tanker fleet, this closure did not have any noticeable effects on freight costs.

The strong increase in oil prices in 1973 gave rise to the deduction in the demand of this product and the annulment of oil tanker orders which surfaced in 1974 and in following years persistently.

3 Interpretation and Conclusions

The ship building industry has a short cycle that starts when there is a depression in the freight market and economic activity begins to grow. As economic activity grows, transport demand increases and freight rates go up, allowing shipowner profits to increase while these answer back demanding more new and second-hand ships. In view of a stronger demand, ship prices increase and the fleet orders peak. In the optimum point of order acceptance, saturation is reached in such a way that, if the activity does not increase, depression is again reached in the freight market as a logical consequence of the imbalance generated by a supply surplus. Therefore, the short cycle is explained by the well known Cobwed theorem. The duration of a full regular short cycle of this kind may last between 3 and 4 years. In this paper the short cycles obtained are 4 years long (see Table 1).

However, short cycles do not explain everything. In reality, more complex economic facts occur. In the short cycle the economic activity and the shipowners were the principal actors, when in the long cycle there also intervene other economic agents as the ship builders themselves (dockyards), the financial middlemen (banks) and the governments of the countries. The long cycle is therefore a consequence of the decisions of all these agents; it may last between 9 and 12 years, and basically coincides with half of the average economic life of a ship. The results obtained provide long cycles of 12.7 years (see Table 1). There may exist even longer cycles, but Lloyd's does not have world fleet data in GRT prior to 1924. Some authors who have used different sources of information, provide evidence for cycles of 25 and 50 years. We can infer from the previous studies that the ship building industry has two phases. A first phase of rising construction, which includes three regular cycles of 3–4 years and another phase of adjustment, fall of freights and breaking up, which also has short cycles of 3–4 years. This scenario corresponds to a first construction phase of 9–12 years and a second adjustment phase of another 9–12 years. Each phase constitutes a cycle of 9–12 years in which are included three short cycles and after two cycles of 9–12 years a long cycle of 18–24 years is generated.

So, what is argued here is that when there is a surge of ship building, as a result of optimistic expectations, the fleet is built and since the average economic life of a ship can be between 18 and 24 years, it is necessary that the long cycle is completed in this time, so the fleet stocks are exhausted.

The attempts by governments, financial middlemen, dockyards and shipowners to avoid the traumatic characteristics of the cycles could only soften the sudden changes of the cycles, but they can not avoid them because the cycles are a logical consequence of the market forces (Massac 1994).

These explanations coincide with the proposals of Hampton (1986), with the exception that this author estimates a long cycle of an approximate duration of 20 years. The explanations provided by Hampton (1986) are based on the long-run evolution of the freights for the 1947–1983 period. The evolution of international freights is one way to analyse the cyclical fluctuations of the maritime transport market. An alternative way to analyse the cycles is to observe the behaviour of the ship building industry, which is the main focus adopted in this paper.

References

Campbell JY, Mankiw NG (1987) Are output fluctuations transitory? Q J Econ 102:857–880

Geweke J (1988) The secular and cyclical behaviour of real GNP in 19 OECD countries, 1957–1983. J Bus Econ Stat 6:479–486

Goodwin TH (1993) Business-cycle analysis with a Markov-switching model. J Bus Econ Stat 11:331–339

Hamilton J (1989) A new approach to the economic analysis of nonstationary time series and the business cycle. Econometrica 57:357–384

Hampton M (1986) Analysis shiping cycles I and II. Seatrade, January and February, 19–23 and 19–23

Harvey AC (1985) Trends and cycles in macroeconomic time series. J Bus Econ Stat 3:216–227

King RG, Plosser CI, Stock JH, Watson MW (1991) Stochastic trends and economic fluctuations. Am Econ Rev 81:819–840

Lloyd's (1994) Lloyd's register of shipping statistical tables. London

Massac GM (1994) Cycles de production et d'exploitation des moyens de transport maritime. J Mar Marchande 16:9–17

Nelson C, Plosser C (1982) Trends and random walks in macroeconomic time series. J Monet Econ 10:139–162

Sarabia JM, Coto-Millán P (1993) Comportamientos pseudo-cíclicos en la evolución de la flota española y mundial: una aproximación empírica. Estudios de Transportes y Comunicaciones 61:59–64

UNCTAD (1968) Review of maritime transport 1967. New York

Part III
Port Economic Impact

A Methodological Discussion on Port Economic Impact Studies and Their Possible Applications to Policy Design

Pablo Coto-Millán, Miguel Ángel Pesquera, and Juan Castanedo Galán

Abstract The main methodologies applied to obtain the economic impact of a port, are discussed in this paper. A systematic classification which allows to evaluate the availability of the resources as well as the advantages, and disadvantages of every methodology, is offered according to the aims. Moreover, with these methodologies, the impact of port industry on the rest of the economy, can be obtained.

1 Introduction

The aim of this work is to carry out a methodological review of port impact studies and their possible applications to policy design. Taking this into account, firstly, the aims and targets of impact studies, are put forward; secondly, the different methodologies used to deal with this issue, are classified and analysed. Not that a simple list of such methodologies is issued, but a systematic classification which allows the evaluation of the availability of the means and the advantages and disadvantages of the possible methodologies, according to the aims. Thirdly, a dynamic view which will undoubtedly be very profitable in the future, is discussed; fourthly, the main criticisms and defences of impact studies, arising from economic literature are offered. Finally, there is a reflection on the possible direction these works may take in the future.

2 Aims and Utility

The main aim of these studies is to obtain the economic impact of the operation of a port for a specific year. The economic impact indicators of the stylized results of such studies are the following: added value, jobs, income and taxes. The main aim

P. Coto-Millán (✉), M. Ángel Pesquera, and J. Castanedo Galán
University of Cantabria, Santander, Spain
e-mail: cotop@unican.es

P. Coto-Millán et al. (eds.), *Essays on Port Economics*, Contributions to Economics,
DOI 10.1007/978-3-7908-2425-4_11,

of some works, such as the US Maritime Administration (1979) and the work by Coto-Millán (1995), is the estimation of the economic impact of ports in U.S.A. and in Spain, respectively. Apart from this, there are further aims such as to understand the correlation between port and other local, regional and national industries, as well as to emulate policies of change on ports and assess the effects of the changes on the economic development of the town, area, region or country.

The studies may be useful for Port Authorities and government regulation Agencies which take decisions on port development, whether the port is in operation or a new project of a port is being planned. It may be interesting for legistators, congressmen and for the public in general, since all of them may be concerned with the economic effects of a national port system. It may be also interesting for port users and for business and institutions related to port industry.

3 Analysis of Different Methodologies

There is no agreement on the opinions about the different methodologies to be applied to port economic impact studies, or on the evaluation of the added value generated by the port. Therefore, an orientative classification of the methodologies applied in the last years, is proposed.

In synthetic terms, there are three types of methodologies to be applied to port economic impact.

Methodology I is based upon the cost aggregation of the different economic agents to develop the transport of goods and services related to the port studied.

Methodology II has its grounds on the added value aggregation of two big groups of economic agents (Port Industry and Port Authority) to study the direct economic impact of ports.

Methodology III is based upon the added value of three big groups of economic agents (Port Industry, Port Authority and Users) to study the total economic impact of ports (both direct and indirect).

3.1 Methodology I

Metholoolgy I was applied, among others, to the study of the port of Baltimore by Hille et al. (1975). In this study, total economic impacts such as the total amount of the costs of the different economic agents were calculated, not grouped under any criteria, but classified into direct, indirect and induced costs. Such costs were defined as follows:

Direct costs are those supported by shipping companies and their representatives.

Indirect costs are payments to companies which depend on the port, as well as the costs of the Administration.

Induced costs are those derived from the consumption of goods and services in house economies which get their income from the services related to the abovementioned direct and indirect costs.

Direct and indirect costs are estimated in this methodology from more or less detailed surveys. Induced costs are estimated from the multipliers obtained from the input–output tables of the region studied.

In this methodology, direct impact results are estimated for direct jobs, direct cost per ton of general cargo, solid bulk and other goods, and for each of the abovementioned direct cost categories. Indirect impacts are estimated in terms of added value and jobs generated by services contracted from business and institutions, which represent the abovementioned indirect costs. Induced impacts are estimated applying an average multiplier to jobs (service sector), and other nine average multipliers, with the same classification as for the direct costs, from which induced costs are obtained; and finally, four more multipliers to indirect costs according to their own classification. Moreover, an approach to the impact is provided in terms of taxes, according to the estimations of the taxes paid by companies and individuals related to direct costs.

The disadvantage of this methodology is the excessive focus on surveys and the fact that very vague and inaccurate induced multipliers are calculated. It seems that certain induced multipliers are obtained for the input–output table, then, the sectors which are more similar to direct and indirect cost categories, are sought and multipliers are simply applied to them. This methodogy is expensive and takes a long time. Its advantage is that it allows to present impact disaggregations per ton of each type of good, such as the total impact of each ton of cars (even of each item), of each ton of goods carried in containers, of each ton of solid bulk,... etc. Furthermore, if the survey is detailed enough, it allows space disaggregations such as the impact for a town, an area and a region.

3.2 *Methodology II*

The Methodology II is based upon the estimation of the added value, and it only estimates direct impacts. Some works which apply this methodology are those realized between 1979 and 1982 by the US Maritime Administration (1979), and those realized in Spain by Fraga and Seijas (1992) and De Rus et al. (1994). In these works, different methods are used to get to estimate the economic impacts of two big groups of economic agents:

1. Port Industry
2. Port Authority

In order to estimate the direct economic impacts of these two groups, the following methods or combinations of methods are used:

1. Manual method of impact estimation per ton.
2. Automatic method of the input–output table direct multipliers:

   (a) Standard multiplier estimation.
   (b) Multiplier estimation and indicative survey use.
3. Detailed survey method.

The services included in Port Industry and Port Authority, will be described and the application and advantages of every method will be subsequently discussed. In Port Industry, ship assistance services are regarded as the supplies and provisions, repairs, auxiliariy services, docking services such as pilotage, towage, and mooring services-, good shipping and storing services, as well as good handling services. Depending on the type of port referred in the study, this concept will be whether broadened or reduced, including the services provided by fishing ships and other services related to fishing sale. It is also possible, if the study requires so, to include the services provided to cruisers, as well as those services derived from repairs, maintenance and watching of the cruisers. Thus, fishing and cruise industries are included in the work by Braun (1990), since they are economically relevant in the port of Cape Canaveral.

As regards to Port Authority, the costs of this institution, such as staff, supplies and contracts with third parties, are considered. The Port Authority usually provides the widest range of facilities and charges a series of tariffs for using them, according to the services. However, it also contracts and pays for the enlargement and improvement of port infrastructures, and port dragging.

The two big groups of economic agents mentioned before, generate direct jobs added value, income and tax, estimated according to the above mentioned methods, and their operative mechanism as well as their advantages and disadvantages, will be analysed below.

The manual method consists of the estimations of the economic impact per ton of a big group of goods (such as: general cargo, solid bulk and oil and its derivatives), calculated through the added value of the port (income obtained from production factors strictly of the port). This approach is made for Broad Brush studies. Such an approach is particularly interesting if the resources available are limited, and its advantage is that it gets short-term results and it can be easily updated. Its main disadvantage is that the magnitudes evaluated may be inaccurate due to particular points of view or prejudices in the sector, which have into topics not verified empirically.

The automatic method consists of the application of the direct multipliers of the region the services provided by companies, these being considered as a proxy of port industry, and to the application of the direct multipliers of the construction sector to the dragging and construction services contracted by the Port Authority. It requires the availability of the input–output tables for the base period and the region studied. It needs specialists in input–output tables and in port operations. With these means, this method takes little time and, in spite of being more expensive, it is much more efficient than the manual method. It can also be complemented with the previous one and/or with a limited survey.

The detailed survey method takes longer and needs specialists in working with a great deal of data and in surveys, which makes it an extremely expensive method.

Nevertheless, it must be said that such a method has been most used in direct impact estimation.

3.3 Methodology III

The Methodology III also is based upon the added value aggregation to estimate direct, indirect and induced economic impact of three big groups of economic agents: Port Industry, Port Authority and Users. Both Port Industry and Port Authority are defined as in the above mentioned second methodology. On the contrary, the group of Users is new and needs to be defined. This group accounts for the economic effects generated by the users of port services. With the aim to estimate impacts, several methods or combinations of methods are used according to the type of economic agent dealt with.

1. Methods to estimate Port Industry impacts:
   (a) Manual method per ton.
   (b) Multiplier automatic method.
      - With standard multiplier values.
      - With a discreet survey.

   (c) Detailed survey method.
2. Method to evaluate Port Authority impact:
   (a) Method based on construction and dragging business survey.
   (b) Automatic method of construction sector multipliers.
3. Methods which calculate the User's impact:
   (a) Limited survey method with manual estimations per ton.
   (b) Automatic method with multipliers.
      - With standard multiplier values.
      - With limited survey.

   (c) Detailed survey method.

Each group of port agents can be estimated with two or three different methods from which different combinations may arise; the choice of a particular combination of methods will be based on the availability of data, the means used in the research and the coherence of the results. The advantages and disadvantages are similar to those described in the second methodology.

Some of the works which used this methodology are: US Maritime Administration (1979), which developed a basic software instrument, the Kit, using this methodology in about forty studies on U.S. ports so far (De Salvo 1994); Coto-Millán and Sarabia (1993); De Salvo and Fuller (1988, 1995); Davis (1983) and Villaverde and Coto-Millán (1997). The works for the port of Seattle can also be mentioned here, as well as the studies by Opuku (1990) and Pinfold (1991), for the ports of New York and New Jersey and Halifax, respectively, and the study by Villaverde and Coto-Millán (1995, 1996) for the port of Santander (Spain).

4 A Dynamic View of Port Impact Studies

There is a static and a dynamic point of view related to time, in order to analyse port impacts. From the static point of view the economic impact of a port is studied with respect to a base year, while from a dynamic point of view this impact is studied comparing several years. In order to apply the latter, it is necessary the knowledge of at least two static studies of the same port and under the same methodology. For this reason, the methodologies applied to static port studies have been previously analysed with respect to a base period, therefore, it is necessary to focus on the dynamic aspects of the port.

There is no doubt that an analysis at a certain point in time, provides a limited point of view. However, it is the existence of two or more impact studies of a port, under the same methodology and in different periods, that allows to obtain a richer dynamic view as regards to data, which is therefore, more valuable to interpret certain facts, and to make predictions. In short, it is a true complementary instrument which in taking political port decisions. Some of these dynamic estimations have been carried out for ports with certain tradition in impact studies. In the former work, the studies made for 1979 and 1984 lead the authors to warn about the possible errors in the predictions made by dynamic port impact studies. The reason for their assumption is that changes in business productivity, regulation and international dollar rate, significantly variate port traffic tonnage. Nevertheless, these authors (Yochum and Agarwal) state that several port impact studies of a given port, provide important data on the change of port economic structure. Such data can be used for the development of adequate econometric models to predict the effect of tonnage and other relevant variables which affect port economy.

As regards to the latter work by Warf and Cox (1989), it presents a dynamic interpretation of two studies of the port of New York in 1977 and 1987, providing additional relevant data, reasonably adjusted to reality. However, with the aim to systematically offer the main attacks to port impact studies, as well as the possible responses arising, the following conclusions have been met.

5 Criticisms and Defences of Port Impact Studies

As in all economic studies which adopt a particular methodology or develop a model, it is necessary to make a series of assumptions in port impact studies. The three abovementioned methodologies are based on manual methods, surveys and input–output tables or combinations of these methods. In Methodology II, if a manual method or detailed survey is adopted, since only direct impacts are estimated, the only possible assumption is that the agents are well identified and that the data they provide is reliable.

In Methodologies I and II, and even in the choice between detailed survey and manual method of Methodology II, it is necessary to estimate indirect impacts. If a detailed survey is applied, indirect impacts can be estimated, assuming that the companies from which services are contracted, or which sell products to the port, have production functions of grade-one homogeneity (that is to say, with constant scale output), with a fixed technology and constant elasticity of substitution. With these three assumptions, the total invoicing of each port supplier of goods and services, can fall into two categories: sales and services for business directly related to port activity, and buying and services for other companies. The invoicing level is considered as output divided into port and non-port output. The amounts of inputs that get the port output, are obtained below in constant amounts, so that indirect jobs, income, tax and added value, can be estimated. It would be necessary to assume again that the agents are well identified and that the data provided is reliable.

With the input–output model, it is possible to calculate impact from direct, indirect and induced multipliers. However, in the input–output analysis the following assumptions are included: production functions with constant technology, lack of scale economies and non-input substitution in the process of production.

These assumptions are restrictive and due to this, the main criticisms naturally arise, as in the case of the economic models. The likelihood of such assumptions can be additionally compared in some cases, while they would be accepted "ad hoc" in others.

The input–output analysis has the Leontief production functions, with the particular characteristic that it can only occur with only one constant amount between the productive factors. Moreover, there is no elasticity of substitution between factors in these functions.

In the surveys, a somewhat more complex production function such as the CES (Constant Elasticity of Substitution) is assumed, with grade-one homogeneity (that is to say, constant scale outputs), and constant elasticity of substitution. Furthermore, when this elasticity is both constant and null, the CES function turns into a Leontief function.

Therefore, the assumptions included both in the survey and in the input–output model, are very similar, and the restrictions which imply that such assumptions, have been the main target of the different criticisms.

The work by Waters (1977) is one of the toughest attacks to the economic impact studies. Waters presents a series of criticisms according to the issues below:

1. Data on jobs, income, wages and other costs of service supplying in relation with the port, are not easily available.
2. The role of the port in the economic development of an area, is not available.
3. The price structure and the concept of a port-town without a port, is inconceivable.
4. Import and export impact on local consumption prices is undetermined.
5. The amount of workers, wages and income related to imports and exports, is uncertain and difficult to determine.

These difficulties and the arduous task of estimating the costs of the services provided by companies related to the port, make this a difficult and unfeasible task, and its results become unreliable. Waters (1977) shows this with a table presenting a relation of 64 possible direct costs to be surveyed, which falls into five big groups: ship expenditure in port services, terminal and port service costs, inland transport, ship crew costs and port services. This data is usually complemented with estimations of values of solid bulk and manufactured goods operated through the port itself or a port in the vicinity. Both cost amounts are considered to be exogenous, and a keynesian multiplier is applied to them to determine their effects on income. The result is a regional estimation based on the costs of the port. The method assumes: (a) that the price level is considered constant; (b) that exports of a region lead to an income growth; (c) that cost changes represent exogenous variables which can generate induced changes on the regional income level through the accelerator impact. To the previous assumptions, the lack of a technological change such as the introduction of containerization, should be added, which in the case of maritime transport means to ignore an important gain of quicker and cheaper costs which are more flexible. All these restrictions lead Waters to reject port impact studies and propose cost-benefit analyses based on surveys and completed with input–output models.

Chang (1978) defended port impact studies based upon the fact that their static condition is not a disadvantage but the aim of the study. This is the main focus of the study, a different point of view to analyse the economic effects of a port for a specific year. Moreover, it is important to carry out short term studies long term effects do not have to be estimated. Port impact studies do not aim to measure the effects of the investment made on port facilities, which is a long term issue, but to calculate the costs of such investment for a specific year. In addition, there may not be significant technological changes or considerable price changing during that year. It should be considered that these studies cannot be used for planning, but this is not the aim of port economic impact studies. However, a comparative analysis of different periods, with economic variables referred to a base period, can solve many of the problems arising from the annual analyses. As it was mentioned before, Waters recommended cost-benefit analyses instead of port impact studies. The former may be useful if the study focuses on the setting-up or the enlargement of a specific port. However, the cost-benefit analysis requires the estimation of the benefits obtained from the investment, and due to this, the port impact studies carried out in recent years can be very useful to identify the sources of the benefit and to check the validity of the benefit predictions through a cost-benefit analysis. Chang (1978) is aware of the problems arising from the multiplier, and proposes an econometric model which allows an approach to the dynamic problem which is: the estimation of the use of the future capacity based on the demand predictions, in order to increase the accuracy of port impact analyses. The traffic demand models have been studied by Coto-Millán (1986, 1988a, b, 1995). The above mentioned econometric models can be used as complement to the traditional port impact studies, so that they would help to take decisions on the future enlargement of a port, and could partly solve some planning problems.

6 Final Reflections

The methodologies presented in this paper can be valid and their possible applications allow different combinations in the use of surveys, manual estimations per ton and input–output multipliers. The manual methods per ton and the surveys provide a wide disaggregation into subsectors, which reveal the importance of these subsectors, and even of the companies or group of companies, in economic terms. Such methods particularly disaggregate the port industry impact for shippers, stevedores, services for ship docking (pilotage, towage, mooring,...), storing services and terminals, supplies,... etc.

The automatic methods cannot provide a sectorial distribution of total added value, jobs, wages and taxes, since all the results are presented in too aggregated terms. However, only with this method is it possible to obtain induced multipliers, since the survey would take us too long.

A good procedure is to develop surveys to estimate direct and indirect impacts and to compare the results obtained applying direct and indirect multipliers. If there are significant differences between the results, it is necessary to compare them to additional cross data (such as that from customs statistics, merchant register offices... etc), and therefore, to check the survey reliability and to make the necessary adjustments. If after this process, the differences are insignificant, the induced multipliers can be applied to both types of data (from the survey and from the multipliers), so that total impacts in intervals are obtained, with insignificant differences [less than 15%, as in Coto-Millán (1995) and between 5% and 15% as Villaverde and Coto-Millán (1995)].

There is another reflection with respect to the dynamic view of the studies. Naturally, as studies of a specific port are performed, some errors and omissions of the previous studies are corrected. These corrections must be taken into account when a dynamic analysis is carried out, so that the methodological differences are minimum, and the data provided is reliable. The different studies become more likely from a dynamic view to the extent that a comparative static analysis is rejected, and an analysis of the "stylized facts" as regards to port impact evolution is realised. There is evidence that such an exposure can provide an advantageous discipline to develop a different and deeper analysis framework than that developed so far.

Finally, it must be said that the use of econometric models on the productive capacity, of the port business productivity and cost and of the port service demand, can complement but will never replace port impact studies. Each of these exposures can give interesting but different results as regards to the economic aspects of a port.

References

Chang S (1978) In defense of port economic impact studies. Transport J 17:79–85

Coto-Millán P (1986) El transporte marítimo en España: 1974–83. Ph. dissertation, Department of Economics Theory, University of Oviedo (Abstract) and European Institute of Maritime Studies (1997)

Coto-Millán P (1988a) Funciones de demanda del transporte marítimo en España. Información Comercial Española 656:125–141

Coto-Millán P (1988b) El transporte marítimo en España (1974–1987): Peculiaridades. Información Comercial Española 658:101–109

Coto-Millán P, Sarabia JM (1993) Análisis de los servicios de transporte marítimo en España: Demanda, Precios, Renta y Series Temporales. Actas de las IX Jornadas de Economía Industrial. Suplemento: Investigaciones Económicas, 201–213

Coto-Millán P (1995) The Conditioned Demands of Spanish Sea Transport 1975–1990, International Journal of Transport Economics, XXII, N°3 325–346. New York

De Rus G et al (1994) Estimación de la Actividad Económica y Estructura de Costes del Puerto de La Luz y Las Palmas. Civitas y Autoridad Portuaria de Las Palmas

De Salvo J, Fuller D (1988) The economic impact of the port of Tampa. Center for Economic and Management Research, Tampa, Florida

De Salvo J (1994) Measuring the direct impacts of a port. Transport J 33:33–42

De Salvo J, Fuller D (1995) The role of price elasticity of demand in the economic impact of a port. Rev Reg Stud 25:13–35

Davis H (1983) Regional port impact studies: a critique and suggested methodology. Transport J 23:61–71

Fraga J, Seijas, JA (1992) El Puerto de Ferrol y su influencia en la economía de la comarca. Junta del Puerto y Ría de Ferrol

Hille J et al (1975) The economic impact of the port of Baltimore on Maryland. Division of Transport Business and Public Policy, College of Business and Management, University of Maryland, College Park, MD

Opuku KA (1990) Economic impact of the port industry on the New York and New Jersey metropolitan region. Port Authority of New York and New Jersey

Pinfold G (1991) Port of Halifax. Economic impact study. Port of Halifax

US Maritime Administration (1979) Port Economic Impact Kit. Office of Port and Intermodal Development, Washington

Villaverde CJ, Coto-Millán P (1995) El impacto del Puerto de Santander en la Economía Cántabra. Autoridad Portuaria de Santander

Villaverde CJ, Coto-Millán P (1996) Impacto Económico Portuario: Metodologías para su análisis y aplicación al Puerto de Santander. Autoridad Portuaria de Santander

Villaverde CJ, Coto-Millán P (1997) Economic impact of Santander port on its Hinterland. Int J Transport Econ 24(2):259–277

Waters RC (1977) Port economic impact studies: practice and assessment. Transport J 16:14–18

An Approach to the Contribution of the Port System

Pablo Coto-Millán, Miguel Ángel Pesquera, and Juan Castanedo Galán

Abstract In this paper, we propose a methodology to estimate the economic impact of the ports of a country. Inspired by traditional literature about the research on port impact, we obtain the estimation of the impact of ports on the Spanish economy in 1993.

Input–output techniques are used to obtain direct, indirect and induced economic impacts for the different types of economic agents: port users, port industry and port authority.

1 Introduction

A series of works have been carried out since the early 1960s in order to evaluate the economic effects of some specific infrastructure such as ports. Thus, the works by Hille et al. (1975) on the port of Baltimore, and the U.S. Maritime Administration (1979), represent the most important background of the research made on the impact of ports. Both studies develop a methodological base which aims to provide a working instrument which can be applied to the most important ports of the different American States. After the first works carried out, this methodology has been applied to different American, European and Canadian ports. Among the most recent works, it is worth mentioning Opuk's work (1990), on the industrial port of the metropolitan region of New York and New Jersey, as well as a work by Pinfold (1991) about the port of Halifax.

There is not much research in Spain, on the economic impact of ports. However, the background of this is represented by Fraga and Seijas (1992), and by De Rus et al. (1994), who carried out some studies developing a methodology of the

P. Coto-Millán (✉), M. Ángel Pesquera, and J. Castanedo Galán
University of Cantabria, Santander, Spain
e-mail: cotop@unican.es

P. Coto-Millán et al. (eds.), *Essays on Port Economics*, Contributions to Economics,
DOI 10.1007/978-3-7908-2425-4_12,

estimation of the economic activity. The work by Villaverde and Coto (1995), carried out more recently, researches on the economic impact through two methods. One of these methods is the estimation of the economic activity with it, the authors mainly lie on De Rus et al. work (1994). The second method proposed is the estimation of the economic impact, mainly inspired by the above mentioned international works. Such method accounts for both direct and indirect economic effects. Our aim in this paper is to estimate some macroeconomic magnitudes of the economic impact of Spanish ports, in terms of confidence intervals. The approach of these magnitudes can determine the decisions taken on infrastructure and transport policy.

The initial hypotheses are the following:

It is assumed that ports can be dealt with in aggregated terms, in the same way as only one port.

All the assumptions underlying the general input–output model, are also considered, especially all those carried out for the Input–Output Tables of the Spanish economy in 1989.

In addition to this, it is assumed that the Spanish production structure and the relations between sectors represented by the input–output table of 1989 (the most recent available), are similar to those of 1993.

Having made these assumptions, it is possible to apply Hille et al. methodology (1975) with the modifications proposed by Villaverde and Coto (1995). However, unlike these works, the economic impact will refer now to all the national ports.

As in the abovementioned works, we will try to determine both the direct and the indirect impacts of three groups of economic agents: port users, port industry and port authority.

The economic impacts to be determined for the abovementioned agents, will refer to:

- Employment
- Household income
- Government income
- Added Value

2 Economic Impact of Port Users

By port users we mean the group of people and/or firms using the ports. Port users are mainly import, export firms and shippers of national trade goods. However, when the economic impact is estimated, only the data corresponding to national exports is used. The reason for this is the use of the Input–Output Tables of the Spanish economy. Using the input–output methodology, imports are referred to as inputs, as well as national trade goods, while exports are referred to as outputs.

In order to estimate the economic impacts of port users, it is necessary to obtain information about the value of national exports. The Subdirección General de Aduanas de España (Spanish Tariff Assistant Management), has provided such information in two ways: an "invoice value" and a "statistical value", the latter being generally known as F.O.B. (free on board) value of goods.

In order to calculate the economic impact of port users, "the ex factory price value" of goods exported through the port, should be used. Although such information is not available, experts of the foreign trade sector have bee consulted and they have provided what is called the contrasted empiric regularity: "The ex factory price value of goods exported through the port, ranges from 75 to 80% of the F.O.B. value".

As regards to a specific port, the information related to all the export operations carried out in a particular year, is more easily available. For instance, in the study by Villaverde and Coto (1995), 3,670 operations were registered, so that it was possible to aggregate these operations in different tariff sectors, and then to establish tables corresponding to the sectors of the Input–Output Tables of the Spanish economy.

When national ports are considered, millions of operations are registered and it is more difficult to classify and deal with so much information. However, the Subdirección General de Aduanas de España, has provided the aggregate value of port exportation, F.O.B., of 2,175,487 million pesetas for 1993. With such data, it is possible to apply the previous hypothesis of 75–80% of the F.O.B. value as an approach to the ex factory price value.

Under the previous assumptions, it is possible to estimate a total impact of port users as Table 1.

Table 1 Total impact of port users (1993)

Impact	Wages (million ptas.)	Government income (million ptas.)	G.N.P. (million ptas.)	Employment
75% impact of F.O.B. Value				
Direct	453,589	26,106	868,019	329,586
Indirect	474,800	83,212	761,964	254,532
Induced[a]	870,467	146,845	1,400,463	466,642
Induced[b]	756,339	146,845	1,400,463	466,642
Total[a]	1.798,856	256,163	3,030,446	1,050,760
Total[b]	1.684,728	256,163	3,030,446	1,050,760
80% impact of F.O.B. value				
Direct	483,828	27,846	925,887	351,559
Indirect	506,453	88,760	812,762	271,501
Induced[a]	928,497	156,635	1,634,226	497,752
Induced[b]	806,761	156,635	1,634,226	497,752
Total[a]	1,918,778	273,241	3,372,875	1,120,812
Total[b]	1,797,042	273,241	3,372,875	1,120,812

[a]Induced average wage = indirect average wage

[b]Induced average wage = rate of direct and indirect average wages

3 Economic Impact of Port Industry

Port industry is known as the group of people and/or firms rendering every kind of service necessary to carry out port operations.

In order to estimate the economic impact of port industry, it is necessary to calculate the difference between F.O.B. value of goods exported, and the estimation of the ex factory price value of goods exported through national ports, which has been previously made. Direct, indirect and induced multipliers corresponding to the sector "service rendered by firms" of the Input–Output Tables of the Spanish economy in 1989, are applied to the value obtained. Economic impact of Port Industry is shown as follow Table 2.

4 Economic Impact of National Port Authority Capital Spending

A third impact corresponding to port authority capital spending, is usually considered in the research on the impact of specific ports. As regards to Spain, the capital spending of all the Spanish ports will be considered.

In order to estimate the economic impact of capital spending of all the Spanish ports, direct, indirect and induced multipliers corresponding to the "construction" sector of the Input–Output Tables of the Spanish economy, should be applied to these. The capital spending of Spanish ports consisted of 45,326 million pesetas for 1992.

With this information it is possible to produce Table 3.

Table 2 Economic impact of port industry (1993)

Impact	Wages (million ptas.)	Government income (million ptas.)	G.N.P. (million ptas.)	Employment
25% impact of F.O.B. value				
Direct	142,495	82,125	427,483	102,248
Indirect	50,036	11,965	116,389	28,281
Induced[a]	175,511	30,457	316,536	96,265
Induced[b]	154,834	30,457	307.166	96,265
Total[a]	368,042	124,547	860,405	226,794
Total[b]	374,365	124,547	860,405	226,794
20% impact of F.O.B. value				
Direct	113,995	65,700	341,986	81,798
Indirect	40,029	9,572	93,111	22,625
Induced[a]	136,252	24,365	253,226	77,012
Induced[b]	121,789	24,365	253,226	77,012
Total[a]	290,276	99,637	688,323	181,435
Total[b]	275,813	99,637	688,323	181,435

[a]Induced average wage = indirect average wage
[b]Induced average wage = rate of direct and indirect average wages

Table 3 Economic impact of Spanish port capital spending (1993)

Impact	Wages (million ptas.)	Government income (million ptas.)	G.N.P. (million ptas.)	Employment
Impact				
Direct	12,737	2,176	25,337	8,113
Indirect	9,926	2,040	19,989	5,303
Induced[a]	21,295	3,581	37,303	11,377
Induced[b]	19,578	3,581	37,303	11,377
Total[a]	43,958	7,797	82.629	24,793
Total[b]	42,241	7,797	82,629	24,793

[a]Induced average wage = indirect average wage
[b]Induced average wage = rate of direct and indirect average wages

5 Summary and Conclusions

Estimating the total impact of National Spanish Ports on the Spanish economy in 1993, the following magnitudes are provided in terms of confidence intervals.

- Between 1,256,988 and 1,372,399 jobs.
- Between 2,002,755 and 2,337,101 million pesetas household income.
- Between 363,597 and 405,585 million pesetas government income.
- Between 3,801,398 and 4,315,909 million pesetas G.N.P.

That is to say, Spanish Ports provide an Added Value ranging from 6.78 to 7.70% of total G.N.P., and generates an amount ranging from 8.20 to 8.95% of Spanish employment in 1993.

References

De Rus G, Roman C, Trujillo L (1994) Estimación de la Actividad Económica y Estructura de Costes del Puerto de La Luz y Las Palmas. Civitas, Madrid

Fraga J, Seijas JA (1992) El Puerto del Ferrol y su influencia en la economía de la comarca. Port Authority and Ría del Ferrol, El Ferrol

Hille J et al (1975) The economic impact of the port of Baltimore on Maryland. Division of Transport Business and Public Policy, College of Business and Management, University of Maryland, Maryland

Opuk KA (1990) Economic impact of the port industry on the New York and New Jersey. Port Authority of New York and New Jersey, New York

Pinfold G (1991) Port of Halifax. Economic impact study. Port of Halifax

U.S. Maritime Administration (1979) Port economic impact kit. Office of Port and Intermodal Development, Washington

Villaverde-Castro J, Coto-Millán P (1995) Impacto del Puerto de Santander sobre la Economía Cántabra. Port Authority of Santander, Santander

The Economic Impact of Ports: Its Importance for the Region and Also the Hinterland

Pablo Coto-Millán, Ingrid Mateo-Mantecón, and José Villaverde Castro

Abstract This study presents the calculations needed to obtain the overall economic impact of the Port of Santander in 2005. Clearly, to obtain this impact we must calculate the direct, indirect en induced impact, both of the Port Community and the Community of Port Users.

This study will show the analyses carried out to obtain the economic impact of the Port of Santander on the economy of the city of Santander, Cantabria and its hinterland in 2005. Significantly, it is the first time that the impact on the city and its hinterland has been calculated, given that previous studies calculated only the impact on the region. This contribution supplements the previous studies carried out in that it quantifies the significance of ports and the companies in their influence area that use their facilities in terms of the creation of value and employment.

1 Introduction

The input–output analysis framework is essentially attributable to two Nobel Prize-winners, Wassily Leontief and Richard Stone, who have contributed enormously to this field of research and influenced its subsequent evolution.

However, it is important to stress that Leontief's closed model (1936), as a mathematical formulation based on the Walrasian models, constitutes the most widely used methodology within synthetic economics' scientific developments as a whole (deductive–inductive).

Likewise, the economic reality being analysed by means of these methodologies requires an in-depth understanding of the interdependence of the various production activities. This means that the process is based on statistics agreed upon by the

P. Coto-Millán (✉), I. Mateo-Mantecón, and J. Villaverde Castro
University of Cantabria, Santander, Spain
e-mail: cotop@unican.es

P. Coto-Millán et al. (eds.), *Essays on Port Economics*, Contributions to Economics, DOI 10.1007/978-3-7908-2425-4_13,

various official bodies, and in many cases this becomes a costly and slow process of data collection that leads to a time lag of 4 or 5 years between the current period and the latest national or regional input–output table that was produced and published. Moreover, a number of authors indicate that the studies can, on occasions, present some weaknesses caused by the statistical handling of the various macroeconomic aggregates and the aforementioned time lag. However, despite these minor obstacles, there is no doubt that, for mesoeconomic analyses (sectorial, territorial or institutional), this methodology must be called upon. This is made evident by the number of articles and contributions referring to the input–output method in the Journal of Economic Literature (JEL) as well as those published in the Economic System Research (ESR) series, which exceeded 600 from 1960 to 2004. This clearly demonstrates how profusely useful this methodology is for applied economics (Fontela and Pulido 2005).

Given that this study constitutes an analysis of the economic impact of port activity, it is worth going over the previous studies that have been carried out in this field.

The first studies on the economic impact of port activity emerged in the United States in the second half of the 1960s. The ports of New York and New Jersey were the first to be taken into consideration. In the 1970s, the first methodological discussions took place, based on the development of the input–output model and its application to the measurement of the impact of ports. The main stances opposing this kind of study were advocated by Robert C. Waters, while those in favour had Semoon Chang as their main champion, and most of Waters' criticisms were dealt with.

It has to be said that the debate is still open and some authors continue to question this type of analysis (Benacchio et al. 2000). However, these criticisms are not conclusive and this argument is supported by the large number of studies in this sector carried out to date. In fact, it has been documented that over 200 studies on the economic impact of ports have applied this methodology in the United States and Europe.

Thus, studies similar to the one presented here have already been carried out, and this methodology has been used to obtain the economic impact of other Spanish ports, which makes it possible to draw comparisons between them. This study adopts the methodology that was used in the investigation carried out in 2001 by Coto-Millán et al., who analysed the economic impact of the Port of Santander in reference to the year 1998. This study will incorporate some modifications that will yield a more detailed analysis of the socioeconomic importance of the Port of Santander, while attempting to confirm that its influence extends beyond the region's physical boundaries. This is the main contribution presented.

In the majority of articles it can be inferred that the function of ports goes beyond their traditional role as mere embarkation and disembarkation points for goods and passengers, and that they have become centres in which a large number of activities are centred. These activities add value to the goods and are fully integrated into the international logistical and intermodal chains. On the basis of this approach, the direct, indirect and induced effects of the ports are identified.

This report presents the calculations needed to obtain the overall economic impact of the Port of Santander in 2005. Naturally, to obtain this impact we must calculate the so-called direct, indirect and induced impact, both of the Port Community and the Community of Port Users. Specifically, the various types of impact under analysis refer to the following aspects:

Direct impact refers to the volume of employment, wages and salaries, sales, Gross Operating Surplus (GOS), taxes and Gross Value Added (GVA) generated by the Port Community and the Port Users Community.

Indirect impact refers to the economic activity deriving from the purchasing and investment relationships that the Port Community and Port Users Community maintain with the rest of the economy.

Lastly, the *induced impact* refers to the effects produced as a result of private consumption deriving from the wages and salaries received by the workers who (in the Port Community or Port Users Community) undertake their activity due to the existence of the Port of Santander.

Thus, this report will present the analyses carried out to obtain the economic impact of the Port of Santander on the economy of the city of Santander, Cantabria and its hinterland in 2005. Significantly, it is the first time that the impact on the city and its hinterland has been calculated, given that previous studies calculated only the impact on the region.

It is worth bearing in mind that, prior to the economic impact study, an exhaustive market study of the companies linked to the Port of Santander was carried out. This study consisted of analysing over 200 companies of several autonomous regions of Spain that are associated with the Port. In this case, the economic impact analysis refers to 2005, since this was the last year for which all of the necessary information exists. This information, in so far as it relates to the companies involved in the port activity, has been extracted from surveys carried out on the companies, as well as the annual reports submitted by them to the Registry of Commerce. The macroeconomic information comes from both the national accounts and the regional accounts published by the National Institute of Statistics (Spanish initials INE).

The structure of this article is as follows: the introduction is followed by a section that summarises aspects of the Port of Santander's goods traffic in a brief analysis that covers the period from 1995 to 2005. Subsequently, the calculation of the direct impact will be presented, followed by the indirect and induced impact, calculated for Santander, Cantabria and the Port of Santander's Hinterland. The article ends with the main conclusions drawn from the work carried out.

2 Analysis of the Port of Santander's Traffic

This analysis of the traffic focuses on a period of ten years, from 1995 to 2005.

Prior to the analysis of the traffic and its evolution, it would be advisable to start by highlighting the significance of the Port of Santander within the state-owned port system in our country. Subsequently, the basic characteristics of the port traffic will be examined and its evolution between the aforementioned years will be analysed.

The traffic study presented in this document is just a small part of the study carried out, given that several databases have been analysed in order to obtain the Port of Santander's main clients, as well as the traffic that they are actually handling due to the use of these port facilities. In order to allocate traffic to a certain client, a detailed study of the customs codes, wharfs and shipping agents that handle the goods was carried out. Equally, the various taxes received by the Port Authority and the fees obtained for occupation of the public domain were analysed. This provided a clear idea of the main agents that work with the port and take advantage of the break-in-bulk point to carry out various operations on the goods, adding value to them throughout the process.

The Spanish state-owned port system is made up of 44 "ports of general interest" (the country's major ports), managed by 28 Port Authorities whose coordination and efficiency control falls to the *Puertos del Estado* ("State Ports", Spanish initials OPPE) public body, dependant on the *Ministerio de Fomento* (the Ministry of Development) and responsible for implementing the Government's ports policy.

The basic legal framework applicable to the Port Authorities is as follows: Law 48/2003, of November 26, on the economic regime and service provision of ports of general interest, alongside the previous laws of 1992 and 1997, provides the Spanish Port System with the instruments needed to improve their competitive position in an open and globalised market, establishing an autonomous management regime for the Port Authorities, which must undertake their activity following business criteria.

Within this framework, the system has evolved towards one in which the management of these major ports adheres to the "landlord" model, in which the Port Authority becomes a supplier of port infrastructure and land and regulates the use of the public domain, while the services are essentially provided by private operators under a system of authorisation or concession (*Anuario estadístico.* Puertos del Estado).

The total goods traffic of the Spanish Ports System in the year 2005 reached a figure of 441,017,185 tons, of which the Port of Santander, as a "port of general interest" managed by the Port Authority of Santander, contributed 6,700,878 tons, or in other words 1.52% of the total traffic. This makes the Port of Santander a medium-sized port within the System.[1] Table 1 shows an ordered list of the port authorities that, through their ports, have handled over four million tons. As can be seen, Algeciras recorded the highest volume of traffic in Spain, followed, at a certain distance, by Barcelona and Valencia. On the Atlantic coast, the most important port is Bilbao, which occupies fourth place nationally. The Port of Santander, with a volume of traffic exceeding 6.7 million tons, occupied sixteenth position in 2005.

[1]Figures taken from the Puertos del Estado's *Anuario estadístico* ("Statistical Yearbook").

Table 1 Total traffic of the Port Authorities handling over four million tons in 2005

Port authority	Tons
B. Algeciras	63,567,017
Barcelona	43,836,845
Valencia	40,862,005
Bilbao	33,237,170
Tarragona	30,986,590
Cartagena	26,770,218
Luz-Palmas	22,768,814
Gijón	21,566,058
Huelva	20,905,260
S.C.Tenerife	18,724,782
A Coruña	13,986,830
Castellón	13,372,753
P.Mallorca	13,027,304
El Ferrol	9,678,537
Almería	9,613,865
Santander	6,636,688
B.Cádiz	5,702,370
Pasajes	5,360,492
Avilés	4,948,704
Seville	4,857,393
Málaga	4,664,642
Vigo	4,252,290
Alicante	3,493,935
Pontevedra	1,849,467
Ceuta	1,543,459
Villa-Arosa	1,184,134
Melilla	801,291
Total tonnage	428,198,913

Source: compiled by author using figures taken from the Puertos del Estado's *Anuario estadístico* for 2005

In 2005, the distribution of the Port of Santander's traffic was as shown in Graph 1:

Firstly, there is a predominance of solid bulk traffic given that, in 2005, 5,139,652 tons were handled, representing 77% of the total traffic of the port. The solid bulk goods include coal and petroleum coke (1,198,009 tons), cereals and flours (619,233 tons), cement and clinker (448,881 tons) and soya cake (306,711 tons). The second most important type of traffic is general goods, which amounted to 1,264,012 tons handled in 2005, 18% of the total, comprising mainly cars (423,928 tons) and paper and wood pulp (272,774 tons). Lastly, the bulk liquid traffic, with 425,652 tons, accounted for 4% of the total and the most abundant goods within this group were chemical products (104,846 tons) and fuel oil (74,572 tons).[2]

[2]Figures taken from the Santander Port Authority Annual Report.

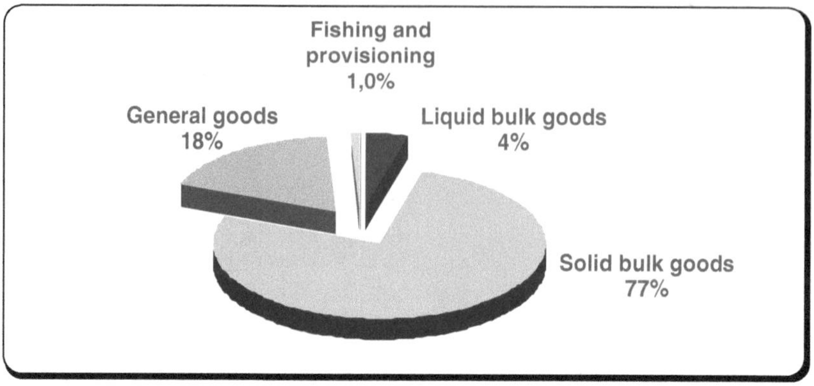

Graph 1 Distribution of the traffic of the Port of Santander. 2005
Source: compiled by author using figures taken from the Santander Port Authority's 2005 Report

Importantly, of the total of 6,700,878 tons of traffic, 1,522,313 tons are goods for exportation (goods loaded), which accounts for 77.3% of the total.

Having analysed the traffic, the economic impact of the Port of Santander can be estimated. The aim is also to assess the influence of the port on the city, on Cantabria and on its hinterland (or influence area).

3 The Direct Impact of the Port of Santander

3.1 Introduction

In this economic impact study, the terminology commonly used in previous studies has been modified and, throughout, the analysis focuses on two communities: the Port Community and the Port Users Community. Basically, the former comprises the companies that provide port or logistical services in the Port Service Zone, while the latter refers to the companies that require and use these services (the port's clients).

The Port Community includes the following agents: the Port Authority, Customs, the State Stevedoring Company (Spanish initials SESTISAN), the Stevedoring Companies, the Shipping Agents, the Tugging Companies, the Mooring Companies, the Pilots, the Bonded Warehouse and Cold Store, the logistical services, the port terminal, intra-port transportation and fishing. In other words, the Port Community comprises the economic agents that carry out operations that are necessary for loading and unloading goods and passengers, both in frontline port activities on the wharves and in supporting activities.

The Port Users Community includes the companies linked to the port's activity via a relationship of dependency, as clients that require services, whether because they need products, raw materials and inputs to carry out their production process, or because they wish to sell their finished products in other parts of the world. The clients have been categorised into various groups, which include: the logistical operators, shipyards, shipowners and dependant industry (production sector) on the one hand, and the road and rail transportation industry on the other. The latter is essential for the modal interchange of the goods by making the port an important and necessary link in the value chain. Thus, the sum of the effects of the companies, as clients, and of the land transport sector will provide the total impact of the Port of Santander's Community of Users.

The variables that will be taken into account to estimate the economic impact of the Port of Santander were mentioned in the introduction: number of employees, sales, salaries and wages, Gross Operating Surplus (GOS), taxes paid and Gross Value Added (GVA). In this study, and with the aim of estimating the economic impact for the year 2005, the Annual Report and Profit and Loss Account for the financial year were requested from the registries of commerce of the various autonomous regions where the companies of the Port Community and the Port Users Community submit their annual accounts (namely the Registries of Commerce of the various provinces of Castilla y León, Madrid, Catalonia and the Basque Country).

Furthermore, a questionnaire was sent requesting various kinds of information from all of the companies included in the study. The questionnaire yielded data on investments, the percentage of activity that is dependant on the services provided by the Port of Santander, staff expenditure, demand, land transport modes used and the relationship with other companies that have been invoiced for purchases, supply or investment in 2005.

Some companies declined to respond to the questionnaire, so some of the aforementioned variables were estimated using their Annual Reports and Profit and Loss Accounts. These were obtained from the Registries of Commerce. In the cases in which the necessary information was unavailable because it was absent from the annual accounts, estimates were made for the different variables using previous impact studies and/or their similarity with other companies in the same sector for which reliable data was available.

The calculation criteria for these figures are shown below and they were applied to all port agents, whether belonging to the Port Community or the Port Users Community:

Wages = wages and salaries + social contributions (allocated in the Profit and Loss Account).

Employment, understood as the wage-earning staff, whether permanent or temporary.

Sales = net turnover + work carried out by the company for fixed assets + other operating income.

Gross Operating Surplus (GOS) = profits before tax (or, preceded by a minus sign, losses before tax) − capital subsidies transferred at the close of the financial year + allocations for amortisation of fixed assets + variation in traffic forecasts and loss of irrecoverable debts + financial and assimilated expenses + variation in forecasts for intangible fixed assets, tangible fixed assets and audit portfolio.

Net taxes = taxes − subsidies

Gross Value Added (GVA) = GOS + Wages

The summaries relating to the estimation of the direct impact of the Port Community and the Port Users Community are now presented below.

3.2 Direct Impact of the Port and Users Community

This section explains various aspects that have been taken into account in the calculation of the direct impact of each agent that makes up the Port Community of the Port of Santander. It is worth mentioning that the companies in question are located in the city of Santander and operate within the port site or very near it.

The Port Community comprises the following agents:

- State Administration
- Shipping companies
- Pilots' Association
- Tugging companies
- Mooring companies
- Stevedoring companies and SESTISAN
- Port terminals
- Intra-port transportation
- Companies that provide logistical services
- Bonded Warehouse
- Fishing and others

A summary of the direct impact of these companies is provided below.

The total direct impact of the Port Community of Santander is presented in the last row of Table 2.

Under the title "Community of Users of the Port of Santander", a set of companies were taken into account, grouped into the following categories:

- Shipowners
- Shipyards
- Dependant Industry Activity
- Logistical Operators
- Land Transport Sector

As with the Port Community, Table 3 lists the values of the various economic figures corresponding to the Port Users Community in its entirety and the last row shows the total impact, equivalent to the sum of each of the figures obtained for each of the aforementioned agents.

Table 2 Port Community. 2005 direct impact

Port community	Wages	Employment	Sales	GOS	Taxes	GVA
Port authority	6,360,677	194	22,936,417	6,981,941	295,355	13,342,618
Customs	937,500	30	0	0	214,025,904	937,500
Shipping agents	2,878,796	74	25,913,115	1,662,458	39,312	4,541,253
Pilots' association	1,218,823	18	1,950,117	195,012	0	1,413,835
Tugging companies	790,022	16	4,587,908	3,647,161	0	4,437,183
Mooring companies	518,550	12	668,623	99,208	167,746	617,758
Stevedoring companies	10,593,088	207	34,678,271	4,313,081	570,228	14,906,169
Port terminals	2,589,036	70	24,453,013	3,524,711	35,553	6,113,747
Intra-port transportation	1,999,641	72	6,183,791	341,298	0	2,340,940
Logistical services	3,481,745	90	87,285,867	5,517,452	64,811	8,999,197
Bonded Warehouse	790,240	21	2,268,385	421,691	119,577	1,211,931
Fishing and others	7,368,240	335	11,791,088	595,262	115,051	7,963,502
Direct impact	39,526,359	1,139	222,716,593	27,299,275	215,433,537	66,825,634

Source: compiled by author. The employment is in units and the rest of the figures are in 2005 euros

Table 3 Port Users Community. 2005 direct impact

Users Community	Wages	Employment	Sales	GOS	Taxes	GVA
Shipowners	3,838,716	109	47,467,374	1,030,957	28,260	4,869,673
Shipyards	3,451,334	57	12,340,828	860,436	−743,166	4,311,770
Dependant Industry	273,323,571	6,608	4,020,159,886	448,515,620	7,664,229	721,839,188
Logistical Operators	7,959,231	209	190,027,198	5,291,292	81,900	13,250,523
Transport	22,368,153	620	72,278,412	7,835,601	63,563	30,203,754
Direct Impact	310,941,003	7,603	4,342,273,700	463,533,905	7,094,786	774,474,908

Source: compiled by author. The employment is in units and the rest of the figures are in 2005 euros

3.3 The Total Direct Impact of the Port of Santander

Once the direct impact of the Port Community and the Port Users Community has been obtained by adding together the relevant figures, we obtain the total economic impact of the Port of Santander in the year 2005. The figures relating to this impact appear in the last row of Table 4.

Once the direct impact has been calculated, the indirect and induced impact of the Port Community will be calculated. This impact will be measured for the city of Santander and for the wider region of Cantabria, as well as for the port's *hinterland* (which goes beyond the regional boundaries).

4 The Indirect and Induced Impact of the Port of Santander

4.1 Methodology

As outlined in the general introduction, the economic impact study estimates two types of effect, called primary and secondary effects. The primary effects, also known as the direct impact, include all of the activities needed for the activity in question, which were calculated in the previous section. The secondary effects, which are those that are now under consideration, refer to the economic activity which takes place within the primary activity's influence area and is therefore economically dependant on it. The secondary impact is, in turn, made up of two types of effect known as indirect and induced effects. At the risk of repetition, it is worth pointing out again that the indirect effects refer to all of the economic activities undertaken in the region where the primary activity takes place which are dependant on this activity through the purchase or sale of goods and/or services. The induced effects are a result of the consumption expenditure deriving from the

Table 4 Direct impact of the Port of Santander in the year 2005

Port of Santander	Wages	Employment	Sales	GOS	Taxes	GVA
Port community	39,526,359	1,139	222,716,593	27,299,275	215,433,537	66,825,634
Users community	310,941,003	7,603	4,342,273,700	463,533,905	7,094,786	774,474,908
Direct impact	350,467,362	8,742	4,564,990,294	490,833,180	222,528,324	841,300,542

Source: compiled by author. The employment is in units and the rest of the figures are in 2005 euros

wages and salaries paid to the people employed directly or indirectly in the primary activity under consideration.

Although there is more than one method for quantifying economic impact, the most commonly used is based on the input–output analysis, using the Input–Output Tables (IOT).

An IOT is a set of countable relationships that reflect the interdependencies that exist in the production system of a given economic area. By definition, the aim is to analyse the branches of activity by adding together the Homogenous Production Units (HPUs), but given that these units cannot be observed directly, the European System of Integrated Economic Accounts (ESA) replaces them with Kind-of Activity Units (KAUs), which are assimilated within the business establishment that undertakes a single activity (Table 5).

On the basis of the KAU's, the economic activity branches are generated – a grouping carried out in accordance with the National Classification of Economic Activities (NCEA). These activities are in turn defined in the National Classification of Goods and Services (NCGS), which distinguish around 6,000 basic

Table 5 Correspondence between IOT 2000 branches and CRE branches

Branches		CNAE-93
1	Agriculture, livestock farming, hunting and forestry	01,02
2	Fishing	05
3	Extraction of energy products, other minerals and oil refining	10, 11, 12, 13, 14, 23
4	Production and distribution of electrical energy, gas and water	40.1, 40.2, 40.3, 41
5	Food, drink and tobacco	15.1–15.9, 16
6	Textiles, clothing, leather and footwear	17, 18, 19
7	Wood and cork	20
8	Paper, publishing and graphic arts	21, 22
9	Chemical industry	24
10	Rubber and plastic	25
11	Other non-metallic mineral products	26
12	Metallurgy and metallic products	27, 28
13	Machinery and mechanical equipment	29
14	Electrical, electronic and optical equipment	30, 31, 32, 33
15	Manufacture of transportation materials	34, 35
16	Assorted manufacturing industries	36, 37
17	Construction	45
18	Trade and repair	50, 51, 52
19	Hospitality industry	55
20	Transport and communications	60.1, 60.2, 60.3, 61, 62, 63, 64
21	Financial intermediation	65, 66, 67
22	Market business and real-estate services	70, 71, 72, 73(p), 74
23	Public administration	73(p)*, 75
24	Education	80(p)
25	Healthcare and social services	85(p)
26	Other activities	90(p), 91.1, 91.2, 91.3, 92(p), 93, 95

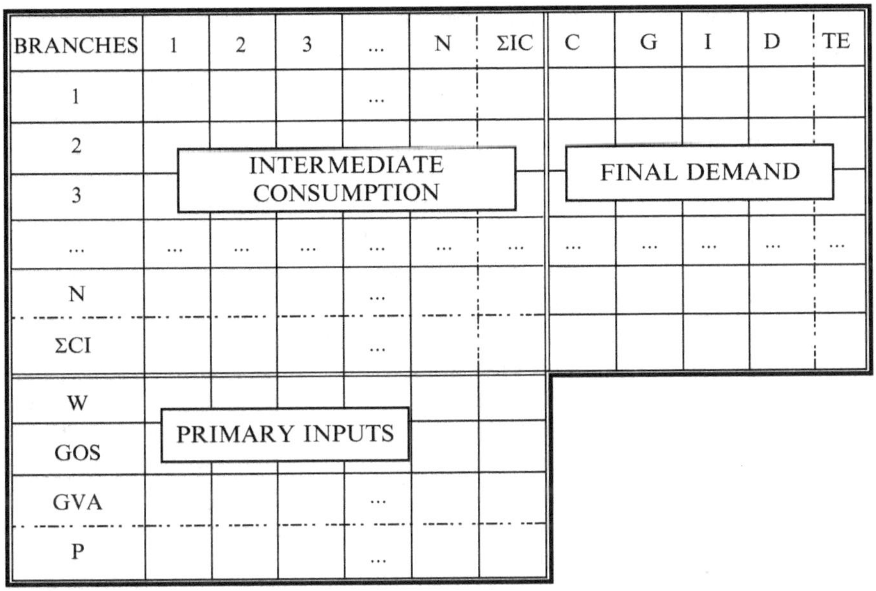

Graph 2 IOT. Adding table
Source: compiled by author

products. Thus, the IOT uses the branches of economic activity in its tables rather than the ideal branches of homogenous production.

The IOTs are structured into three large blocks: intermediate consumption, final demand and primary inputs. This is shown in Graph 2:

The intermediate consumption block (top left) shows the consumption generated through the production process between the various goods and services in terms of resources and employment, or in other words, as the inputs and outputs of a branch. Assuming the existence of *n* branches of activity, the reading in columns indicates, in each cell, the quantity of inputs that are acquired from the branch of activity located in the corresponding row. If the table is read in rows, the cells will contain the outputs or final products that the activity shown in the row sells to the branch of activity located in the corresponding column, to be used as factors of production by the branch. The last row of the table indicates the total amounts acquired as inputs by each branch from the rest of the branches of economic activity, while the last column shows the amount of outputs that each branch of activity sells to the rest of the branches of the economy. The sums of the totals row and column must tally or, in other words, the Intermediate Consumption (IC) totals must be the same whether they are being obtained by rows or calculated by columns.

The final demands block contains the various destinations to which the branches of activity direct their production. In this case, the rest of the destinations which each branch directs its production to are shown; that is, the outputs for final consumption in the home (C), outputs for public consumption (G), outputs for gross capital formation (I), the sum of the final destinations excluding intermediate consumption (D) and the total sum of the final destinations (TE).

The final block, for primary inputs, contains the primary factors of production. If we analyse the rows, the first coincides with the last block of the Intermediate Consumption; that is to say, the Intermediate Consumption of each branch from the rest of the branches of the economy (IC). W contains remuneration from the employment factor, i.e. the wages and salaries of the workers. The GOS represents remuneration from the capital factor contained in the Gross Operating Surplus. GVA is the Gross Value Added of each branch, a figure which, in general terms, is provided by adding together the factors' incomes (W + GOS). And lastly, P represents the value of the production of each branch as the sum of the value of the intermediate consumption plus the Value Added (IC + GVA).

The values in row P for each branch tally with those shown in the Total Employment (TE) column or, in other words, the total value of the production of each branch must be identical to the value of the employment used for this production.

An overall analysis of all of the blocks allows us to interpret the various rows and columns of each branch of activity. The column of a given branch of activity represents its cost structure, since it shows both the intermediate consumption and the remuneration for the factors needed for production. The row for a given branch represents the destinations of the production, whether intermediate or final.

4.2 The Input–Output Demand Model and the Gross Value Added Model

Using the Walrasian general equilibrium as a starting point, Leontief[3] attempts to restrict and simplify this model by means of a series of suppositions, the most important of which is the elimination of all of the price effects in the replacement of inputs; there are no limited factors and the technical production coefficients are fixed. These coefficients represent the use that any branch of activity makes of another per unit of production. Thus, if the technical coefficient is represented by a_{ij}, it is interpreted as the use that branch j makes of products from branch j (x_{ij}) per unit of product j (P_j); that is, $a_{ij} = x_{ij}/P_j$.

The establishment of this supposition (the non-replaceability of the inputs by prices or technology) gives rise to the three central hypotheses of the model:

- Homogeneity: each product is supplied for just one branch of activity.
- Proportionality: the amount of input used depends on the level of production of each branch. This presupposes the existence of constant returns to scale.

[3]The early input–output tables produced for the United States economy, as well as the corresponding model, were devised by the economist Wassily Leontief in 1936.

- Additivity: there are no external economies or diseconomies or, in other words, the total effect of various types of production constitutes the sum of the individual effects.

Leontief's production function or fixed proportion function for obtaining a product in which two factors are used, X and Y, tells us that there is only a combination of productive factors suitable for producing each level of output.

The basic demand model, on the assumption that a_{ij} is the data and that the final output demand (D) is determined exogenously, allows the calculation of the production levels (P) needed to meet that demand. The model's set of equations is obtained from the IOT and the following is applicable:

$$\begin{aligned} x_{11} + x_{12} + x_{13} + \cdots + x_{1n} + D_1 &= P_1 \\ x_{21} + x_{22} + x_{23} + \cdots + x_{2n} + D_2 &= P_2 \\ \vdots \quad + \vdots + \vdots + \cdots + \vdots + \vdots &= \vdots \\ x_{n1} + x_{n2} + x_{n3} + \cdots + x_{nn} + D_n &= P_n \end{aligned} \tag{1}$$

where each row assumes that the sum of the destinations of the production of a branch to the rest of the branches (x_{ij}) and to the final demand (D), constitutes the value of its production (P). Given that $a_{ij} = x_{ij}/P_j$ we can replace in (1):

$$\begin{aligned} a_{11}P_1 \ + a_{12}P_2 + a_{13}P_3 + \cdots + a_{1n}P_n + D_1 &= P_1 \\ a_{21}P_{12} + a_{22}P_2 + a_{23}P_3 + \cdots + a_{2n}P_n + D_2 &= P_2 \\ \vdots + \vdots \quad + \vdots \quad + \cdots + \quad \vdots \quad + \vdots &= \vdots \\ a_{n1}P_1 + a_{n2}P_2 \ + a_{n3}P_3 + \cdots + a_{nn}P_n + D_n &= P_n \end{aligned} \tag{2}$$

Reordering expression (2), it follows that:

$$\begin{aligned} (1 - a_{11}P_1) - a_{12}P_2 \quad - a_{13}P_3 - \cdots - a_{1n}P_n \quad &= D_1 \\ -a_{21}P_1 \quad + (1 - a_{22}P_2) - a_{23}P_3 - \cdots - a_{2n}P_n \quad &= D_2 \\ \cdots\cdots\cdots\cdots\cdots\cdots\cdots\cdots\cdots\cdots\cdots\cdots & \\ -a_{n1}P_1 \quad - a_{n2}P_2 \quad - a_{n3}P_3 - \cdots + (1 - a_{nn}P_n) &= D_n \end{aligned} \tag{3}$$

In matrix terms, the demand model could be represented by the following equation:

$$(I - A)P = D \tag{4}$$

where I is the identity matrix (all of the elements are zero except those corresponding to the main diagonal, which are equal to the unit), A is the technical coefficients matrix, P is the production column vector and D is the final demand column vector. The matrix $(I - A)$ is called the Leontief matrix.

However, the most common reformulation of the model is obtained by pre-multiplying both members of the (4) by $(I - A)^{-1}$, so that:

$$P = (I - A)^{-1}D \tag{5}$$

This expression reflects in a simple manner the validity of the model when it comes to resolving the problem posed: the necessary production of each branch (vector P) to meet a final demand (vector D) determined exogenously, given a productive structure shown by Leontief's inverse matrix $(I - A)^{-1}$.

The economic impact model for the Gross Value Added is explained below. Taking the IOT shown in the intermediate consumption block, analysing the branches of activity by rows, we can extract the following set of equations:

$$\begin{array}{c} x_{11} + x_{12} + x_{13} + \cdots + x_{1n} + Y_1 = X_1 \\ x_{21} + x_{22} + x_{23} + \cdots + x_{2n} + Y_2 = X_2 \\ \vdots + \vdots + \vdots + \cdots + \vdots + \vdots = \vdots \\ x_{n1} + x_{n2} + x_{n3} + \cdots + x_{nn} + Y_n = X_n \end{array} \tag{6}$$

where each row assumes that the sum of the destinations of the output of a branch to the rest of the branches (x_{ij}), plus the vector Y_i, which represents the spending on consumption and investment related to the activity considered in the case of the analysis of indirect effect, and the production aimed at private consumption in the case of the induced effects analysis, will be equal to a certain amount of production X_i. If we replace x_{ij} with its expression in technical coefficient terms, the following applies:

$$\begin{array}{c} a_{11}X_1 + a_{12}X_2 + a_{13}X_3 + \cdots + a_{1n}X_n + Y_1 = X_1 \\ a_{21}X_1 + a_{22}X_2 + a_{23}X_3 + \cdots + a_{2n}X_n + Y_2 = X_2 \\ \vdots + \vdots + \vdots + \cdots + \vdots + \vdots = \vdots \\ a_{n1}X_1 + a_{n2}X_2 + a_{n3}X_3 + \cdots + a_{nn}X_n + Y_n = X_n \end{array} \tag{7}$$

The technical coefficients are constant in accordance with the hypotheses of the demand model. In other words, each branch will keep constant the proportion of its production that goes to the other branches, whatever the overall level of the production that has been reached. This hypothesis allows the use of the national or regional IOTs to obtain the technical coefficients and subsequently incorporate them into the set of simultaneous equations (7). Operating in a similar way to the previously presented case, the following expression is obtained:

$$X = (I - A)^{-1}Y \tag{8}$$

Using the IOTs, and dividing the value added of each branch by the value of the output generated, we obtain the value added coefficients:

$$
\begin{aligned}
\alpha_1 &- \frac{GVA_1}{X_1} \\
\alpha_2 &= \frac{GVA_2}{X_2} \\
&\vdots \\
&\vdots \\
\alpha_n &= \frac{GVA_n}{X_n}
\end{aligned}
\tag{9}
$$

where GVA_i is the Gross Value Added of ouput branch i, and α_i is the coefficient of the corresponding value added. The model assumes that these coefficients remain constant, much like the technical coefficients. In other words, each branch generates the same percentage of Gross Value Added regardless of the output level that the branch reaches. The IOT allows us to calculate the value of these coefficients so that they can subsequently be incorporated into our analysis.

To find the value of the value added in the set of simultaneous equations, the following applies:

$$
\begin{aligned}
GVA_1 &= \alpha_1 X_1 \\
GVA_2 &= \alpha_2 X_2 \\
&\vdots \\
&\vdots \\
GVA_n &= \alpha_n X_n
\end{aligned}
\tag{10}
$$

and replacing X_i with its value expressed in (8), we obtain the value added model:

$$
GVA = A_{GVA}(I - A)^{-1}Y \tag{11}
$$

where GVA is the Gross Value Added column vector for each branch and A_{GVA} is the diagonal matrix of the value added coefficients.

According to the expression obtained, given some predetermined indirect or induced effects (Y), a production structure (represented by Leontief's inverse matrix $(I - A)^{-1}$) and an economic structure (determined by the coefficient matrix A_{GVA}), we obtain the Gross Value Added (GVA) that the primary activity under consideration generates in the various branches of economic activity.

4.3 The Indirect and Induced Impact of the Port of Santander

This study uses the input–output method as the basis for the analysis. As mentioned in the general introduction, this methodology has a fairly long tradition in our country, especially when applied to the analysis of the economic activities of ports. The methodology has been used to carry out a large number of studies for various ports of Northern Spain, including (TEMA 1994), Santander (Villaverde and Coto-Millán 1996) and Gijón.

The methodology applied in this economic impact study is based on the premise that the Port of Santander's activity is considered a primary activity.

Having calculated the direct effects in the previous section, this section will estimate the indirect and induced effects following the methodology already described. Given that the Port of Santander is located in Cantabria, and that there are no recent IOTs for the region, a regionalisation of the latest national IOT, referring to the year 2000, has been used. Furthermore, we also make use of information from the Spanish Regional Accounting (Spanish initials CRE); this information is complemented by the purchasing and investment data for the Port of Santander, as well as the information corresponding to the consumption spending deriving from direct and indirect wages and salaries (of the individuals who work in activities related to the port).

In particular, the estimation of the indirect and induced effects is made following an analytical process which comprises three stages. In the first stage, the indirect and induced impact vectors are estimates. In the second, the input–output methodology is applied to calculate a value-added vector. And in the third and final stage, the indirect and induced impact is calculated taking into account the impact vectors, the added-value vectors and the indices for all the relevant figures (employment, sales, salaries and GOS) calculated in relation to the GVA.

As mentioned above, the IOT used in this report refers to the year 2000. This table considers the national economy divided into 73 branches of activity coinciding with the divisions of the CNAE-93 (the Spanish National Classification Economic Activities). This breakdown of branches was also used for the technical coefficient matrices and Leontief's inverse matrix. In order to simplify the interpretation of the results, the original IOT has been reduced to 26 branches of activity in accordance with the breakdown used in the Spanish Regional Accounting (CRE). The CRE considers 27 branches of activity. Given, however, that the last branch ("Homes that use domestic staff") has a value of zero in all cells of the intermediate consumption block of the IOT, we decided to subsume it into Branch 26 ("Other services and social activities").

The next step in our analysis consists of regionalising the national IOT. Although there are various methods for carrying out this operation, the "modified coefficients of localisation" method has been chosen. To this end, again in relation to the year 2000, we calculated the corresponding "localisation coefficients" for all 26 branches of activity, both for Cantabria and for Spain.

This method consists of applying the formula indicated below for each of the 26 branches of activity:

$$\text{SLQ} = C_I = \frac{GVA_i^R / GVA^R}{GVA_i^N / GVA^N}$$

Where: SLQ: Simple Location Quotient.
GVA_i^R: GVA of Cantabria of sector i
GVA^R: GVA of Cantabria total
GVA_i^N: GVA of Spain of sector i
GVA^N: GVA of Spain total

With these coefficients, a matrix of 26 rows by 26 columns is constructed, in which all of the cells except those corresponding to the main diagonal in which the aforementioned localisation coefficients appear have a value of zero. When the localisation coefficient exceeds the unit, according to the convention adopted the value is equal to one, hence, in these cases, the term "modified localisation coefficients".

By multiplying this matrix by the national technical coefficients, the matrix of regional technical coefficients is obtained. Likewise, by multiplying this matrix by the distribution coefficients matrix, the distribution coefficients matrix for Cantabria is obtained.

Thus, as a kind of compendium of the information detailed above, we have the following matrices: Total IOT for Spain, used to obtain the technical coefficient matrices, distribution coefficient matrices and Leontief's inverse matrix. Equally, the national IOT is used to calculate for Cantabria the technical coefficient matrices, distribution coefficient matrices and Leontief's inverse matrix by means of using the modified localisation coefficients. All of these are based on the 26 branches of activity system. The results obtained are shown in Table 1 of the appendix. With these coefficients, a matrix of 26 rows by 26 columns is constructed, in which all of the cells except those corresponding to the main diagonal in which the aforementioned localisation coefficients appear have a value of zero. By multiplying this matrix by the national technical coefficients matrix, the regional technical coefficients matrix is obtained (Table 2 of the appendix, appearing at the end of the article). Finally, the regional inverse matrix $(I - A^R)^{-1}$, which is the one used in the calculations relating to the indirect and induced effects, is obtained.

The impact vectors have also been calculated. And, to determine the elements corresponding to these vectors, certain figures must be broken down into sectors, specifically:

1. To calculate indirect effects, the purchases and investments made by the Port Community and the Users Community must be broken down into sectors; the vector thus obtained is called the "indirect impact vector".
2. To calculate the induced effects, private consumption must be broken down into sectors; the vector thus obtained is called the "induced impact vector".

Given that we do not have our own information corresponding to the autonomous region of Cantabria, the Port of Santander's purchases are broken down into sectors by assuming that the "Maritime Transport" branch, in which the activity undertaken by the Port of Santander is included, has a similar distribution to the national equivalent; this breakdown is obtained from the national IOT.

As far as breaking down the investments is concerned, we proceeded in the same way as with the purchases, albeit making use, initially, of the IOT's column relating to gross capital formation.

To calculate the purchases and investments, this study took into account the following expressions:

Purchases = operating consumption + other operating costs − taxes (figures obtained from the Profit and Loss Account).

Investment = tangible and intangible fixed assets for the year (obtained from the report that accompanies the Registry of Commerce accounts, in the tangible and intangible fixed assets sections).

The figures for the of the Port Community's purchasing (150,695,017 euros) and investments (15,736,555 euros) have been obtained by applying the aforementioned expressions.

Furthermore, the induced effects vector is calculated by breaking down into sectors the internal private consumption obtained with the information provided by the national IOT.

Once the indirect and induced impact vectors have been obtained, it is possible to calculate the indirect and induced effects, although firstly, it is necessary to estimate, with the information provided by the CRE, the ratios of wages and salaries, employment and GOS in relation to the GVA for each branch of activity. Lastly, by multiplying the value added at factor cost vector, deflated to 2005 euros, by the value added vector obtained in the previous step, the induced impact values are determined for the various economic figures considered in this study.

Taking into account the results obtained both for the direct impact and for indirect and induced impact, having applied the methodology cursorily explained, Table 6 presents a summary of them. Naturally, the total impact of the Port Community is calculated by adding together the direct, indirect and induced impact of the aforementioned community.

Table 6 Total impact of the Port community 2005

Port community	Wages	Employment	Sales	GOS	Taxes	GVA
Direct impact	39,526,359	1,139	222,716,593	27,299,275	215,433,537	66,825,634
Indirect impact	40,037,430	1,974	356,216,437	45,209,012	9,144,265	87,388,484
Induced impact	14,088,135	842	85,438,275	13,674,681	2,909,366	27,803,770
Total impact	93,651,925	3,955	664,371,305	86,182,968	227,487,168	182,017,887

Source: compiled by author. The employment is in units and the rest of the figures are in 2005 euros

In the light of these results, we can indicate that the impact of Santander's Port Community is of 3,955 jobs. The calculations also reveal that these jobs are salaried with 93.6 million euros and that the GOS and GVA were 86.2 million and 182 million euros, respectively.

To understand the significance of these figures and in particular the ones referring to employment and GVA, we must compare them with Cantabria's figures for 2005, in which the number of people employed was 255,200 and the GVA generated reached 10.16534 billion euros. Thus, the Port Community employment represents around 1.6% of the total employment and 1.8% Cantabria's GVA. This means that, for each 100 jobs created in the region, 1.6 are created by the Port Community, and that for each 100 euros of GVA generated in the region, almost 2 euros (1.8) are provided by Santander's Port Community. Furthermore, the results indicate that quality employment is being created, given that the salaries received by the Port Community workers are competitive.

4.4 The Port of Santander's Impact on the Region. Conventional Approach

Having calculated the economic impact of the Port Community, we can proceed with the broken down calculation of the overall economic impact of the Port of Santander and its influence on various geographical areas. This analysis will allow us to study the influence of the port on various regions and this is an innovation compared to the studies that have previously been carried out.

In particular, we analyse the economic impact of the Port of Santander on the city of Santander, Cantabria, and subsequently on the rest of the regions of its hinterland or influence area.

To calculate the economic impact of the port on Santander, we have assumed that all of the Port Community's activity takes place in this city. We also took into account the companies of the Port Users Community which carry out their activity in Santander. Therefore, the selection criterion for the companies was, as it is throughout the study, the geographical area in which the companies related to the Port carry out their activity.

The first aspect analysed was the direct impact of the companies that carry out their activity in Santander, for the purpose of which (as mentioned above) the annual accounts were requested from the Registry of Commerce, following which we proceeded to calculate the percentage of the companies' activity that is carried out with the port through surveys and estimates based on other studies.

Thus, adding together the figures for the direct impact of the Port Community and Port Users Community companies, calculated as indicated, the direct impact of the port on the city of Santander is obtained. These figures appear in Table 7:

Once we know these variables, which represent the port's direct impact on the city, figures for purchases (409,349,883 euros) and investments (20,260,864 euros) can also be obtained, and the input–output method can be used to calculate the

Table 7 Direct impact of the port on the city of Santander

Santander	Wages	Employment	Sales	GOS	Taxes	GVA
Users community	28,394,565	633	322,119,980	37,803,374	298,809	66,197,938
Port community	39,526,359	1,139	222,716,593	27,299,275	215,433,537	66,825,634
Direct impact	67,920,924	1,772	544,836,573	65,102,648	215,732,346	133,023,572

Source: compiled by author. The employment is in units and the rest of the figures are in 2005 euros

port's indirect and induced impact on the city of Santander. The results obtained are shown in Table 8.

Bearing in mind that the city of Santander's disposable income amounts to 38% of Cantabria's, it can be assumed that Santander's GVA represents approximately 38% of Cantabria's GVA. Consequently, we are saying that, in absolute figures, the city of Santander's GVA in 2005 was of around 3.825 billion euros.

Comparing the previous figure with the GVA generated by the port in the city of Santander, it follows that the port has created approximately 11% of the Cantabrian capital's GVA.

As far as employment is concerned, and assuming that there is a direct relationships between the number of people who live in a population centre and the volume of employment, it would be possible to estimate this figure for Santander. We can assume that Santander's employment represents approximately 33.8% of the total employment of Cantabria, which implies that the figure for total employment in Santander in the year 2005 was of around 86,260 workers. This means that the impact of the port on employment in the city is of around 9.6% of the total. In other words, for each 100 jobs generated in Santander, almost ten depend, in a direct, indirect or induced manner, on the activity generated by the port.

Having analysed the influence of the port on the city of Santander, we can study its influence on the wider region. Given that the city of Santander belongs to Cantabria, the variables of the study for Cantabria will encompass both the results obtained for the city and those corresponding to the rest of the companies that are located in Cantabria (but outside of Santander) in the part of their activity related economically speaking to the Port of Santander.

Taking as reference the employment figure for Cantabria in 2005 of 255,200 workers and the GVA of 10.16534 billion euros, it follows from the figures in Table 8 that the employment generated by the Port of Santander represents 4.5% of the region's total employment, while the GVA accounts for 6.1% of Cantabria's total. In other words, for each 100 euros of wealth created in Cantabria, 6 are due to the Port of Santander. And, for each 100 jobs created in Cantabria, 4.5 are due to the economic activity associated with the Port of Santander.

4.5 *The Port of Santander's Economic Impact on the Hinterland*

As for the hinterland, we have calculated the direct, indirect and induced impact of the Port of Santander on the regions of Castilla y León, Catalonia, Madrid and the

Table 8 Total impact of the port on the city of Santander

City of Santander	Wages	Employment	Sales	GOS	Taxes	GVA
Direct impact	67,920,924	1,772	544,836,573	65,102,649	215,732,346	133,023,571
Indirect impact	97,940,009	4,756	925,835,615	115,433,437	22,876,229	218,619,972
Induced impact	29,368,526	1,754	178,107,053	28,506,628	6,064,946	57,960,527
Total impact	195,229,459	8,282	1,648,779,241	209,042,714	244,673,521	409,604,070

Source: compiled by author. The employment is in units and the rest of the figures are in 2005 euros

Basque Country. These regions have been chosen because, following the market study, they proved to be the places where the companies that carry out their activity in relation to the Port of Santander are located. In this first estimate, we chose to continue using the same IOT that was used for Cantabria. Naturally, this introduces distortions into the final results, which will be rectified in forthcoming additions to this study. Moreover, the application of the IOTs of each autonomous region will constitute the contribution made in forthcoming analyses.

As indicate above, as an innovation this study includes an analysis of the impact of the Port of Santander on its influence area beyond and city and the region in which it is located. If this aspect was taken into account in previous studies, for simplicity's sake the employment and GVA of the hinterland were imputed as though they had been generated in Cantabria. On this occasion, however, the regions where the Port Users Community companies are located is specified and the impact of the port on these regions is calculated. The results are subsequently added together to obtain the overall influence on the hinterland and to thereby make it possible to calculate the total impact of the Port of Santander, which comprises the impact on the city of Santander, the impact on Cantabria and the impact on the rest of the hinterland.

The impact of the Port of Santander on *Castilla y León* is attributable to a long-standing relationship that exists between the companies located in this autonomous region and the port.

In this case, as ever, the study started by obtaining the direct impact of the companies of the Port Users Community that are located in Castilla y León. To do so, we made use of the figures provided by the corresponding registries of commerce, as well as by a market study that allowed us to extract the percentages of the activity of these companies that results from their relationship with the Port of Santander (Table 9).

In addition to the Port Users Community companies located in Castilla y León, in this case we also took into account the land transport that is associated with the traffic of goods to and from the Port of Santander, since this is seen as another link in the value chain and one which depends on the decision made by the companies to choose the Port of Santander as the gateway for its products and provisioning. The first line of Table 3.6 shows the results corresponding to the variables for the study for Castilla y León.

Table 9 Total impact of the Port of Santander on Cantabria

Cantabria	Wages	Employment	Sales	GOS	Taxes	GVA
Direct impact	98,405,893	2,548	770,371,424	141,517,528	215,320,083	239,923,421
Indirect impact	133,381,983	6,466	1,269,075,899	157,935,423	31,230,997	298,463,516
Induced impact	41,042,023	2,452	248,901,622	39,837,535	8,475,660	80,998,866
Total impact	272,829,899	11,465	2,288,348,945	339,290,486	255,026,741	619,385,803

Source: compiled by author. The employment is in units and the rest of the figures are in 2005 euros

Once the direct impact has been calculated, the purchases (2,570,760,672 euros) and investments (4,186,934 euros) are taken into account and the input–output method described above is applied to calculate the indirect impact. Subsequently, once the direct and indirect wages and salaries are known, the induced impact is calculated. The figures corresponding to this impact are summarised in the second and third lone of Table 10.

In the light of these results, it is worth underlining the great importance of the Castilla y León companies in the total impact of the Port of Santander. In relative terms, they account for a volume of employment equivalent to 15.82% of Cantabria's employment and a GVA equivalent to 20.61% of the autonomous region's GVA.

As in the previous case, firstly we calculated the variables corresponding to the companies that, although located in *Catalonia*, have a relationship with the Port of Santander, given that they use it for import and export. The direct impact is thereby obtained, the figures for which appear in the first line of Table 11.

Once the variables for the direct impact of the Port of Santander on Catalonia have been calculated, the indirect and induced impact is calculated in using the same method that was mentioned for Castilla y León. In this case, the purchases made by Catalan companies related to the Port of Santander stood at 303,687,276 euros and investments amounted to 1,040,097 euros. The last line of Table 11 shows the results for the total impact of the Port of Santander on Catalonia.

We then calculated the direct impact of the Port of Santander on *Madrid*. In this regard, it is worth mentioning that, as everybody knows, Madrid is home to a large proportion of our country's companies and logistical centres. In this case we have taken into account the companies that belong to Santander's Port Users Community and that carry out their activity in Madrid and we have obtained the percentage of their activity that results from goods channelled through the Port of Santander. The corresponding direct impact appears in the first line of Table 12.

Bearing in mind these variables and the purchases and investments of the companies that materialised in relation to their activity at the Port of Santander (the purchases in this case were 486,612,710 euros and the investments were 500,856 euros), the indirect and induced impact was obtained (second and third lines of Table 12). The total impact, shown in the last line of the aforementioned table, is the second most significant behind Castilla y León.

Lastly, the economic impact of the Port of Santander on the *Basque Country* was calculated. The direct impact is shown in the first line of Table 3.9 and, as can be seen, it is particularly significant, which is fairly logical since in the Basque Country there are two major ports, Bilbao and Pasajes.

Taking the purchases (704,104 euros) and investments (250,756 euros) of the companies located in the Basque Country that use the infrastructures of the Port of Santander, we can obtain the indirect and induced impact, which is shown in the second and third lines of Table 13.

To summarise, it is important to present the information on the economic impact of the Port of Santander on its hinterland overall, for the purpose of which the results obtained for the direct, indirect and induced impact for Castilla y León, Catalonia, Madrid and the Basque Country were added together. These overall

Table 10 Total impact of the Port of Santander on Castilla y León

Castilla y León	Wages	Employment	Sales	GOS	Taxes	GVA
Direct impact	226,970,367	5,553	2,956,436,497	323,940,058	6,955,312	550,910,426
Indirect impact	555,872,024	26,547	5,585,624,475	684,586,599	132,920,762	1,270,276,365
Induced impact	138,615,688	8,280	840,642,514	134,547,639	28,625,769	273,566,277
Total impact	921,458,079	40,380	9,382,703,486	1,143,074,296	168,501,842	2,094,753,068

Source: compiled by author. The employment is in units and the rest of the figures are in 2005 euros

Table 11 Total impact of the Port of Santander on Catalonia

Catalonia	Wages	Employment	Sales	GOS	Taxes	GVA
Direct impact	6,796,819	173	321,646,312	12,590,977	45,360	19,387,796
Indirect impact	65,928,317	3,151	660,850,937	81,049,957	15,749,730	150,514,562
Induced impact	12,877,234	769	78,094,699	12,499,317	2,659,300	25,413,985
Total impact	85,602,370	4,093	1,060,591,948	106,140,251	18,454,390	195,316,343

Source: compiled by author. The employment is in units and the rest of the figures are in 2005 euros

Table 12 Total impact of the Port of Santander on Madrid

Madrid	Wages	Employment	Sales	GOS	Taxes	GVA
Direct impact	18,104,442	463	515,673,979	12,753,989	207,569	30,858,431
Indirect impact	105,079,265	5,017	1,056,746,419	129,488,008	25,134,763	240,203,976
Induced impact	21,811,791	1,303	132,278,812	21,171,665	4,504,391	43,046,862
Total impact	144,995,498	6,783	1,704,699,210	163,413,662	29,846,723	314,109,269

Source: compiled by author. The employment is in units and the rest of the figures are in 2005 euros

Table 13 Total impact of the Port of Santander on the Basque Country

Basque country	Wages	Employment	Sales	GOS	Taxes	GVA
Direct impact	189,842	5	862,081	30,627	0	220,469
Indirect impact	272,336	14	1,993,796	269,345	58,200	556,198
Induced impact	81,836	5	496,302	79,435	16,900	161,509
Total impact	544,014	24	3,352,179	379,407	75,100	938,176

Source: compiled by author. The employment is in units and the rest of the figures are in 2005 euros

results, as a summary of the impact of the Port of Santander on the rest of the hinterland (excluding Cantabria), are presented in Table 14.

Table 15 is presented as a summary, in relative terms, of the importance of the volume of employment and the GVA of the various autonomous regions (ARs) under analysis, calculated as though they were produced in Cantabria's economy:

5 Conclusions

The main results of the study on the economic impact of the Port of Santander on the three geographical areas under analysis (city, autonomous region and hinterland) are presented below. A summary of these results is shown in Table 16:

Table 14 The impact of the Port of Santander on the hinterland

Hinterland impact	Wages	Employment	Sales	GOS	Taxes	GVA
Direct impact	252,061,469	6,194	3,794,618,870	349,315,652	7,208,241	601,377,121
Indirect impact	727,151,942	34,729	7,305,215,627	895,393,909	173,863,455	1,661,551,102
Induced impact	173,386,549	10,357	1,051,512,326	168,298,056	35,806,361	342,188,633
Total impact	1,152,599,960	51,280	12,151,346,823	1,413,007,617	216,878,057	2,605,116,855

Source: compiled by author. The employment is in units and the rest of the figures are in 2005 euros

Table 15 Relative impact of the Port of Santander on the hinterland

ARs	Employment (%)	GVA (%)
Castilla-León	15.82	20.61
Madrid	2.65	3.09
Catalonia	1.6	1.92
Basque Country	0.0094	0.0092

Source: compiled by author

It is worth highlighting that in the city of Santander the impact of the port represents approximately 11% of the capital's total GVA; or in other words, for each 100 euros of value added generated by the economic activity of Santander, almost 11 euros are associated with the activity of the Port Community and the Community of Users that make use of the Port of Santander's infrastructures. As far as employment is concerned, the port is responsible for 9.6% of city jobs.

As for the information yielded from the study for Cantabria, we can confirm that 4.5% of Cantabria's employment results from the Port of Santander's influence in the region, and 6.1% of the GVA results from the business activity carried out by companies associated with the port activity.

It is worth highlighting that the Port of Santander's influence in the hinterland regions is very important, since just 19.2% of the added value associated with the activities that are dependant on the Port of Santander are found in Cantabria, which clearly shows that the scope of the port service goes beyond the regional boundaries. Hence it is a public infrastructure of general interest to the State, as mentioned previously.

Furthermore, if we take into account the employment figure that would be produced if the total impact (with the hinterland) were calculated as though it were produced in Cantabria, it follows that 24.6% of Cantabria's employment and almost 32% of its GVA would result from the Port of Santander. Evidently, as indicated above, the impact of the hinterland, due to its magnitude and importance, must be assigned to the autonomous regions in which the impact actually materialises. Its impact cannot be computed in the Cantabrian economy, which is why in future applications all of the above will be taken into account, and using the input–output tables for each region, the impact of the Port of Santander will be assigned and measure appropriately.

To finish, Table 17 shows a comparison of results which clearly demonstrates the differences arising from the adoption of this new system of company designation and impact estimation, taking into account, as mentioned earlier, the geographical area in which the companies carry out their main activity connected to the Port of Santander.

As can be seen, in this case, with the new methodology applied in this study, greater detail on the geographical influence area of the Port of Santander has been achieved. This study, like nearly all studies, can be improved, and this is the intention for the future, introducing new information that will be provided by the use of input–output tables for the autonomous regions involved, allowing us to assess the influence of the port with greater accuracy and observe its influence on these regions.

Table 16 Total impact of the Port of Santander

Total impact	Wages	Employment	Sales	GOS	Taxes	GVA
(1) City of Santander impact	195,229,459	8,282	1,648,779,241	209,042,714	244,673,521	409,604,070
(2) Cantabria impact (without city)	77,600,440	3,183	639,569,705	130,247,773	10,353,219	209,781,732
(3) Cantabria impact (1) + (2)	272,829,899	11,465	2,288,348,946	339,290,487	255,026,740	619,385,802
(4) Rest of Hinterland impact	1,152,599,960	51,280	12,151,346,823	1,413,007,617	216,878,057	2,605,116,855
Total impact (3) + (4)	1,425,429,859	62,746	14,439,695,767	1,752,298,103	471,904,798	3,224,502,658

Source: compiled by author. The employment is in units and the rest of the figures are in 2005 euros

Table 17 Comparison of results with previous studies

	Year	% Regional employment	% Regional GVA
Port community impact (1)	2005	1.6%	1.8%
Port and users community impact (2)	2005	4.5%	6.1%
Hinterland impact (3)	2005	20.1%	25.9%
Total impact with Hinterland (2) + (3)	2005	24.6%	32%
Impact of port industry	1998	1.8%	2%
Impact of dependant industry	1998	13%	17.5%
Total impact	1998	14.8%	19.5%
Impact of Port industry	1993	2%	2%
Impact of dependant industry	1993	10%	10%
Total impact	1993	12%	12%

Source: compilied by author using previous economic impact studies. (Coto-Millán et al. 2001; Coto and Villaverde 1995)

Appendix

Table 1 Original and modified localisation coefficients (LC)

Branches	Cantabria GVA	Spain GVA	Cantabria	Spain	Original LC	Modified LC
1	317,715	23,498,000	0.04511	0.04118	1.09543	1
2	33,777	1,486,000	0.00480	0.00260	1.84154	1
3	58,952	4,478,000	0.00837	0.00785	1.06658	1
4	119,094	11,324,000	0.01691	0.01985	0.85206	0.85206
5	177,874	14,119,000	0.02526	0.02475	1.02068	1
6	33,182	7,443,000	0.00471	0.01305	0.36119	0.36119
7	26,729	2,501,000	0.00380	0.00438	0.86586	0.86586
8	46,620	9,366,000	0.00662	0.01642	0.40327	0.40327
9	178,680	9,691,000	0.02537	0.01699	1.49378	1
10	85,316	4,661,000	0.01211	0.00817	1.48297	1
11	100,019	7,976,000	0.01420	0.01398	1.01596	1
12	412,546	16,062,000	0.05858	0.02815	2.08091	1
13	114,891	7,375,000	0.01631	0.01293	1.26213	1
14	108,701	7,302,000	0.01544	0.01280	1.20607	1
15	102,411	11,504,000	0.01454	0.02016	0.72124	0.72124
16	24,081	5,415,000	0.00342	0.00949	0.36029	0.36029
17	713,760	47,584,000	0.10135	0.08340	1.21526	1
18	603,616	63,955,000	0.08571	0.11209	0.76466	0.76466
19	658,260	43,407,000	0.09347	0.07608	1.22862	1
20	378,279	41,755,000	0.05371	0.07318	0.73398	0.73398
21	275,488	26,279,000	0.03912	0.04606	0.84932	0.84932
22	1,067,586	84,768,000	0.15159	0.14857	1.02035	1
23	361,247	35,513,000	0.05130	0.06224	0.82413	0.82413
24	366,044	27,941,000	0.05198	0.04897	1.06138	1
25	384,398	28,986,000	0.05458	0.05080	1.07442	1
26	293,149	26,171,000	0.04163	0.04587	0.90750	0.89648
Total	7,042,415	570,560,000	1	1	1	1

Table 2 Regional technical coefficients matrix 2000

Ramas	1	2	3	4	5	6	7	8	9	10	11	12
1	0.07015	0.00314	0.00002	0.00019	0.27376	0.01736	0.06759	0.01681	0.00185	0.00329	0.00004	0.00003
2	0.00000	0.00000	0.00000	0.00000	0.00305	0.00000	0.00000	0.00000	0.00001	0.00000	0.00000	0.00000
3	0.01256	0.03472	0.31589	0.23392	0.00159	0.00091	0.00551	0.00191	0.06176	0.00336	0.04308	0.03266
4	0.01073	0.00600	0.01051	0.12898	0.00844	0.00917	0.01314	0.01810	0.01210	0.01810	0.03596	0.01659
5	0.11529	0.04303	0.00001	0.00011	0.18715	0.01276	0.00002	0.00137	0.00375	0.00004	0.00001	0.00000
6	0.00020	0.00620	0.00008	0.00016	0.00052	0.09366	0.00093	0.00017	0.00111	0.00462	0.00071	0.00043
7	0.00189	0.00439	0.00142	0.00005	0.00493	0.00163	0.23042	0.00801	0.00058	0.00178	0.00614	0.00287
8	0.00043	0.00034	0.00037	0.00245	0.00623	0.00146	0.00706	0.10417	0.00445	0.00234	0.00385	0.00084
9	0.03606	0.00580	0.00824	0.01204	0.00717	0.04194	0.02876	0.03429	0.19237	0.14337	0.03667	0.03299
10	0.00526	0.00887	0.00096	0.00024	0.01221	0.01258	0.01037	0.00691	0.01245	0.13095	0.00643	0.00587
11	0.00019	0.00023	0.00061	0.00033	0.01221	0.00007	0.00144	0.00001	0.00295	0.00211	0.07929	0.00336
12	0.01048	0.00954	0.00668	0.01746	0.01480	0.00542	0.02117	0.02584	0.00323	0.02396	0.03919	0.22739
13	0.00738	0.00037	0.00672	0.02732	0.00450	0.00832	0.01952	0.01104	0.01242	0.01705	0.04436	0.02633
14	0.00014	0.00173	0.00061	0.01413	0.00039	0.00064	0.00086	0.00075	0.00113	0.00211	0.00281	0.00365
15	0.00011	0.03488	0.00029	0.00008	0.00043	0.00024	0.00029	0.00003	0.00009	0.00029	0.00072	0.00064
16	0.00000	0.00012	0.00012	0.00020	0.00001	0.00010	0.00013	0.00108	0.00009	0.00005	0.00011	0.01576
17	0.00500	0.00047	0.00231	0.00953	0.00292	0.00140	0.00108	0.00203	0.00120	0.00083	0.00777	0.00206
18	0.03023	0.02163	0.00326	0.02486	0.02598	0.03298	0.03939	0.02784	0.00923	0.01934	0.02617	0.02141
19	0.00044	0.00037	0.00039	0.00090	0.00060	0.00067	0.00099	0.00061	0.00368	0.00120	0.00073	0.00124
20	0.00888	0.04715	0.02093	0.01367	0.03349	0.02554	0.03283	0.03077	0.02954	0.02703	0.09009	0.02966
21	0.00887	0.01116	0.00542	0.01300	0.00764	0.00798	0.00621	0.00750	0.00597	0.00762	0.00839	0.00741
22	0.00490	0.02021	0.01959	0.05721	0.06168	0.04404	0.02515	0.05805	0.05982	0.05315	0.07788	0.03867
23	0	0	0	0	0	0	0	0	0	0	0	0
24	0.00085	0.00143	0.00096	0.00174	0.00126	0.00175	0.00122	0.00151	0.00169	0.00164	0.00148	0.00250
25	0.00470	0.00227	0.00040	0.00081	0.00126	0.00055	0.00130	0.00035	0.00079	0.00089	0.00073	0.00009
26	0.00122	0.00135	0.00113	0.00104	0.00153	0.00087	0.00063	0.00967	0.00148	0.00198	0.00149	0.00064

13	14	15	16	17	18	19	20	21	22	23	24	25	26
0.00037	0.00001	0.00000	0.00135	0.00320	0.00158	0.01106	0.00022	0.00004	0.00022	0.00152	0.00110	0.00081	0.01241
0.00001	0.00000	0.00000	0.00000	0.00000	0.00001	0.00761	0.00002	0.00000	0.00000	0.00007	0.00025	0.00041	0.00001
0.00232	0.00111	0.00073	0.00095	0.01223	0.00519	0.00646	0.04383	0.00179	0.00175	0.00390	0.00638	0.00757	0.00519
0.00579	0.00592	0.00536	0.00538	0.00269	0.01832	0.00584	0.01235	0.00480	0.00626	0.01768	0.01024	0.00750	0.01212
0.00021	0.00000	0.00000	0.00003	0.00000	0.00086	0.16595	0.00107	0.00002	0.00018	0.00430	0.00528	0.00975	0.00627
0.00042	0.00041	0.00267	0.00780	0.00020	0.00114	0.00344	0.00073	0.00002	0.00071	0.00094	0.00027	0.00179	0.00250
0.00104	0.00173	0.00084	0.08548	0.01541	0.00097	0.00311	0.00169	0.00001	0.00033	0.00005	0.00003	0.00006	0.00392
0.00106	0.00156	0.00063	0.01912	0.00080	0.00200	0.00155	0.00201	0.00481	0.01610	0.00610	0.00429	0.00240	0.00988
0.00930	0.01148	0.00924	0.02400	0.01097	0.00555	0.01576	0.00438	0.00042	0.00625	0.00298	0.00238	0.04677	0.01452
0.01059	0.02142	0.03867	0.01700	0.01092	0.00287	0.00142	0.00576	0.00013	0.00174	0.00022	0.00032	0.00099	0.00686
0.00402	0.00586	0.00441	0.00350	0.10704	0.00111	0.00330	0.00265	0.00005	0.00049	0.00027	0.00029	0.00183	0.00141
0.12000	0.07499	0.08387	0.17097	0.06636	0.00317	0.00233	0.00305	0.00043	0.00373	0.00157	0.00109	0.00035	0.00658
0.05160	0.01015	0.01517	0.01146	0.01345	0.00407	0.00443	0.00612	0.00023	0.00173	0.01286	0.00128	0.00068	0.00740
0.03295	0.13991	0.01808	0.01383	0.03642	0.00239	0.00368	0.01771	0.00164	0.00650	0.00177	0.00583	0.03632	0.01039
0.00297	0.00104	0.16917	0.00175	0.00006	0.02730	0.00038	0.01036	0.00019	0.00144	0.00977	0.00046	0.00021	0.00071
0.00054	0.00018	0.00010	0.01261	0.00064	0.00029	0.00196	0.00117	0.00051	0.00252	0.00086	0.00165	0.00048	0.00508
0.00316	0.00064	0.00069	0.00285	0.22337	0.01150	0.00943	0.00716	0.01071	0.05383	0.00892	0.01265	0.00732	0.02097
0.01675	0.00886	0.00862	0.02849	0.03091	0.03333	0.03550	0.01837	0.00165	0.00630	0.00647	0.00339	0.03114	0.01657
0.00058	0.00048	0.00074	0.00086	0.00201	0.00307	0.00080	0.02116	0.00707	0.00556	0.00549	0.00647	0.00856	0.00977
0.01492	0.01103	0.01392	0.02073	0.01480	0.04883	0.00954	0.15420	0.02113	0.02723	0.02445	0.00746	0.00981	0.01832
0.00615	0.00406	0.00459	0.00604	0.00905	0.01810	0.01087	0.01021	0.14169	0.02650	0.00786	0.00371	0.00568	0.00824
0.04144	0.04608	0.02815	0.04028	0.04565	0.12921	0.04965	0.07782	0.08166	0.08707	0.06286	0.03152	0.05849	0.07942
0	0	0	0	0	0	0	0	0	0	0	0	0	0
0.00047	0.00064	0.00080	0.00053	0.00031	0.00184	0.00104	0.00101	0.00099	0.00117	0.00073	0.00198	0.00106	0.00067
0.00038	0.00022	0.00059	0.00080	0.00000	0.00445	0.00192	0.00181	0.00067	0.00250	0.00099	0.00095	0.03946	0.00200
0.00051	0.00048	0.00123	0.00089	0.00016	0.00288	0.00509	0.00243	0.00270	0.01876	0.00240	0.00104	0.00588	0.10030

References

Alcaide Inchausti A (1970) Wassily Leontief: Análisis económico input–output. Editorial Ariel, Barcelona

Alonso-Carrera J, Freire-Serén MJ (2001) Infraestructuras públicas y desarrollo económico de Galicia. Universidad de Vigo, Vigo, p 39

Benacchio M, Ferari C, Haralambides HE, Musso E (2000) On the economic impact of ports: local vs. national costs and benefits. International workshop Genoa – June 8-10, 2000. Special interest group on maritime transport and ports member of the WCTR Society

Carsten F, Mattoo A, Cristina Neagu I (2002) Trade in international maritime services: how much does policy matter? World Bank Econ Rev 16(1):81–108

Castillo-Manzano JI, Coto-Millán P, Pesquera MA, López-Vapuesta L (2004) Comparative analysis of port economic impact studies in the Spanish port system (1992–2000). Chapter 19, 297–316, Essays on microeconomics and industrial organisation, 2nd edn. Physicxa-Verlag, Springer, Heidelberg

Chang S (1978) In defense of port economic impact studies. Transport J 17:79–85

Coto-Millán P (1999a) A methodological discussion on Port Economic Impact Studies and their possible applications to policy design. Chapter 7, 115–128. Maritime Transport Applied Economics, Civitas

Coto-Millán P (1999b) Port economic impact: methodologies and application to the Port of Santander. Chapter 8, 131–154. Maritime Transport Applied Economics, Civitas

Coto-Millán P (1999c) An approach to the contribution of the port system in the Spanish economy. Chapter 9, 157–164. Maritime Transport Applied Economics, Civitas
Coto-Millan P, Banos-Pino J, Rodriguez-Alvarez A (2000) Economic efficiency in Spanish ports: some empirical evidence. Marit Pol Manag 27(2):169–174
Coto-Millán P, Gallego JL, Villaverde J (2001) Crecimiento y Desarrollo Portuario. Aplicación al Puerto de Santander, Autoridad Portuaria de Santander
Coto-Millán P, Villaverde Castro J, Mateo Mantecón I (2008) Impacto Económico del Puerto de Santander en la Ciudad, en Cantabria y en otras Regiones Españolas. Autoridad Portuaria de Santander
Davis CH (1983) Regional port impact studies: a critique and suggested methodology. Transport J 17:61–71
Dietzenbacher E, Lahr ML (eds) (2004) Leontief and input–output economics. Cambridge University Press, New York, p 418
Dollar D, Alejandro M, Ximena C (2002) Maritime transport costs and port efficiency. Policy Research Working Paper Series 2781, The World Bank
Fernández JI, Martí ML, Puertas R (2004) Metodología para la medición de impacto económico de los puertos: el caso del puerto de Castellón. I Congreso Internacional de Transporte: los retos del transporte en el siglo XXI: Castellón de la Plana, 4–6 mayo 2004, pp 403–420.
Fontela E, Pulido A (2005) Técnicas de la investigación en el análisis input–ouput. Revista Asturiana de economía-RAE, N° 33, pp 9–29
Leontief W (1984) Analisis economico input–output. Biblioteca de Economía, Barcelona, p 359
Leontief W (1986) (1986): Input–output economics, 2nd edn. Oxford University Press, New York, p 436
Leontief W, Strout, A (1998) Multiregional input – output analysis. International Liberary of Critical Writings in Economics, London
Leontief W (1966) Input–output economics. Oxford University Press, New York
Llano Verduras C (2004) Economía sectorial y espacial: El comercio interregional en el marco input–output. Instituto de estudios Fiscales, Madrid
Martínez E, Gutiérrez P, López LJ, Martín FJ (1999) El impacto económico de los puertos de Santa Cruz de Tenerife sobre la provincia. Hacienda pública Española n° 148:175–185
Memorias Anuales de los años 1995 a 2005. Autoridad Portuaria de Santander.
Ministerio de Fomento. Ley 27/1992, de 24 de noviembre, de Puertos del Estado y de la Marina Mercante
Ministerio de Fomento. Ley 62/1997, de 26 de diciembre, de modificación de la Ley 27/1992, de Puertos del Estado y de la Marina Mercante
Ministerio de Fomento. Ley 48/2003, de 26 de noviembre, de régimen económico y de prestación de servicios de los puertos de interés general
Ministerio de Fomento (2005) Anuario estadístico de Puertos del Estado del año
Pulido A, Fontela E (1993) Análisis input–output. Modelos, datos y aplicaciones. Pirámide, Madrid
Villaverde J, Coto-Millán P (1995) El impacto económico del Puerto de Santander en la economía cántabra. Autoridad Portuaria de Santander
Villaverde J, Coto-Millán P (1996) Impacto económico portuario: Metodologías para su análisis y aplicación al Puerto de Santander. Autoridad Portuaria de Santander
Villaverde J, Coto-Millán P (1998a) Port Economic Impact: methodologies and application to the Port of Santander. Int J Transport Econ 25(2):159–179
Villaverde J, Coto-Millán P (1998b) Guest Editor's introduction. Int J Transport Econ 25(2):109–112
Villaverde J, Coto-Millán P (2000) Nuevos enfoques en la demanda de transportes en España. Estudios de Construcción y Transportes, n° 86, pp 99–110
Villaverde J, Coto P, Aza R, Baños J, Canal JF (2004) Impacto de los puertos de Avilés y Gijón en la economía asturiana. Papeles de Economía Española. Comunidades Autónomas. Principado de Asturias, N° 20, pp 207–219
Waters RC (1977) Port impact studies: practice and assessment. Transport Rev 16:14–18
Yochum GR, Agarwal VB (1987) Economic impact of a port on a regional economy. Growth Change 18(3):74–87

The Effect of Port Infrastructures on Regional Production

Pablo Coto-Millán, José Baños Pino, and Ingrid Mateo-Mantecón

Abstract This study highlights the impact that investment in a project to expand the Port of Santander would have, and the effects of this investment on economic growth and employment in Cantabria. To this end a methodology based on estimating an aggregate production function has been used which, by applying econometric techniques of cointegration, detects a stable relationship between the regional output, employment, human capital, the supply of private capital and the port infrastructure. This approach measures the magnitude of the effects of an increase in the capital endowment of the port on private production, being complementary to results obtained using other economic research tools, such as input–output studies or social assessment projects through cost-benefit analysis.

1 Introduction

The per capita income level of an economy depends, basically, on the levels of accumulation of its factors of production, including investment in material capital, both public and private, as well as the accumulation of human capital. Within material capital, large public infrastructures stand out for their importance, since it is usually argued that the lack of them constitutes an obstacle to regional economic development.

For this reason, in recent years the question of what allocation of public capital should be promoted from economic policy has become more relevant, and moreover, what type of infrastructures should be selected in order to prompt greater regional economic growth. Thus, investment in transport and communication

P. Coto-Millán (✉) and I. Mateo-Mantecón
University of Cantabria, Santander, Spain
e-mail: cotop@unican.es

J. Baños Pino
University of Oviedo, Oviedo, Spain

P. Coto-Millán et al. (eds.), *Essays on Port Economics*, Contributions to Economics,
DOI 10.1007/978-3-7908-2425-4_14, © Springer-Verlag Berlin Heidelberg 2010

infrastructures has become, through the EU structural funds, one of the main economic instruments of the European Union (EU) to contribute to real convergence among member states.

Traditionally Spain, and above all some of its regions, had been deprived of adequate planning in its infrastructures, which caused a considerable delay with respect to those installed in other EU countries. The need to expand and modernize existing infrastructures, especially those regarding transport, responds to the idea that, due to the peripheral situation of the Spanish economy, a better allocation of these can lead to an increase in productivity of companies and, as a result, in the competitiveness of their products. Consequently, one would expect that the economic effects of investment in infrastructures would be significant and quantitatively important.

As Aschauer (1989c) showed, the empirical evaluation of the economic effects of public funding in infrastructures is obtained from estimating the degree of response in regional output with respect to the public funding. This requires estimating a production function, in which the public funding is incorporated as an additional factor of production to the work and private funding.

It is precisely in this context that the present research is based, with the aim of determining, based on a growth model and using a rigorous econometric methodology, the elasticity of output response in Cantabria set against the investment in public infrastructure, and more specifically, against the projected investments in the expansion of the Port of Santander. Ultimately, the aim of this study is to calculate the percentage change of the gross added value of Cantabria, both in the short term and long term, due to the increase in spending in the port infrastructure and, in addition, to contrast what the effects will be on the creation of regional employment.

1.1 Economic Effects of Investment in Infrastructures

Traditionally, and from the viewpoint of an aggregate function of production, only the effect of the stock of private capital on the productivity of the countries or regions was investigated. However, from the work of Aschauer (1989c), the importance of public infrastructures on the productivity of the private sector should not be ignored. In this case, the theoretical framework used consists of expanding the traditional arguments of the production function and appraising the output elasticities with respect to the various types of capital. Although this approach has been used in many studies under very restrictive simplifying assumptions, it has led the debate on measuring the aggregate macroeconomic effects of infrastructures (see, for example, the views of Gramlich (1994), Draper and Herce (1994) or de la Fuente (1996a)).

Infrastructures are all those capital goods that make up the base of socioeconomic activity, in the measure in which they determine or condition the productive potential of different parts of the territory and the geographical location of the factors of mobile production. From this definition it follows that the development

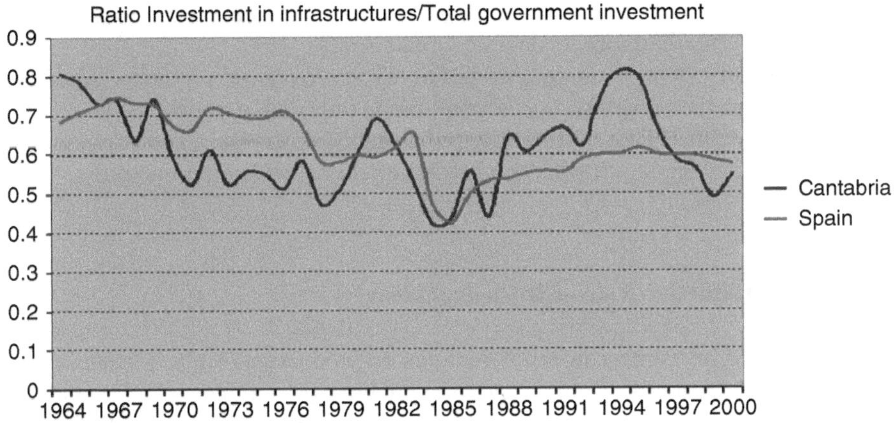

Graph 1 Ratio Investment in infrastructures/Total government investment

of a region and the interregional inequality of incomes is closely related to the provision of infrastructures. The lack or shortage of roads, motorways, ports, railways, and communication, water and energy infrastructures, among others, hinders the operation of the economic system and has a negative effect on employment and the standard of living of citizens.

Infrastructures are financed from public expenditure, which is made up of total public investment and public consumption, which is to say that not all public expenditure is directed to production. Aschauer (1987) shows that an increase in public investment spending has a more stimulating effect on domestic production than equal increases in public consumption expenditure. For the specific case of Spain, Argimón et al. (1993) confirms this result.

Moreover, the amount of public investment allocated to infrastructures makes up the highest proportion. Graph 1 shows the relative growth of investment in infrastructures over total public investment carried out by the Public Administrations in Cantabria and the whole of Spain since 1964.

It can be seen that both series have similar profiles, showing two clear trends: a decrease until 1985 and an increase from that time on. Nevertheless, Cantabria is below the national average in the first part of the time period considered, reaching a difference of almost twenty percent, due to the pronounced fall in the amount produced between 1970 and 1979. From 1985 it seems, however, that investment in infrastructures in Cantabria recovers, surpassing the national average.

The expansion of public expenditure recorded in Spain in recent years has been justified, to a large extent, by fulfilling redistributed objectives. Thus, in the period 1975–1991, the costs of redistributing income and wealth account for 60% of the increase in net total interest costs. Nevertheless, despite its positive effects, public expenditure is not a free good for society. The relief of poverty or the public provision of pensions or unemployment insurance has two cost elements. Firstly, that associated to the diversion of resources which would be used in directly productive private activities. Secondly, government intervention tends to disrupt pricing signals and, therefore, economic dynamism.

Below, the effect of accumulating public capital over the productivity of the private sector of the economy is analyzed, paying special attention to the role of public infrastructures. These, especially those related to transport and communications, are the ones that a priori have a greater effect on private productivity.

1.1.1 The Productive Role of Infrastructures

The influence of investment in infrastructures on production can be carried out by means of an aggregate function, considering public capital as a new input. Without losing generality, a Cobb–Douglas type production function would have:

$$Y_t = A_t \times N_t^{\alpha} \times K_t^{\beta} \times G_t^{\gamma} \tag{1}$$

where Y is production, A a parameter of technological efficiency, N employment, K the stock of private capital, and G the stock of public capital, while the sub-index t represents time. In (1), γ would be the contribution of public capital to the growth of productivity.

Then, the importance of infrastructures would be reflected in the estimated value of γ. In particular, it would be necessary to check whether this coefficient is statistically different from zero. In this sense, the aggregate estimates of γ for the U.S.A. are between 0.3 for Munnell (1990) and 0.5 for Aschauer (1989a), values which are considerably reduced with the estimate broken down by states or regions in the 1970s and 1980s. The above-mentioned empirical work referring to the aggregate case of Spain [Argimón et al. (1993), Bajo and Sosvilla (1993), Dolado et al. (1994), De la Fuente (1994), Flores et al. (1994) or García-Fontes and Serra (1994)] reproduces this pattern of estimates, with a minimum of 0.2 and a maximum of 0.9.

Before drawing conclusions about the budgetary policies that are derived from the important role that infrastructures seem to have, it is necessary to pay some attention to two issues on which the infrastructure–productivity–welfare relationship rests. One is the possibility of reverse causation and the other is the crowding out effects on private investment that public projects might cause. Although the existence of a long-term relationship between production and public capital implies that this causes that in a statistical sense, a connection in the opposite direction cannot be ruled out a priori. The directional contrasts of this relationship, carried out in Argimón et al. (1993), clearly establish that public capital is what causes increases in output, and not the reverse.

With reference to the displacement effect, it is known that there are two general conditions which can produce this. Firstly, public spending may expel private investment due to resource constraints. If the economy is in a situation of full employment, that is to say when potential production has been reached, an increase of public expenditure will be displaced to private investment. Secondly, the displacement effect may occur when interest rates rise substantially due to an increase in public debt and reduce private investment sensitive to them.

In addition, empirical findings on the connection between the size of the stock of public capital and the growth of output based on a unidirectional lineal relationship may not be valid if this were to pass through indirect channels, as it would be that which relates public expense with private investment and this with growth. In any case, empirical type deficiencies raise the possibility of reverse causation discussed earlier. Since private investment is the engine of growth, any crowding out effect that public expenditure had on private investment would damage the possibilities of growth and expansion of consumption levels. The bulk of empirical literature agrees in pointing out the existence of a significant negative effect of public consumption on growth, while the effects of public investment tend to be positive although less strong. Aschauer (1989b) shows that the direct "crowding-out" effect that public investment has is smaller than the stimulating effect or "crowding-in" associated to the role of public capital as a factor of production used by the private sector and its complementarity with private capital.[1]

The aggregate results clearly suggest that a public investment policy that increases the stock of public capital from carrying out the best infrastructure projects, as well as not crowding out private investment, can have very positive effects on long-term income and welfare.

In any case, for (1) a high, positive, and statistically significant γ coefficient cannot justify a policy of indiscriminate expansion of infrastructure investment. For one thing, the past does not prefigure the future relationship between public capital and the level of production. In other words, the productive efficiency of public capital in the past does not guarantee that future investments will also be productive. In fact, infrastructures are characterised by their specificity, in the sense that, for example, it can be very beneficial to build a motorway network without this having to be the expansion of this network.

Nevertheless, as productive as the capital stock of an economy can be, the specific selection of projects – public or private – should always seek endorsement from appropriate assessment techniques, especially those based on cost-benefit analysis. At any rate, it should be emphasised that the coefficients of the production function should also contain information about the optimality of private and public capital allocation from an aggregate perspective. This information is undoubtedly more consistent than any that can be derived from a simple international comparison of public capital/private capital or public capital/output ratios [see, for example, Mas et al. (1994)].

[1]Public and private capital is complementary when the marginal productivity of private capital increases it making it the quantity of public capital available.

1.1.2 Overview of Time-Series Empirical Evidence

Existing econometric studies on the determinants of national or regional productivity start with a supposition that there is a stable relation between aggregate output, on the one hand, and the resources of productive factors (work and different types of capital) and the level of technical efficiency on the other. This relationship, already discussed, is often represented by means of a Cobb–Douglas aggregate production function: $Y_t = A_t \times N_t^{\alpha} \times K_t^{\beta} \times G_t^{\gamma}$.

The availability of time cutting data on the variables $Y_{i,t}$, $N_{i,t}$, $K_{i,t}$ and $G_{i,t}$, has facilitated the emergence of relatively recent works that have addressed the issue of estimating the productivity of production factors and their relation to expenditure in public infrastructures.

Ratner (1983) offers probably the first estimate of an aggregate production function in which public capital is included as input. After adjusting the stocks of public and private capital by the rate of use of the productive capacity, this author estimates a production function at annual data levels of the U.S. private sector and finds that the public capital coefficient is positive and significant, although quite reduced (0.056), as can be seen in the first column of Table 1.

Aschauer (1989a) estimates a very similar specification, with longer time series for the same country, obtaining a higher coefficient (0.39) (second column). This author breaks down the total public capital stock into different categories and concludes that the types of equipment with a greater impact on productivity are the transport, energy supply and water cycle infrastructures.

Aschauer's work had a great impact in its day and can be considered the starting point of literature on the subject. Perhaps the main reason for this success is that his findings appeared at a time when both economic analysts and the public agents were looking, without much luck, for the causes of the worrying decline in the rate of growth recorded by most industrial economies in the second half of the seventies. In this context, Aschauer's work was very appealing inasmuch as it offered a plausible diagnosis of the problem and a simple prescription for economic policy: increase public investment in infrastructures.

It was not long, however, before there was criticism. Many authors agree that the public capital coefficient obtained by Aschauer may be too high, suggesting that such results are due to different econometric problems. One difficulty is that Aschauer's findings could reflect a problem of reverse causality. In this interpretation, public capital would be a superior good, and the correlation between this and the increase in productivity would only reflect the tendency of governments to invest more in periods of rapid growth.

Aschauer (1989a), nevertheless, is conscious of this possibility and offers evidence against it. Firstly, he states that his results do not change substantially when delayed public capital values are used. Furthermore, if there is a problem of reverse causality, a strong correlation between the increase in productivity and many components of investment and public consumption could be expected, whereas such correlation exists only in connection with the expenditure devoted to productive infrastructures.

Table 1 Estimated time series production functions

	Ratner (1983)	Aschauer (1989a)	Serra 1 (Gramlich 1994)	Serra 2 (Gramlich 1994)	Bajo and Sosvilla (1993)	Argimón et al. (1993)	Mas et al. (1993)
α	0.710 (12.62)	0.35 (4.85)	0.59 (10.70)	0.55 (2.91)	0.39	0.22 (10.86)	0.290 (4.52)
β	[0.234]	[0.26]	0.40 (15.64)	0.45 (5.29)	[0.42]	[0.18]	0.613 (9.48)
γ	0.056 (2.70)	0.39 (16.23)	0.27 (11.69)	0.18 (1.02)	0.19	0.60 (7.74)	0.290 (4.52)
Variables on:	Levels	Levels	Levels	Differences	Cointe-gration	Cointe-gration	Cointe-gration
Country	U.S.A.	U.S.A.	Spain	Spain	Spain	Spain	Spain
Period	1949–73	1945–85	1969–88	1969–88	1964–88	1964–89	1964–89

Notes: Statistics t in brackets under each coefficient, except for Bajo and Sosvilla, which use Wald statistics proposed by Phillips and Hansen
The coefficients which appear in square brackets are only indirectly estimated, from assumptions on the degree of returns to scale. Ratner (1983) assumes that the production function presents constant returns on prívate and public capital and work. Aschauer (1989a) makes the same assumption after checking it

A second and probably more serious objection, is that Aschauer's analysis does not allow for other possible productivity determinants. As Holtz-Eakin (1994) observes, American post-war data essentially contains only one observation: the simultaneous fall in investment and the growth in productivity. It is possible, however, that this was a coincidence. The decline in the rate of American public investment at the beginning of the seventies is due, to a large extent, to two factors: the end of construction of the interstate motorway network, and the stabilization and subsequent decline of population of a school age (Gramlich 1994). In turn, the fall of the productivity growth rate could reflect the fast increase in energy prices because of the so-called oil crisis or some other factor. Whatever the case, the coincidental timing of the two processes could be by chance and the apparent significance of public capital should be such that it serves as a *proxy* for any other omitted variable that would actually be the cause of the problem. In this regard, the work of Ford and Poret (1991), which extends Aschauer's analysis to eleven member countries of the OECD, is aimed in this direction, finding a significant correlation between public investment and productivity in only half of the cases. A second sign in the same direction is the significant difference between the results of Aschauer and Ratner. Both work with very similar data, except that Ratner's sample finishes in 1973. Therefore, the high coefficient obtained by Aschauer could be due to the influence of the most recent observations which already include the crisis of the seventies.

The third criticism of Aschauer's findings concerns econometric issues which are especially important in time-series studies. Some studies suggest that the findings of this author could be an example of "spurious regression" identified by Granger and Newbold (1974). These authors argue that in many cases, the apparently good results from regressions in levels between non-stationary variables (i.e., that they show a trend) are not reliable, suggesting carrying out regressions in differences to obtain consistent estimators. That is, it would mean taking, for each variable x, its first difference, $\Delta x_{i,t} = x_{i,t+1} - x_{i,t}$. Thus, reconstructing the (1) expression in logarithmic terms for the period $t+1$ and subtracting the same equation in t, a function is obtained that relates the growth rate of the product in each period ($\Delta Log\ Y_{i,t}$), with the accumulation rates of the various factors of production and the rate of technical progress ($\Delta Log\ A_{i,t}$)[2]

$$\Delta Log\ Y_{i,t} = \Delta Log\ A_{i,t} + \alpha \times \Delta Log\ N_{i,t} + \beta \times \Delta Log\ K_{i,t} + \gamma \times \Delta Log\ G_{i,t} \qquad (2)$$

Equation (1) is a specification in levels of the production function, while (2) would be the same function specified in first differences.

When making the estimate in first differences of all the variables, the results tend to be less favourable to the hypothesis that infrastructures have a substantial impact on productivity – as seen for example upon comparing the third and fourth columns

[2]Remember that the increase in logarithm between two consecutive periods is approximately equal to the percentage increase of the variable.

of Table 1. At any rate, a simple estimate in differences presents its own problems. Munnell (1992) observes that such a specification tends to give results of little credibility on the contribution of the stock of private factors to growth.

Nevertheless, recent advances in time series econometrics suggest that the results of an estimate in levels could be far more reliable than at first thought. In particular, the estimate in the levels of a relationship between non-stationary series is consistent when the variables are co-integrated, which is to say, when there is a linear combination of these that is stationary. In fact, the estimators of ordinary least square thus obtained will be, in large samples, better than normal, although it is also certain that their distributions will not be standard, for which the usual inference contrasts lose their validity.

Table 1 summarizes the results of three works relating to the Spanish case using the methodology of cointegration (columns 5–7). Bajo and Sosvilla (1993) use the total public capital variable contained in the MOISEES database model (Molinas et al. 1990; Corrales and Taguas 1991), while Argimón et al. (1993) and Mas et al. (1993) work with series based on a narrower concept of productive public capital or infrastructures. The results of these studies tend to agree in the sense that the existence of cointegration is accepted in most cases, allowing the hypothesis of spurious regressions to be rejected and ensuring the consistency of the estimators. The coefficients obtained for the public capital variable have high values and seem to be significant. Nevertheless, the application of new techniques does still not allow the question to be considered settled, since the estimated coefficients vary a lot from one study to another.

It is precisely in this context of cointegration analysis in which the empirical contribution of this research work is presented, where also, the stock of port infrastructures from public production capital will be broken down, as an additional explanatory variable. Given the importance of the econometric methodology employed, in the following section the main characteristics of this technique will be briefly discussed.

2 Analysis of Data

In this section the characteristics of the data series used in the model are defined, and the order of integration of the different time series is analyzed. Subsequently, the possibility that there is a long term and stable relationship between the above is studied, interpretable in terms of an aggregate production function. In other words, the possibility that there is cointegration is investigated.

2.1 *Data: Sources and Description*

The variables are considered to be regional production, employment, human capital, private and public capital. All the data refers to the Cantabrian economy and covers the period 1964–2000.

The level of regional production has been measured using the Gross Added Value (GAV) at the cost of the factors. This series, as well as the employment figures, comes from the Spanish national income publication and its provincial distribution, published by the BBVA Foundation. The data on human capital, measured by the proportion of those employed that at least have secondary education, has been obtained from the work Mas et al., of the Instituto Valenciano de Investigaciones Económicas (IVIE), "Capital Humano, series históricas 1964–2001".

The series on private and public capital comes from "El stock de capital en España y sus comunidades autónomas", prepared by the IVIE and commissioned by the BBVA Foundation. The series on private capital refers to net capital and only includes the set of long term, tangible, reproducible assets; consequently, intangible assets, inventories, assets under construction, land and natural resources are excluded. In the case of public capital, the so-called social public capital has been excluded, which would include education and health. The remaining capital would be included, called productive public capital, which would basically cover infrastructures relating to the areas of transport and communications: ports, roads, airports, railway networks, etc. State, autonomous and local administration infrastructures have been included.

In what follows, the abbreviations used to refer to the different variables are:

Production, regional GAV (VAB) = Y
Total employed = N
Private capital = K
Productive public capital, excluding port infrastructures = GK
Public capital in port infrastructures = GP
Human capital = H

All the variables are expressed in thousands of Euros from 1986, except the employment levels, which are in thousands of workers. Furthermore, to avoid possible problems of heterocedasticity in the estimates, all the series are transformed into logarithms.

2.2 Univariate Analysis

The univariate properties of the series can be analyzed from three points of view: graphs of the series, functions of autocorrelation and partial autocorrelation, and nonstationarity tests. In this research, determining the order of integrability of each series has been carried out following the methodology for unit root tests. More concretely, the tests used are those proposed by Phillips and Perron (1988), which are strong against heterocedasticity and whose null hypothesis is that the variable contains a unit root, and the Kwiatkowski et al. (1992) test, which maintains stationarity as a null hypothesis.

2.3 Contrast of Unit Roots in the Normal Frequency[3]

Fuller (1976) and Dickey and Fuller (1979, 1981) consider the problem of contrasting unit roots in autoregressive processes. In a simple Data Generating Process (DGP), an AR(1) that lacks deterministic components,

$$y_t = \rho y_{t-1} + e_t; \quad t = 1, 2, \ldots, T, \tag{3}$$

where it is supposed that $e_t \sim$ NIID $(0, \sigma^2)$ e y_0 is fixed, the contrast on the existence of a unit root in y, that is, that the null hypothesis is $\rho = 1$ set against the alternative that $\rho < 1$,[4] is carried out by means of the statistical t of the least-squared estimate of (3). Though it can be calculated more simply from the following reparametrization of (4):

$$\Delta y_t = \phi y_{t-1} + e_t \tag{4}$$

with $\Delta y_t = (y_t - y_{t-1})$, and $\phi = \rho - 1$, for which our interest would lie in contrasting H_0: $\phi = 0$ on the alternative H_1: $\phi < 0$. The t statistic of the estimator, ϕ, τ, under the null hypothesis is not based on a standard distribution, and its asymptotic distribution, obtained by simulation, as well as the critical values of τ, can be found in Fuller (1976). But it should be noted that these critical values would only be valid if they satisfy the assumptions of non-existence of trends and constants in the (3). For this reason, then, three possible DGPs are considered:

$$y_t = \rho y_{t-1} + e_t; \quad t = 1, 2, \ldots, T, \tag{5}$$

$$y_t = \alpha + \rho y_{t-1} + e_t; \quad t = 1, 2, \ldots, T, \tag{6}$$

$$y_t = \alpha + \beta t + \rho y_{t-1} + e_t; \quad t = 1, 2, \ldots, T, \tag{7}$$

where in whatever case it is supposed that e_t follows an IID process. If $\rho < 1$, as an absolute value, then (5) would be a stationary AR(1) process, while (6) would be a stationary AR(1) process around a constant, and (7) a stationary AR(1) process around a linear trend.

Furthermore, it should be pointed out that in (3), Dickey and Fuller supposed that the data generating process (DGP) was an AR(1). Assuming that the DGP is an AR (p), Dickey and Fuller defined a new augmented statistical process (ADF) based on the following regression:

[3]Reviews of the literature on stationarity and integrity can be found in Dolado et al. (1990) or Banerjee et al. (1993), among others.

[4]Then the alternative hypothesis would be that y_t follows a stationary process AR(1).

$$\Delta y_t = \alpha + \varphi y_{t-1} + \sum_{i=1}^{p} \delta \, \Delta y_{t-i} + u_t$$

where p is the optimum order of the y_t lags, sufficiently long to ensure that the u_t residuals are white noise.

As an alternative to the inclusion of lagged values to eliminate the serial correlation in (3), Phillips (1987) and Phillips and Perron (1988) suggest some non-parametric adjustments from the statistics of Dickey–Fuller that generalize the specification of the data generating process, allowing u_t to follow a quite general process and, especially, any ARMA(p, q) model. These statistics are represented by $Z(\tau_\mu)$, and their asymptotic distribution is identical to that tabulated by Fuller (1976). Then it would be a matter of comparing the critical value of $Z(\tau_\mu)$ with the critical value $\tau_{\mu(T,\alpha)}$ and reject the null hypothesis on the existence of a unit root if $|Z(\tau_\mu)| > |\tau_{\mu(T,\alpha)}|$.

Some economic time series can be characterized as second-order integrated processes, for which the presence of more than one unit root can be checked. In this sense, Dickey and Pantula (1987) and Pantula (1989) develop a simple sequential process to test the hypothesis of multiple unit roots, consisting of starting by considering the greatest number of possible unit roots and decrease this number by one each time that the null hypothesis is rejected, completing the process when the null hypothesis is accepted.

The proposed specific process for checking the null hypothesis of existence of d unit roots against the alternative $d - 1$ roots is a simple application of the Phillips–Perron tests, but considering initially the greatest possible number of unit roots.

Moreover, a series of investigations (Dickey and Fuller (1981), or Schwert (1987, 1989) among others) have shown the weaknesses in normal processes of checking unit roots. Specifically, the tests of Dickey–Fuller, Phillips–Perron or Sargan–Bhargava are low-powered if the sample of available data is not sufficiently large.[5] Therefore, some authors have proposed checking the null hypothesis of stationarity on the alternative of a unit root, which is to use reverse Dickey–Fuller or Phillips–Perron tests. The basic idea is that with the standard tests the null hypothesis would only be rejected if extreme values are obtained, or, in other words, these tests could often not distinguish series with a unit root against an AR(1) series with an autoregressive coefficient very close to the unit.

Tests of this last type, which are based on the null hypothesis of stationarity against the alternative unit root hypothesis, have been developed by Park (1990), Khan and Ogaky (1992), Kwiatkowski et al. (1992) or Leybourne and McCabe (1994). But, given that up till now the new processes have also suffered problems of low power,[6] in this research it has been believed convenient to use a combination of both types of contrasts, using the new contrasts in combination with the standard

[5]In these studies, it is shown that for the simple sizes generally available standard tests may have low power against persistent alternatives.

[6]See the study of Kwiatkowski et al. (1992).

ones. The results, therefore, will depend on the combination of findings from both types of processes, in accordance with the suggestion carried out in Kwiatkowski et al. (1992). On the other hand, it should also be pointed out that the power of these tests is sensitive to how the parameters are set.

Specifically, Kwiatkowski ct al. (1992) propose a Lagrange statistical process (LM), denoted henceforth as KPSS, to contrast the null hypothesis of the stationarity of a series, that can be considered as the sum of a deterministic trend, a random walk and a stationarity error term. In particular this starts from the DGP:

$$y_t = \delta t + r_t + \varepsilon_t$$

$$r_t = r_{t-1} + \mu_t$$

where ε_t and μ_t is uncorrelated white noise. The initial value of the random walk is considered fixed, for which it acts as a constant or average to which the series returns. The null hypothesis of stationarity implies that the variance of noise in the random walk, $\sigma_\mu{}^2$, is zero, which is equivalent to $\sigma_r{}^2 = 0$. Under the null hypothesis, y_t is stationary round a constant (if $\delta = 0$) or round a trend (if $\delta \neq 0$).

In practice, to perform the KPSS test, the regression of y_t is carried out on a constant (η_τ test) or a constant and a trend (η_μ). The remains of this regression, e_t, are used to calculate the Lagrange multiplier:

$$LM_{KPSS} = \frac{\sum_{t=1}^{T} S_t^2}{s_\varepsilon^2} \tag{8}$$

where s_ε^2 is the estimated variance of ε_t[7] and $S_t = \sum_{i=1}^{t} e_i$. The distribution of LM statistics is not standard and the critical adequate values are found in Kwiatkowski et al. (1992).

Moreover, Kwiatkowski et al. (1992) bear in mind the case when t presents an autocorrelation and problems of heterocedasticity. In this assumption, in expression (8), the denominator would be replaced by the long-term variance estimator ε_t in accordance with the $s^2(k)$ process recommended by Newey and West (1987), where k is the number of lags.[8]

---

[7]The sum of the remainder squared divided into $T - 1$, with T being the sample size.

[8]Of course, the value of the test depends on the number of lags chosen. Kwiatkowski et al. (1992) propose a value of, at most, eight. In this regard see also Phillips (1987) and Phillips and Perron (1988).

2.4 Results of the Univariate Analysis

The results of applying the Phillips and Perron (1988) (PP) tests and Kwiatkowski et al. (1992) (KPSS) tests to the series under analysis are shown in Tables 2 and 3, respectively, both for the first differences and the levels of the variables. The results obtained show that, for all the series analyzed, the existence of two unit roots can be rejected, with the exception of the variable stock of private capital (Log K), where the P–P test does not allow rejection of the null hypothesis where I(2) is at 5%, even though the KPSS test shows us that the variable is I(1). Furthermore, it can be confirmed that all the series are integrated in the primary order, I(1), or stationary in the first differences. Keeping in mind the greatest theoretical suitability of the KPSS test for quite short time periods, from the univariate analysis of the variables it is suggested taking all the series, production, employment, human capital, private

Table 2 Unit root and stationarity contrasts

First differences of the series				
Variables in logarithms	PP $Z(t\tilde{\alpha})$	KPSS η_μ	KPSS η_τ	Conclusión
Log Y	−3.09*	0.16		I(0)
Log N	−4.67**	0.24		I(0)
Log K	−1.77	0.36		I(1) ó I(0)
Log GK	−4.16**		0.08	I(0)
Log GP	−3.21*	0.21		I(0)
Log H	−3.94	0.33		I(0)

Notes: *and **denotes significance at 5% and 1%, respectively
$Z(t\tilde{\alpha})$ is the Phillips–Perron test for the null hypothesis of non-stationarity [see Perron (1988)]
η_μ and η_τ are the Kwiatkowski et al. (1992) tests for the null hypothesis of stationarity, when there is only one constant or a constant and a trend, respectively

Table 3 Unit root and stationarity contrasts

Series on levels				
Variables in logarithms	PP $Z(t\tilde{\alpha})$	KPSS η_μ	KPSS η_τ	Conclusión
Log Y	−1.25	0.72*		I(1)
Log N	−1.51	0.48*		I(1)
Log K	−2.90	0.65*		I(1)
Log GK	−1.95	0.72*		I(1)
Log GP	−1.32		0.16*	I(1)
Log H	−1.61	0.71*		

Notes: *denotes significance at 5% and 1%, respectively
$Z(t\tilde{\alpha})$ is the Phillips–Perron test for the null hypothesis of non-stationarity [see Perron (1988)]
η_μ and η_τ are the Kwiatkowski et al. (1992) tests for the null hypothesis of stationarity, when there is only one constant or a constant and a trend, respectively

capital, public capital without port infrastructures and public capital in ports, in the first differences.

3 Impact of Investment in Port Infrastructures on the Output and Regional Employment

3.1 Introduction

After the univariate analysis of the series to be included in the model, it is necessary to study the possible long term relationships between the variables. Being non-stationary series, relationships of cointegration could appear between them which, if not identified and incorporated into the model, would cause the estimates to be inconsistent. In other words, it is a matter of determining whether or not there are any linear stationary relationships between non-stationary variables (integrated in the primary order), which allows a contrast of the possible equilibrium relationships between these variables, as suggested by economic theory.

The first part of this section succinctly outlines the methodology of cointegration that will be used to estimate the production function and to evaluate, thus, the elasticity of output regarding the public capital in port infrastructures. As is known, this type of analysis not only allows a statistical characterisation of the time series, but also a contrast of the common trends between them. The final part of the epigraph focuses on the results of the estimate.

3.2 Basic Concepts of the Cointegration Analysis

The concept of cointegration provides an adequate framework to contrast the significant long-term relationships between non-stationary time series.[9]

If an economic model is considered to be made up of a group of series integrated in the primary order, there is a danger of making an incorrect inference on the existing econometric relationship between them, falling victim to the so-called problem of "spurious regression" (Granger and Newbold 1974). Therefore, the possible existence of long term relationships between a group of non-stationary series, which are of interest from the perspective of economic theory, should be associated to the number of cointegration relationships between them. In this sense, it is said that n non-stationary series x_{it}, each one of which follows a non-stationary process I(1), are cointegrated, CI(1,1), if a linear combination of the same exists

---

[9]A variable is integrated in order d if its difference in order d admits a stationary and invertible ARMA representation. In that case x_t is I(d). In particular, the concept of cointegration has often been applied to series that are integrated in the primary order, I(1).

given by $z_t = \sum_{I=1}^{n} \beta_i x_{it}$ which is a stationary process [I(0)].[10] This cointegration relationship between the variables is interpreted as a long term equilibrium, since the situations of imbalance are stationary and, therefore, transient.

To avoid the problem of spurious regression of series with trends it was common practice to estimate regressions in the differences of the variables. Nevertheless, this type of modelling was shown to be incapable of describing long term behaviour. For this reason, different authors have treated the subject of stationarity using error correction models, which, in principle, allow a consistent economic interpretation on including not only the variables in differences but also in levels. Engle and Granger (1987) rigorously established the link between the concepts of cointegration and error correction models (ECM), showing that the cointegrated variables can always be represented in terms of ECM and vice versa.

3.2.1 Single Equation Cointegration Models

Since Granger (1981) introduced the concept of cointegration, there have been various processes suggested in literature to test this property. The best-known one applies the standard tests of unit roots on the remainder of a static regression estimated by normal least squares. It would mean, therefore, testing the hypothesis $\lambda = 1$ in the following regression:

$$\hat{u}_t = \lambda \hat{u}_{t-1} + \varepsilon_t$$

where $\hat{u}_t$ is the remainder of the static least squared estimate of the variables of interest in levels. For this purpose various tests can be used, such as the Dickey–Fuller (CRDF) or the Augmented Dickey–Fuller (CRADF). However, the distribution of $\hat{\lambda}$ is not the same as in the univariate case, so it depends heavily on the number of variables in the cointegration regression.

Additionally, Engle and Granger (1987), also proposed an estimation process in two phases to determine the parameters of the ECM, showing that after estimating the cointegration vector by COM, the remaining parameters of the ECM can be estimated consistently by introducing the remainder of the static regression delayed a period in the ECM. That is, in the first phase the static regression between the levels of the variables[11] would be estimated using COM, being tested if the remainder of that regression is I(1). If the null hypothesis is rejected which is I(1), then the short term dynamics can be estimated in the second phase.

Nevertheless, the two-stage Engle and Granger (1987) process presents various problems. So, on the one hand, the estimate in the first phase would not be efficient without considering all the information contained in the ECM model. In addition,

---

[10]If the variables are I(d) and there is a linear combination of these which is I(d–b), it is said that they are cointegrated to the order (d, b), CI(d, b).

[11]Remember that these variables are not stationary.

another possible difficulty is that it implicitly assumes that the cointegration relationship is unique, and if the cointegration rank were greater that one, this would be no more than the result of a linear combination of the existing cointegration relationships.[12] Therefore, when there are more than two variables, the possible existence of multiple cointegration should be considered, since in general there could be (n − 1) cointegration relationships between n variables. In this case the long term model should consist of a complete system of cointegration equations. On the other hand, it is necessary to point out that tests on the estimated parameters cannot be carried out in the first phase of the Engle and Granger process, on following those unknown distributions. Finally, it should also be mentioned that problems of simultaneousness may arise upon estimating the error correction model completely, which would require using a process of instrumental variables.

Stock (1987) has shown that the estimators of cointegration relationships, obtained by means of static regressions, tend towards their true values at a greater rate than normal with stationary series. Nevertheless, despite this asymptotic property of "super consistency", these estimators can have significant biases for hypothesis testing when finite samples of finite size are used. Finite sample bias depends on the generating process of the data, as various works have shown, based on the experiments of Monte Carlo. One of the solutions to this problem consists of estimating the cointegration vectors in complete dynamic specifications, although corrections could also be used to avoid these biases in finite samples.

In particular, the non-parametric correction proposed by Phillips and Hansen (1990) and Hansen (1992) could be considered, which is robust to the problems of autocorrelation and simultaneousness, and allows the use of standard inference in the hypothesis tests on linear restrictions of the parameters.

Another way to estimate and test cointegration relationships is the direct estimation of each ECM equation by non-linear least squares. This estimate would be accurate given the non-linearity of the parameters of each equation representing the error correction mechanism. Then, if none of the speed of adjustment parameters turns out to be significant in estimating the ECM, the variables would not be cointegrated. Consequently, the test on the significance of those parameters becomes a cointegration test. But once again the inference on those parameters is not standard, and its critical values are included in Banerjee et al. (1993) and in Banerjee et al. (1995).[13] Also, Kremers et al. (1992) and Banerjee et al. (1993) show that this estimation of the ECM and the test on the significance of the adjustment coefficient is a more powerful test than those presented by Engle and Granger (1987). This is explained because the Dickey and Fuller test on the remainder of the cointegration relationships imposes a common factor restriction on the Data Generating Process (DGP), which, of course, need not be met.

[12]The only cointegration vector would be normalized on any of the parameters, but its economic interpretation would be very doubtful.

[13]In this estimate, under the null hypothesis of non-conitegration, the term which goes with the ECM coefficent is I(1), while the dependent variable is I(0).

On the other hand, if in the DGP only one of the equations contains an error correction term, and the rest do not, there is no problem of simultaneousness, that is, a variable is endogenous and the rest are weakly exogenous, being efficiently able to estimate the ECM equation pertaining to the endogenous variable by non-linear least squares.

3.2.2 Cointegration in Multi-Equation Models

All the previous single equation methods present the problem that, in reality, multiple cointegration relations can exist. To solve this problem, Johansen (1988, 1991) and Johansen and Juselius (1990) have proposed, specifically, a consistent solution to simultaneously estimate the space of cointegration vectors, using a process of maximum verisimilitude. This method, as well as allowing inference on the number of cointegration vectors, makes it possible to carry out tests on various types of relationships between the parameters of interest.

Whatever, it is necessary to point out that Johansen's methodology is not free from problems, and may not be useful when the calculated cointegration vectors do not have an easy economic interpretation. This becomes especially evident when the cointegration rank is high.

In short, Johansen's method starts by considering a vector x_t that includes the p variables of the model ($x_t = x_1, \ldots, x_p$), all of them I(1). The data generating process would be the following vector autoregression (VAR):

$$x_t = \sum_{i=1}^{k} \prod_i x_{t-i} + \varepsilon_t; \quad t = 1, \ldots, T \tag{9}$$

where Π_i represents a matrix of parameters ($p \times p$), and ε_t the error term that follows an independent Gaussian p-dimensional process, with zero average and covariance matrix Ω [$\varepsilon_t \sim$ NIID $(0, \Omega)$]. The null hypothesis of interest is that the subspace dimension of the cointegration vectors of is r, using a maximum verisimilitude process for checking that is based on the reparametrization of (7) as an error correction model in the following way:

$$\Delta x_t = \Gamma_1 \Delta x_{t-1} + \Gamma_2 \Delta x_{t-2} + \cdots + \Gamma_{p-1} \Delta x_{t-k+1} + \Pi x_{t-k} + \varepsilon_t$$

$$\text{With} \quad \Gamma_i = -I + \sum_{i=1}^{k-1} \Pi_i \quad \text{and} \quad \Pi = -I + \sum_{i=1}^{k} \Pi_i$$

and where I is the identity matrix.

As all the x_t vectors in differences are composed of variables I(0) the p linear combinations of variables I(1), $\Pi\, x_{t-k}$, should also be I(0), either because there are some cointegration vectors or because Π is a matrix of zeros. Therefore, the cointegration test proposed by Johansen becomes a test on the rank r of the Π matrix, coinciding this with the number of lineally independent cointegration vectors which exist between the p variables in x_t If $r = p$, the Π matrix has a complete rank, for which x_t is stationary. Of the other hand, if $r = 0$ then there is no cointegration and the model should be estimated in differences. In another case, if $0 < r < p$ is fulfilled, r cointegration vectors would exist. Then it would be possible to define a β matrix, $(p \times p)$, so that its columns are the r cointegration vectors, such that $\beta' x_{t-k}$ be I(0). Also it would be feasible to carry out the following breakdown of Π.

$$\Pi = \alpha\beta'$$

for which:

$$\Delta x_t = \Gamma_1 \Delta x_{t-1} + \Gamma_2 \Delta x_{t-2} + \cdots + \Gamma_{p-1} \Delta x_{t-k+1} + (\alpha\beta') x_{t-k} + \varepsilon_t$$

where α represents the weighting matrix with which each cointegration vector enters each one of the Δx_t equations.

Johansen, in order to determine how many eigenvectors represent significant cointegration relations, proposes tests based on the verisimilitude ratio. A first test, the trace test, verifies the null hypothesis that there is a minimum of r cointegration vectors set against the more general alternative that there is p, $r \leq p$. The critical values of the trace test are tabulated in Johansen (1988) and Osterwald-Lenum (1992). To verify if the vector of constants enters in a restricted manner or not in the model, Johansen (1992) suggests carrying out the sequence of the test starting with the assumption that that only appears in the error correction term. Each time that the null hypothesis is rejected it would be checked under the assumption that the constant is not restricted. If this hypothesis is also rejected a higher cointegration rank would be checked again under the assumption that the constant is restricted, and so on till the first time the null hypothesis is not rejected.

The estimation of the system with Johansen's techniques also shows us if it is possible to maintain the hypothesis of weak exogeneity, by checking that the cointegration vector does not enter the equations of the remainder of the regressors.

3.3 *Effects of Investment in Port Infrastructures on the Production of Cantabria*

In this section various estimates of a Cobb–Douglas aggregate production function for the Cantabrian economy are presented, obtained through the application of cointegration analysis techniques and using annual data covering the period 1964–2000. The strategy of the cointegration analysis, or search for stationary linear relationships between the variables, is performed in two phases. In the first phase the production function is appraised by ordinary least squares, applying the non-parametric correction developed by Phillips and Hansen (1990). Furthermore, it will be checked whether or not the existence of constant performances to scale in production is accepted.

In the second phase a dynamic model will be appraised, both by non-linear methods (error correction models) and maximum verisimilitude (Johansen's techniques of cointegrated autoregressive vectors).

The starting point of the analysis is given by an aggregate production function for regional economy, in which public capital appears as a factor of production different than privately owned capital. Specifically, as mentioned in Sect. 1.1, it is assumed that productive technology can be included through a Cobb–Douglas type function: $Y_t = A_t \cdot N_t^{\alpha} \cdot H_t{}^{\delta} \cdot K_t^{\beta} \cdot G_t^{\gamma}$, where Y represents production, A is a parameter of technological efficiency, N the level of employment, H the human capital, K the stock of private capital, and G the stock of public capital.

In this production function all the variables in levels will be included, bearing in mind the results obtained from the analysis on the order of integrability of the variables, discussed previously. Also, to mitigate possible problems of heterocedasticity, the variables are transformed into logarithmic terms. As is well known, this transformation allows direct identification of the coefficients α, δ, β, and γ as elasticities of output against the productive factors of work, human capital, private capital and public capital, respectively.

Moreover, public capital has been divided into two components: the regional port infrastructures, denoted by GP, and the remainder of public productive infrastructures, called GK. In this way the equation to be appraised would be adapt to the expression: $Y_t = A_t \cdot N_t^{\alpha} \cdot H_t{}^{\delta} \cdot K_t^{\beta} \cdot GP_t^{\gamma_1} GK_t^{\gamma_2}$, with γ_1 and γ_2 now representing the percentage of output responses against a percentage change in the resources of productive port and non-port infrastructures, respectively.

Table 4 includes the results of this production equation, taking the regional gross added value at factor cost as output (GAV), and applying the correction proposed

Table 4 Long term static equation

Production function to be appraised:

$$Log(VAB_t) = Log(A) + \mu \cdot t + \alpha \cdot Log(N_t) + \delta \cdot Log(H_t) + \beta \cdot Log(K_t) + \gamma_1 \cdot Log(GP_t) + \gamma_2 \cdot Log(GK_t)$$

Variable	Coefficient	t-Student
Log(A)	−5.997	−6.299
t (trend)	−0.002	−0.089
Log(N)	0.720	11.061
Log(H)	0.205	8.551
Log(K)	0.695	16.928
Log(GP)	0.161	4.548
Log(GK)	−0.080	−1.163

R^2 adjusted = 0.998

1. *Standard equation error* = 0.010

ADF test of cointegration on the remainder of the equation in levels:

CRADF = −8.85, which is significant at 5%

by Phillips and Hansen (1990). As can be seen, the CRADF test (−8.85) allows the rejection, at normal levels of significance, of the null hypothesis of non-cointegration, for which the relationship that is established between the variables studied, GAV, stock of private capital (K), level of employment (N) and values of the public infrastructures (GP and GK), can be considered as a long term equilibrium.

Nevertheless, it should be noted that the productive non-port public capital variable (public capital less the resources destined for health, education and port infrastructures) is not significant, as is the trend variable. The remainder of the input coefficients show appropriate signs and are statistically significantly. This would mean that the stock of port infrastructures would better capture the influence of the productive public capital on the regional output than the remaining infrastructures (roads, railways or airports). Then in what follows both the variable trend and all the different public capital from the port infrastructures are eliminated from the estimates.

Therefore, in Table 5 the results to appraise the aggregate production function for the Cantabrian economy are collected, during the period 1964–2000, using the level of employment, human capital, private capital and stock of port infrastructures of the region as inputs. Now all the variables show the correct signs and are statistically significant, besides which the CRADF test (−8.69) accepts the existence of a stable long term relationship between them. The estimated elasticity of private production with respect to the stock of public capital in port infrastructures is 0.17. In addition, the elasticity of output against private capital is 0.64, while the elasticity against the work factor is 0.71, reducing the percentage response to 0.17 in the case of human capital. These values are in line with those found in Spain by other authors who have used time series.

The next step of the study focuses on the possible existence of constant performances to scale with regard to all the private factors. As can be seen in Table 5, the

Table 5 Long term static equation

Production function to be appraised:

$$Log(VAB_t) = Log(A) + \alpha \cdot Log(N_t) + \delta \cdot Log(H_t) + \beta \cdot Log(K_t) + \gamma \cdot Log(GP_t)$$

Variable	Coefficient	t-Student
Log(A)	−5.503	−8.675
Log(N)	0.714	10.972
Log(H)	0.175	10.958
Log(K)	0.644	21.775
Log(GP)	0.105	10.958

R^2 adjusted = 0.998

1. *Standard equation error* = 0.010

ADF test of cointegration on the remainder of the equation in levels:

CRADF = −8.85, which is significant at 5%

Wald test on the existence of *constant performances of scale* in the private factors: $\alpha + \beta = 1$: 45.10 (Probability of accepting the null hypothesis: 0.0001)

Table 6 Results of the Johansen–Juselius process, Long term equation

Production function to be appraised:
$Log(VAB_t) = Log(A) + \alpha \cdot Log(N_t) + \delta \cdot Log(H_t) + \beta \cdot Log(K_t) + \gamma \cdot Log(GP_t)$

Eigenvalue	VAR(p)	Ho	H1	Trace test	Critical value at 95%[a]
0.664	4	$r = 0$	$r \geq 1$	82.38	69.82
0.436	4	$r \leq 1$	$r \geq 2$	44.16	47.86
0.310	4	$r \leq 2$	$r \geq 3$	24.09	29.79
0.265	4	$r \leq 3$	$r \geq 4$	11.11	15.49
0.009	4	$r \leq 4$	$r \geq 5$	0.34	3.84

Eigenvalue	VAR(p)	Ho	H1	λ-maximum test	Critical value at 95%[a]
0.664	4	$r = 0$	$r \geq 1$	38.21	33.87
0.436	4	$r \leq 1$	$r \geq 2$	20.07	27.58
0.310	4	$r \leq 2$	$r \geq 3$	12.98	21.13
0.265	4	$r \leq 3$	$r \geq 4$	10.77	14.26
0.009	4	$r \leq 4$	$r \geq 5$	0.34	3.84

The Cointegration Vector, normalized over Log(VABt) results in:
$Log(VABt) + 5.56\, Log(A) - 0.685\, Log(N_t) - 0.203\, Log(H_t) - 0.631\, Log(K_t) - 0.163\, Log(GP_t)$

[a]The critical values of the test have been extracted from Osterwald-Lenum (1992)

estimated values for the coefficients of the inputs, α and β add up to 1.35. The Wald test on the null hypothesis: $\alpha + \beta = 1$, which is distributed as a chi-square with one degree of freedom, should be rejected, since the probability of accepting this null hypothesis of constant performances to scale is 0.0001%.

In the second phase of the analysis applied, and for the purpose of confirming the results obtained, we proceeded to appraise the production function following the methodology of Johansen and Juselius. In this case, the results of the contrast are shown in Table 6. Both the trace and the λ-maximum tests show the existence of a cointegration relationship between the GAV, private capital, employment, human capital and port public capital variables. Once the corresponding cointegration vector has been normalized, it is estimated that the long term elasticity of regional output with relation to the port public capital is equal to 0.163, that of private capital 0.63, that of employment 0.69 and that of human capital 0.20, all very similar values to those obtained in the preceding static regression.

The incorporation of the vector error correction as an explanatory variable in all the equations of the VAR model assumes that this is only significant in determining changes in regional output. Consequently, and also applying Engle and Granger's Representation Theorem, a single equation model in the form of the error correction mechanism constitutes, in our case, the most rigorous and flexible structure for appraising the production function. This modelling will allow, additionally, the appraisal of elasticities both in the short and long term and the calculation of the corresponding dynamic adjustment toward the equilibrium.

In this way the relationship of short and long term has been jointly appraised using non-linear least squares. The results are presented in Table 6, showing the coefficients that correspond to the short term (variables in the first differences) and long term (variables in levels) adjustments. Once again, and this time by virtue of the t-statistic of the ECM coefficient (−5.01), the existence of a cointegration relationship between the variables studied cannot be rejected.

The coefficients obtained for the long term adjustment of the different productive factors does not differ significantly from those estimated in previous static relationships. Thus, the long term elasticities set against the employment (α), human capital (δ), private capital (β) and port public capital (γ) are estimated at 0.60, 0.18, 0.62 and 0.095, respectively. In turn, the short term response of the regional GAV before a change in the endowment of public capital is more reduced, calculated at 0.052 (coefficient Ω). This would signify that a 10% increase in public infrastructures would cause an increase of 0.52% of the regional GAV to the following year in which they started operating. Besides, the ECM coefficient (θ) indicates that every year almost 91% of the imbalances produced in the previous period are corrected.

In the final part of Table 7 some statistical validation tests are also presented: N is a normality test of the remainder of the model; LM is an autocorrelation test of

Table 7 Dynamic equation of the Error Correction Mechanism

$$\begin{aligned}\Delta(Log\ VAB_t) = &\sum_{j=0}^{p}\Theta_j\Delta(Log\ N_{t-j}) + \sum_{j=0}^{p}\Psi_j\Delta(Log\ H_{t-j})\\ &+\sum_{j=0}^{p}\Omega_j\Delta(Log\ GP_{t-j}) + \sum_{J=0}^{p}\Phi\ \Delta(LogK_{t-j})\\ &+\sum_{j=1}^{p}\Gamma_j\Delta(Log\ VAB_{i,t-j}) - \theta\ [(Log\ VAB_{t-1})\\ &- Log\ A - \alpha\cdot(Log\ N_{t-1}) - \delta(LogH_{t-1})\\ &-\beta\,(LogK_{t-1}) - \gamma(Log\ GP_{t-1})]\end{aligned}$$

Variable	Coefficient	t-Student
MCE (coef. θ)	−0.906	−5.016
Log(A)	−3.801	−5.002
$Log(N_{t-1})$	0.602	8.481
$Log(H_{t-1})$	0.188	12.467
$Log(K_{t-1})$	0.625	23.797
$Log(GP_t)$	0.096	3.892
$\Delta[Log(N_t)]$	0.461	4.978
$\Delta[Log(H_t)]$	0.165	2.941
$\Delta[Log(K_t)]$	1.001	6.295
$\Delta[Log(GP_t)]$	0.052	3.075

R^2 adjusted = 0.865
Standard error of the equation = 0.008
Banerjee et al. cointegration test (1993): −5.01* (t-Student over the ECM), which is significant at 5%
N (2) = 0.926
LM(4) = 0.907
ARCH(4) = 0.083
DW = 1.802

Table 8 Accumulated effects on the regional GAV (in percent) for each percentage point increase in planned investment in the expansion of the Port of Santander

Years	Accumulated impact of investment in port infrastructures
2005	0.052433
2006	0.091532
2007	0.095197
2008	0.095540
2009	0.095572
2010	0.095575
2011	0.095576
2012	0.095576
2013	0.095576
2014	0.095576
2015	0.095576

Source: Compiled by author

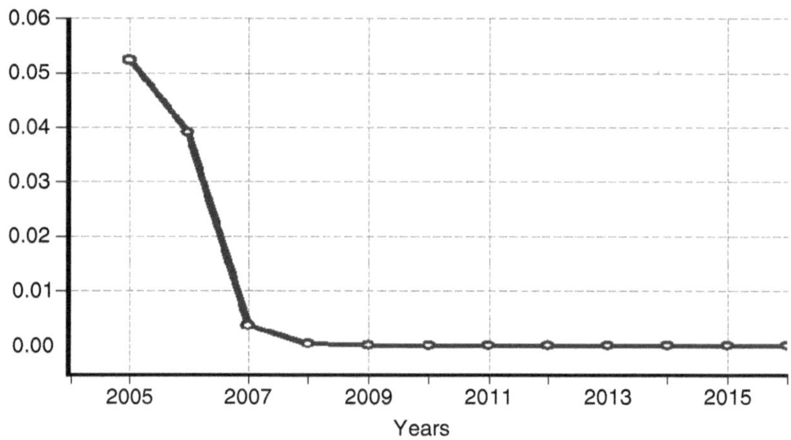

Graph 2 Elasticity of response of the regional GAV for each percentage point increase in planned investment in the expansion of the Port of Santander
Source: Compiled by author

the remainder, and ARCH is a test on the conditional autoregressive heterocedasticity of the remainder. None of these tests show any problems of poor specification of the model obtained.

Based on the estimates obtained for the dynamic equation shown in Table 8, the response of the regional GAV to a percentage change in the resources of public port infrastructures has been calculated. The dynamic model to resolve, made up of an equation in finite differences, is: $\text{Log}(\text{VAB}_t) = (1 - 0.906).\ \text{Log}(\text{VAB}_{t-1}) + 0.052\ \text{Log}(\text{GP}_t) + (0.086 - 0.052)\ \text{Log}(\text{GP}_{t-1})$.

The effect, over time, of a one percent change in the resources of public infrastructures on the Cantabrian GAV is shown in Graph 2. In the same graph it can be seen how the regional output reacts not only during the same year in which the investment is made. That is, Graph 2 summarizes the effects caused in the long

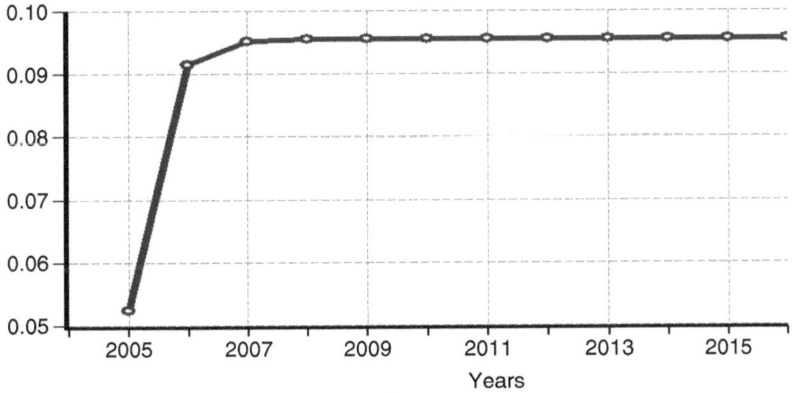

Graph 3 Accumulated response of the regional GAV (in percent) for each percentage point increase in planned investment in the expansion of the Port of Santander
Source: Compiled by author

term on production in Cantabria as a consequence of a one percent increase in the resources of the port infrastructures, considering that this occurred in 2005.

The accumulated elasticities of the regional GAV with respect to the increases in public capital like those predicted in the expansion project for the Port of Santander are represented in Graph 3. In this case the accumulated long-term effects of a 1% increase in the stock of public infrastructures are shown, a result of adding over time the response of regional output in each period.

All the effects that are shown refer to a broad time period, with the purpose of ensuring that the consequences of an increase in the endowment of capital for a year would not be diluted, in a manner that could affect the evolution of other variables. In every case, and as can be seen in Table 8 the adjustment is practically complete in the 5 years following the date when new infrastructures come into operation, reaching, as has already been mentioned, an accumulated effect on the regional GAV of 0.095 percent for each percentage point increase in stock of the productive infrastructures of the Port of Santander.

3.4 *Effects of Investment in Port Infrastructures on Employment in Cantabria*

In a market economy companies determine the level of employment that they want to hire from a process of maximization their profits. In a formal way, the problem of companies would be to achieve the greatest difference between its income and costs, that is, to maximize a certain function Π,

$$\Pi = P \cdot Y - (W \cdot N + CU_K \cdot K + CU_{KG} \cdot KG)$$

where P represents the price level, Y the output, N the employment, K the stock of private capital, KG the productive public capital, W the salary by worker and CU_K and CU_{KG} the unit price or cost of use of private and public capital, respectively.

In this optimization exercise of it is considered that the level of output is derived according to an aggregate production function, which relates the total quantity of goods and services produced with the use made of the work and capital factors, the latter being of a public or private character. That is, a technological relationship would exist of the type Y = Y(L, K, KG) that acts as a constraint in the objective profit function of benefits Π.

Under conditions of perfect competition and in the absence of adjustment costs, companies will resolve the problem above by choosing the level of employment according to an expression which equates the real salary (W/P) with the differential of the production function regarding the work factor. This condition, which is interpreted as an aggregate labour demand, states that companies demand labour to the point in which the increase in income derived from contracting an additional unit of the factor equals the increase in costs that would suppose, or in economic terminology, companies will choose the level of employment so that the marginal product of this input coincides with the real salary.

If it is confirmed that in the production function of the regional economy there is explicitly as much private productive capital as public, the differential of this function regarding the work factor will depend again on these elements. Consequently, deriving the production function that was used in the previous section of this study, $Y_t = A_t \cdot N_t^{\alpha} \cdot H_t{}^{\delta} \cdot K_t^{\beta} \cdot GP_t^{\gamma}$, with regard to employment and equating the result to the real salary, the following expression is obtained:

$$\frac{\partial Y_t}{\partial N_t} = \alpha \cdot A_t \cdot N_t^{\alpha-1} \cdot H_t{}^{\delta} \cdot K_t^{\beta} \cdot GP_t^{\gamma} = W/P$$

Clearing the employment level we arrive at the aggregate work demand function:

$$N_t = \left[\frac{\alpha A_t H_t{}^{\delta} K_t{}^{\beta} KG_t{}^{\gamma}}{W/P}\right]^{\frac{1}{(1-\alpha)}}$$

Now taking logarithms, in order to convert the multiplicative equation into linear, and with the parameters previously estimated for the function of long term production (denoted by ∧), would give:

$$Log\, N_t = \frac{1}{1-\hat{\alpha}}\left[Log\,\hat{\alpha} + Log\,A + \hat{\delta}\,Log H_t + \hat{\beta}\,Log\,K_t + \hat{\gamma}\,Log\,GP_t - Log\left(\frac{W_t}{P_t}\right)\right]$$

Then, in the long term equilibrium, and considering the adjustment processes, the multiplier of public investment on employment created could be calculated as:

$$\frac{\partial Log\,N_t}{\partial Log\,GP_t} = \frac{1}{1-\hat{\alpha}} \cdot \hat{\gamma} = 0.023$$

This means that, in the long term, for each percentage point of increase in the real stock of productive public capital, in real terms, employment would increase by 0.023%. The dynamic impact multiplier of public investment on employment, accumulated over time, is included in Table 9.

4 Summary and Conclusions

In this work, the impact which new investment in the expansion project of the Port of Santander would have on economic growth and employment in Cantabria has been studied. To this end a methodology based on estimating an aggregate production function has been used which, by applying econometric techniques of cointegration, detects a stable relationship between the regional output, employment, human capital, the supply of private capital and the port infrastructure. This approach measures the magnitude of the effects of an increase in the capital endowment of the port on private production, being complementary to results obtained using other economic research tools, such as input–output studies or social assessment projects through cost-benefit analysis.

The methodology employed in this work makes it possible to evaluate the impacts that will result once the investment planned for the Port of Santander has been carried out, quantifying its effects from the moment when the new supply of port infrastructures can be used by the private sector, either to generate more economic activity or to increase its productive efficiency. In this sense, it is assumed that the economic impacts will not be completely produced in the same year in which the investment is made, but are spread out over time, thus considering the short term as well as the long term effects. It should be noted that the year 2005

Table 9 Accumulated effects on employment in Cantabria (in percent) of an increase of a percentage point in the endowment of port capital

Years	Accumulated impact of investment in port infrastructures
2005	0.01258052
2006	0.02196196
2007	0.02284118
2008	0.02292358
2009	0.02293130
2010	0.02293202
2011	0.02293209
2012	0.02293209
2013	0.02293209
2014	0.02293209
2015	0.02293209

has been taken as the start of the port expansion. If the process were to be delayed, the effects would in turn displace themselves over time. In that case, it would be necessary to substitute the year used as the start for the real start year in all the analysis. Also it has been taken into account that the investments projected by the port authority and the transport operators have identical considerations, that of enlarging of the stock of the public port infrastructures.

Among the results obtained the following stand out:

1. The increase in the stock of port capital will cause a growth in production and in regional employment. The magnitude of these effects is very considerable: the GAV of Cantabria will respond, in the long term, with a 0.095 percent increase for each percentage point of increase, in real terms, of the port infrastructures, while the level of employment will grow by 0.023 percent.
2. In the specific case of the planned expansion of the Port of Santander, the final results of its impact on production and employment are obtained by multiplying the coefficients of the estimated elasticities (0.09525 and 0.023) by the percentage variation that represents the new contributions of capital in port infrastructures in relation to the regional aggregate. It should be noted in this conclusion that what matters is not only the value of the coefficient of the estimated elasticity, but also the representativeness percentage of the capital invested, in terms of the public capital of Cantabria. This detailed explanation can be seen in Table 10, where the values of the variables are referred to 2005.

Consequently, the following conclusions from the study should be highlighted:

(a) In the long term and considering that the impact of the new infrastructures will be prolonged until the year 2022, the GAV of Cantabria will increase by 6.34 percent with regard to its value in 2005. This means that, on average, a sustained annual increase of 0.36% in regional production would be achieved from 2005 until 2022. Both the annual impacts as well as those accumulated over time and their breakdown due to the investments of the port authority and transport operators can be seen in Graphs 4 and 5, respectively.
(b) More than ninety percent of all the accumulated effects are produced after 5 years has elapsed from the time when the new resources of port public capital enter operation. Detail of these impacts is offered in Table 11.
(c) With respect to employment, by adding the effects of the investment of the port authority and transport operators, this gives a total generation of 3,654 jobs at the end of the considered period, as can be analyzed in Table 12.

Table 10 Representativeness of the planned investment in the expansion of the port on the stock of port public capital of Cantabria in 2005

Planned public investment in the expansion of the Port of Santander (a)	136,597
Stock of public port infrastructures in Cantabria (b)	183,843
% (a)/(b)	0.74

Source: Compiled by author. Monetary figures in thousands of constant euros, 1986

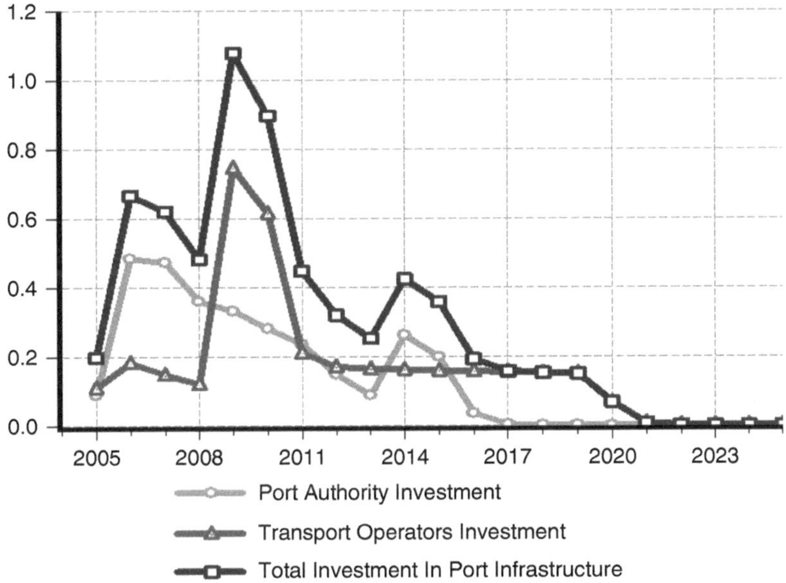

Graph 4 Annual response in the GAV of Cantabria (in percentage) on the planned investment in the expansion of the Port of Santander

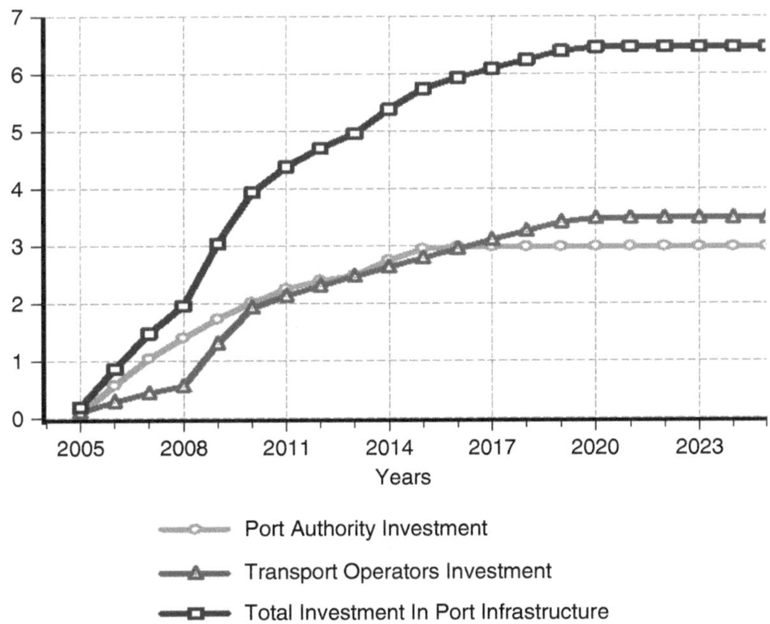

Graph 5 *Accumulated* response in the GAV of Cantabria (in percentage) on the planned investment in the expansion of the Port of Santander

Table 11 *Accumulated* effects on the GAV of Cantabria (in percentage) on the planned investment in the expansion of the Port of Santander

Years	Accumulated impact port authority investment	Accumulated impact transport operators investment	Accumulated impact total investment
2005	0.0882	0.1101	0.1983
2006	0.5711	0.2924	0.8635
2007	1.0428	0.4403	1.4831
2008	1.4018	0.5613	1.9631
2009	1.7335	1.3062	3.0398
2010	2.0145	1.9206	3.9351
2011	2.2499	2.1315	4.3814
2012	2.3987	2.3022	4.7010
2013	2.4863	2.4670	4.9533
2014	2.7479	2.6291	5.3770
2015	2.9465	2.7887	5.7351
2016	2.9816	2.9459	5.9275
2017	2.9849	3.1008	6.0857
2018	2.9852	3.2534	6.2387
2019	2.9853	3.4038	6.3891
2020	2.9853	3.4713	6.4566
2021	2.9853	3.4776	6.4629
2022	2.9853	3.4782	6.4635
2023	2.9853	3.4783	6.4635
2024	2.9853	3.4783	6.4635
2025	2.9853	3.4783	6.4635

Table 12 *Accumulated* effects on employment in Cantabria (by numbers) on the planned investment in the expansion of the Port of Santander

Año	Accumulated impact port authority investment	Accumulated impact transport operators investment	Accumulated impact total investment
2005	50	62	112
2006	323	165	488
2007	589	249	838
2008	792	317	1.110
2009	980	738	1,718
2010	1,139	1,086	2,224
2011	1,272	1,205	2,477
2012	1,356	1,301	2,657
2013	1,405	1,395	2,800
2014	1,553	1,486	3,040
2015	1,666	1,576	3,242
2016	1,685	1,665	3,351
2017	1,687	1,753	3,440
2018	1,688	1,839	3,527
2019	1,688	1,924	3,612
2020	1,688	1,962	3,650
2021	1,688	1,966	3,653
2022	1,688	1,966	3,654
2023	1,688	1,966	3,654
2024	1,688	1,966	3,654
2025	1,688	1,966	3,654

(d) The greatest rate of job creation would occur between the years 2006 and 2011, motivated, above all, by the investment of the transport operators, as can be seen in Graphs 6 and 7.

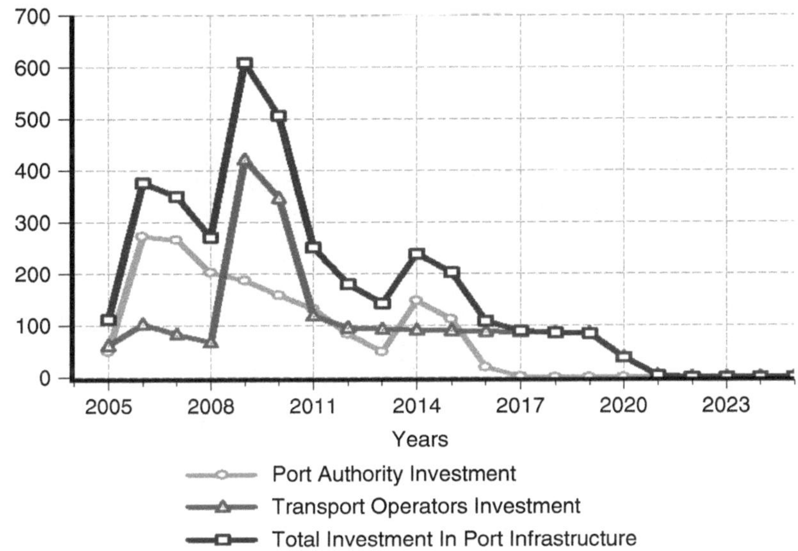

Graph 6 Annual creation of regional employment from the planned investment in the expansion of the Port of Santander

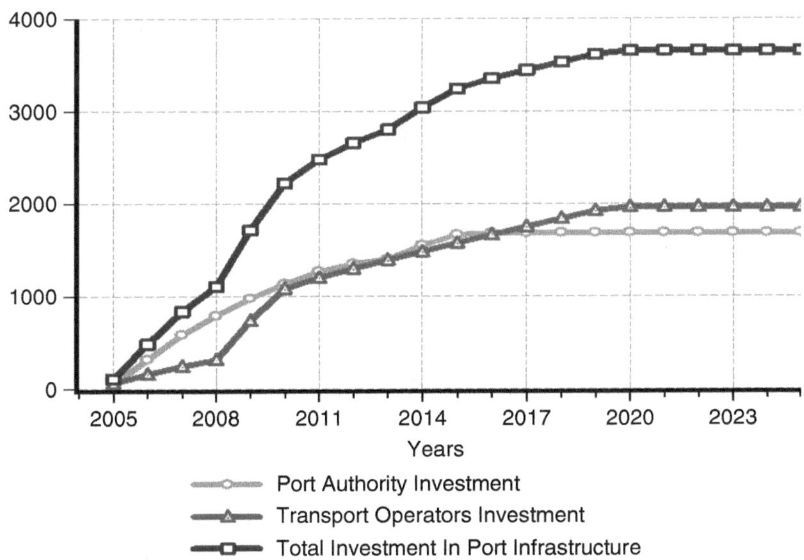

Graph 7 *Accumulated* creation of regional employment from the planned investment in the expansion of the Port of Santander

All these conclusions should be considered with the limitation that it is assumed that the new infrastructures will be accompanied by new port traffic, so as to justify the investment, as the corresponding cost-benefit analysis would show.

References

Argimón I, González-Páramo JM (1997) Efectos de la inversión en infraestructuras sobre la productividad y la renta de las CC.AA.: Especial referencia al transporte por carretera en Galicia. In: Pérez Touriño E (ed) Infraestructuras y desarrollo regional: Efectos económicos de la Autopista del Atlántico. Ed. Civitas, Madrid

Argimón I, González-Páramo JM, Roldán JM (1993) Productividad e infraestructuras en la economía española, Documento de Trabajo 9313. Banco de España, Madrid

Argimón I, González-Páramo JM, Martín MJ, Roldán JM (1994) Productividad e infraestructuras en la economía española. Moneda y Crédito 198:207–238

Aschauer DA (1987) Is government spending stimulative? Federal Reserve Bank of Chicago staff memoranda

Aschauer DA (1989a) Is public expenditure productive? J Monet Econ 23:177–200

Aschauer DA (1989b) Does public capital crowd-out private capital? J Monet Econ 24:171–188

Aschauer DA (1989c) Public investment and productivity growth in the Group of Seven. Econ Perspect 13(5):17–25

Bajo O, Díaz C, Montávez MD (2002) Optimal endowments of public investment: an empirical analysis for the Spanish regions, Documento de Trabajo E2002/4. Fundación Centro de Estudios Andaluces, Sevilla

Bajo O, Sosvilla S (1993) Does public capital affect private sector performance? An analysis of the Spanish case, 1964–1988, Papeles de Trabajo, 1/93. Instituto de Estudios Fiscales, Madrid

Banerjee A, Dolado J, Galbraith J, Hendry DF (1993) Co-integration, error-correction and the econometric analysis of non-stationary data. Oxford University Press, Oxford

Banerjee A, Dolado J, Mestre R (1995) On the power of co-integration test: dimension invariance vs common factors. Queen's Institute for Economic Research, Discussion Paper, 922

Corrales A, Taguas D (1991) Series macroeconómicas para el periodo 1954–88. Un intento de homogeneización. In: Molinas C, Zabalza A (eds) La economía española. Una perspectiva macroeconómica. A. Bosch, Barcelona, pp 583–646

De la Fuente A (1994) Capital Público y Productividad. In: Esteban JM, Vives X (eds) Crecimiento y convergencia regional en España y en Europa, vol II. Instituto de Análisis Económico-CSIC, Barcelona, pp 479–505

De la Fuente A (1996) Infraestructuras y productividad: un panorama de la evidencia empírica. Información Comercial Española 757:25–39

Dickey DA, Fuller WA (1979) Distribution of the estimators for autoregressive time series with a unit root. J Am Stat Assoc 74:427–431

Dickey DA, Fuller WA (1981) Likelihood ratio statistics for autoregressive time series with a unit root. Econometrica 49:1057–1072

Dickey DA, Pantula SG (1987) Determining the order of differencing in autoregressive processes. J Bus Econ Stat 5:455–461

Dolado J, Jenkinson T, Sosvilla-Rivero S (1990) Cointegration and unit roots. J Econ Surv 4:249–273

Dolado JJ, González Páramo JM, Roldán JM (1994) Convergencia económica entre las provincias españolas: Evidencia empírica (1955–1989). Moneda y Crédito 198:81–119

Draper M, Herce JA (1994) Infraestructuras y crecimiento: un panorama. Revista de Economía Aplicada 2(6):129–168

Engle RF, Granger CWJ (1987) Co-integration and error correction: representation, estimation and testing. Econometrica 55:251–276

Flores R, Gracia M, Pérez T (1994) Efectos de la inversión en infraestructuras sobre la economía española. Universidad Complutense de Madrid, mimeo

Ford R, Poret P (1991) Infraestructure and private sector productivity. OCDE Econ Stud 17:63–88

Fuller WA (1976) Introduction to statistical time series. Wiley, New York

Fundación BBVA (varios años) Renta nacional de España y su distribución provincial. Bilbao

García-Fontes W, Serra D (1994) Capital Público, Infraestructuras y crecimiento. In: Esteban JM, Vives X (eds) Crecimiento y convergencia regional en España y en Europa, vol II. Instituto de Análisis Económico-CSIC, Barcelona, pp 451–478

Granger CWJ (1981) Some properties of time series data and their use in econometric model specification. J Econom 16:121–130

Granger CWJ, Newbold P (1974) Spurious regressions in econometrics. J Econom 2:111–120

Gramlich EM (1994) Infrastructure investment: a review essay. J Econ Lit 32(3):1176–1196

Hansen BE (1992) Efficient estimation and testing of cointegrating vectors in the presence of deterministic trends. J Econom 53:87–121

Hendry DF (1990) Econometric modelling with cointegrated variables: an overview. In: Engle RF, Granger CWJ (eds) Long-run economic relationships. Oxford University Press, Oxford, pp 51–63

Holtz-Eakin D (1994) Public sector capital and the productivity puzzle. Rev Econ Stat 76 (1):12–21

Johansen S (1988) Statistical analysis of cointegration vectors. J Econ Dyn Control 12:231–254

Johansen S (1991) Estimation and hypothesis testing of cointegration vectors in Gaussian vector autoregressive models. Econometrica 59:1551–1580

Johansen S, Juselius K (1990) Maximum likelihood estimation and inference on cointegration: with applications to the demand for money. Oxf Bull Econ Stat 52:169–210

Khan JA, Ogaky M (1992) A consistent test for the null of stationarity against the alternative of a unit root. Econ Lett 49:7–11

Kremers J, Ericsson N, Dolado J (1992) The power of co-integration tests. Oxf Bull Econ Stat 54:325–348

Kwiatkowski D, Phillips PCB, Schmidt P, Shin Y (1992) Testing the null hypothesis of stationarity against the alternative of a unit root: how sure are we that economic time series have a unit root? J Econom 54:159–178

Leybourne SJ, McCabe BPM (1994) A consistent test for a unit root. J Bus Econ Stat 12:157–166

Mas M, Maudos J, Pérez F, Uriel E (1993) "Capital público y productividad de la economía española", mimeo. Instituto Valenciano de Investigaciones Económicas, Valencia

Mas M, Maudos J, Pérez F, Uriel E (1994) Capital público y productividad en las regiones españolas. Moneda y Crédito 198:163–192

Mas M, Pérez F, Uriel E (eds) (2003a) El stock de capital en España y su distribución territorial. Fundación BBVA, Bilbao

Mas M, Pérez F, Uriel E, Serrano L, Soler A (2003b) Capital humano, series históricas 1964–2001. BanCaja, Valencia

Molinas C, Ballabriga C, Canadell E, Escribano A, López E, Mestre R, Sebastián M, Taguas D (1990) MOISEES: Un modelo de investigación y simulación de la economía española. A. Bosch, Barcelona

Munnell AH (1990) How does public infrastructure affect regional economic performance? N Engl Econ Rev 30:11–32

Munnel AH (1992) Mortgage lending in Boston: interpreting HMDA data. Working Papers 92–97. Federal Reserve Bank of Boston, Boston

Newey WK, West KD (1987) A simple, positive semi-definite heteroskedasticity and autocorrelation consistent covariance matrix. Econometrica 55:703–708

Osterwald-Lenum M (1992) A note with fractiles of the asymptotic distribution of the maximum likelihood cointegration rank test statistics: four cases. Oxf Bull Econ Stat 54:461–472
Park J (1990) Testing for unit roots and cointegration by variable addition. In: Fomby T, Rhodes F (eds) Advances in econometrics Cointegration, spurious regressions and unit roots. JAI Press, Grenwich
Pantula SG (1989) Testing for unit roots in time series data. Econom Theory 5:265–271
Phillips PCB (1987) Time series regression with a unit root. Econometrica 55:277–301
Phillips PCB, Hansen BE (1990) Statistical inference in instrumental variables regression with I(1) processes. Rev Econ Stud 57:99–125
Phillips PCB, Perron P (1988) Testing for a unit root in time series regression. Biometrica 75:335–346
Ratner J (1983) Government capital and the production function for US private output. Econ Lett 13:213–217
Sadei [Sociedad Asturiana de Estudios Económicos e Industriales] (2003) Datos y cifras de la economía asturiana 2002. Servicio de Publicaciones del Principado de Asturias, Oviedo
Schwert GW (1987) Effects of model specification on tests for unit roots in macroeconomic data. J Monet Econ 20:73–103
Schwert GW (1989) Tests for unit roots: a Monte Carlo investigation. J Bus Econ Stat 7:147–159
Stock J (1987) Asymptotic properties of least-squares estimators of cointegrating vectors. Econometrica 55:1035–1056

Part IV
Regulation and Economic, Technical and Allocative Efficiency

Bootstrapped Technical Efficiency of African Seaports

Carlos Pestana Barros, Albert Assaf, and Ade Ibiwoye

Abstract This paper analyzes the efficiency of a representative sample of African seaports using a bootstrapped DEA – Data Envelopment Analysis approach. Bias and confidence intervals are estimated for the efficiency scores. The paper finds that the original efficiency scores are biased. The bootstrapped results indicated that Nigerian seaports are the most efficient, followed by Mozambique and Angola. Discussions of the results as well as related policy implications are provided.

1 Introduction

The aim of this paper is to analyse the technical efficiency of major African seaports located in Nigeria, Angola and Mozambique. A bootstrapped DEA model is applied using input and output data over the period 2004–2006. The performance of seaports, which motivates the present study, has been traditionally measured by alternative frontier models, such as ratio analyses between inputs and outputs (Song and Cullinane 2001), simple DEA models, (Panayides et al. 2009), or regression frontier models (Barros 2005). The use of ratio analysis is nowadays uncommon, but traditional DEA and regression frontier models are still popular methods.

The major aim of the present research is to analyse the efficiency of African seaports. In contrast to research on Asian, USA and European seaports (Cullinane

C. Pestana Barros (✉)
Instituto Superior de Economia e Gestão, Technical University of Lisbon and UECE (Research Unit on Complexity and Economics), Lisbon, Portugal
e-mail: cbarros@iseg.utl.pt

A. Assaf
Department of Hospitality and Tourism Management, Isenberg School of Management, University of Massachusetts, Amherst, MA, USA

A. Ibiwoye
Department of Actuarial Science, University of Lagos, Lagos, Nigeria

P. Coto-Millán et al. (eds.), *Essays on Port Economics*, Contributions to Economics, DOI 10.1007/978-3-7908-2425-4_15, © Springer-Verlag Berlin Heidelberg 2010

et al. 2004, 2005, 2006; Coto-Millán et al. 2000), few studies have addressed the performance of African seaports. The focus on Nigeria, Angola and Mozambique is also new in the seaport context. This paper also improves on related research in the area by using the DEA-bootstrap approach, which has several advantages over the simple DEA and SFA methods, traditionally used to estimate the efficiency of seaports. More details about the advantages of the bootstrap methods are provided in the methodology section.

The remainder of this paper is organized as follows. Section 2 presents the contextual setting. Section 3 reviews the literature. Section 4 details the methodology. Section 5 presents the data and the results. Section 6 concludes.

2 Contextual Setting

Productivity analysis on Middle Eastern and African seaports has been previously studied by Al-Eraqi et al. (2008) and Barros et al. (2009), using operational and financial data. West African seaports were, however, absent from these studies. The sample used in this paper is extended to cover Nigerian, Angolan and Mozambique seaports.

Table 1 presents some characteristics of the seaports analysed.

The table includes the main seaports of Angola, Mozambique and Nigeria. The seaports in Angola and Mozambique are mainly old and public seaports, built by Portugal during the colonial era. The Nigerian seaports analysed are country seaports and do not include private seaports such as the Bonny Island oil export terminal explored by Shell. The Nigerian Ports Authority is the main regulator of the major seaports in Nigeria. The Port of Apapa is the primary outlet for Nigeria exports. It is the country's biggest port, handling a wide range of commodities and containing facilities specialized in handling wheat, oil, cement, fish, dry cargo, and containers. Our sample also includes specialised seaports such as the RoRo Port – Lagos which is located along with Tin Can Island Port – Lagos, about 7 km west of the city. Tin Can Island Port merged with Roro Ports – Lagos in 2006, after the private operators – Port and Terminal Multiservices Ltd. (PTML) – took over the terminals. The RoRo Port – Lagos handles only ships designed to carry rolling-stock cargo which does not require cranes to be loaded on or off. Tin Can Island is a generic seaport. The Container terminal – Lagos is also located in Apapa tin Can island complex and claimed to be the most modern multi-purpose terminal in Africa, providing the highest standards in the region for vehicle handling, storage and logistics and it is operated by the Grimaldi group.

The Federal lighter terminal – Onne and Federal ocean terminal – Onne located in Port Harcourt are also a complex specialised in oil and gas. The other Nigerian seaports are regional seaports. Intels is the private concessionaire which operates Port Harcourt, Onne and Warri, seaports. The Calabar seaport is now publicly managed but there is plan for it to be privatised.

Table 1 Sample characteristics

Country	Name	Ships arrived and departed	Tons embarked and disembarked	Containers	Max depth of berths (m)	Total area (m^2)	Number of cranes	Number of employees
Angola	Luanda	275	305,715	91,050	9.5	150,000	10	1,074
Angola	Lobito	220	244,515	72,823	10.5	11,000	8	859
Angola	Namibe	130	145,172	43,236	10.5	10,000	4	510
Angola	Cabinda	107	119,553	35,606	12.5	40,000	4	420
Angola	Soyo	161	179,330	53,409	12.5	39,000	6	630
Angola	Port Amboim	77	85,395	25,433	20	4,638	4	300
Angola	Namibe	130	145,172	43,236	10.5	10,000	4	510
Angola	Cabinda	107	119,553	35,606	12.5	40,000	4	420
Mozambique	Maputo	330	367,200	109,362	12.8	158,203	12	1,290
Mozambique	Beira	412	458,858	136,660	11	31,406	15	1,612
Mozambique	Chinde	156	96,286	18,024	9.5	27,124	13	592
Mozambique	Mocimboa de Praia	293	33,800	90,448	10.5	29,979	14	538
Mozambique	Mozambique Island	268	21,429	55,295	12.5	35,689	17	831
Mozambique	Nacala	237	75,429	11,812	9	25,696	12	318
Mozambique	Pembane	231	79,714	12,871	11.5	32,834	16	685
Mozambique	Quelimane	293	38,000	30,448	10.5	29,979	14	538
Nigeria	Apapa Port – Lagos	750	8,570,348	188,480	9	145,534	34	1,306
Nigeria	Tin Can Island Port – Lagos	403	1,849,227	44,465	11.5	54,400	12	1,103
Nigeria	RoRo Port – Lagos	104	1,183,912	26,141	11.5	190,000	8	183
Nigeria	Container Terminal – Lagos	206	1,637,171	23,891	10.5	6,800	6	200
Nigeria	Port Harcourt – Port Harcourt	246	1,203,054	15,356	7.8	53,804	29	958
Nigeria	Delta Ports – Warri	550	1,266,240	42,335	11.5	2,367	4	432
Nigeria	Calabar Port – Calabar	80	176,055	27,563	11	35,000	4	286
Nigeria	Federal Lighter Terminal – Onne	226	631,718	15,032	5.7	1,185	7	300
Nigeria	Federal Ocean Terminal – Onne	55	61,200	18,227	13	200,000	2	215
	Average	252.64	818,666.20	51,650.80	11.06	57,158.13	11.11	660.00
	St. Dev	161.56	1778,915.39	44,926.15	2.59	62,774.38	7.88	402.66

3 Efficiency or Performance Studies

Following Farrell (1957), researchers applying frontier estimation techniques represent the production technology by a bounding function that reflects the best-practice production.

Three scientific methods to analyse efficiency quantitatively are the ratios analysis, stochastic econometric frontier (SFA) and data envelopment analysis (DEA). All methods have strengths and weaknesses (Tongzon 1995).

3.1 Research Based on DEA

There are several DEA models available to analyse efficiency, such as the DEA – CCR model from Charnes et al. (1978), DEA – BCC model from Banker et al. (1984), DEA cross-efficiency model from Doyle and Green (1994), DEA Super-Efficiency model from Andersen and Petersen (1993), DEA threshold model from Thanassoullis (1999), and DEA modified Super-efficiency model from Lovell and Rouse (2003).

Most of these methods have so far been applied to the seaport industry. Roll and Hayuth (1993) were the first to introduce DEA to the port sector. They presented a hypothetical numerical example of 20 ports using manpower, capital cargo as inputs and cargo throughput, service level, consumer satisfaction, and ship calls as outputs. The authors concluded that redeveloped ports are capable of receiving larger vessels and increase its throughput.

In a similar study, Martinez Budria et al. (1999) analyzed the efficiency of 26 Spanish ports using the DEA – BCC model. In their study, ports were divided into three groups namely "high complexity ports", "medium complexity" and "small complexity". The results showed that high complexity ports have the best efficiency.

In another European application, Barros (2003a) studied the technical and allocative efficiency of five Portuguese seaports. His study aimed to evaluate the effectiveness of the incentive regulation policy on increasing production efficiency of ports. Based on the results, it was concluded that the incentive regulation policy should be revised to further enhance efficiency. In a follow-up study, Barros (2003b) used the DEA – Malmquist and Tobit model to analyse the technical efficiency and technological change of ten Portuguese seaports from 1990 to 2000. The author concluded that state owned ports and ports with some private control did not improve their total productivity over the period 1990–2000, mainly due to the lack of technological improvements.

Studies on Portuguese seaports also include Barros and Athanassiou (2004) who applied the DEA – CCR and BCC to estimate the relative efficiency of two Greek ports and four Portuguese ports from 1998 to 2000. The broader purpose of this study was to implement international benchmarking procedures in order to identify areas of potential improvements and hence implementing a strategy that leads to improved performance with context to the European Seaport Policy. Their results showed that scaled economies and privatization contribute to higher performance.

Some other studies include Cullinane et al. (2005), which assessed the efficiency of 57 international, Cullinane et al. (2006), which assessed the efficiency of 69 European ports, and Cullinane et al. (2005), which also assessed the efficiency of European seaports. However, all these cited papers have adopted the traditional DEA models. Innovative DEA models have been adopted by Barros and Peypoch (2007), Managi (2007) and Lozano and Villas (2009) who used the Luenberger indicator, and Al-Eraqi et al. (2009) who used the Simar and Wilson (2007) DEA bootstrap procedure. A critical survey of DEA models in the seaport context is provided in Table 2.

3.2 Research Based on Stochastic Frontier Models

The traditional stochastic frontier model was initially proposed by Aigner et al. (1977) and Meeusen and van den Broeck (1977). Many varieties of this basic model have also appeared in the literature (Kumbhakar and Lovell 2000). Pre-eminent extensions are the time variant inefficiency model, time invariant inefficiency with fixed effects and the time invariant random effects model (Battese and Coelli 1992; Cornwell et al. 1990; Pitt and Lee 1981). Contemporary stochastic frontier models are the random frontier model (Greene 2004, 2005); and the latent frontier model, (Orea and Kumbhakar 2004; Alvarez et al. 2004).

In the port context, Liu (1995) was the first to use the stochastic frontier approach. The author tested the impact of ownership on the performance of UK ports, and concluded that publicly owned ports are less efficient than those privately owned. He also noticed that the geographical location of seaports has a major influence on efficiency. The study also showed that ports on the west coast of the UK are almost 10% less efficient than those located in the east coast.

In a more recent study, Notteboom et al. (2000) collected data on a sample of 36 European container terminals located in the Hamburg – le-Havre range and the western Mediterranean region. As a process of benchmarking these ports, the analysis also included four other Asian terminals. The authors realized that most ports in the Hamburg – le-Havre had an efficiency level close to 80%, while Belgium ports were considered to be the most efficient and Italian ports amongst the lowest. Smaller terminals in larger ports were also found to be more efficient than smaller terminals in smaller ports.

The impact of port size on efficiency was also assessed by Coto-Millán et al. (2000) in their study on Spanish ports. The authors showed that smaller ports were more efficient than larger ports, mainly due to the autonomy of larger ports. Asian seaports were also analysed by Cullinane et al. (2002), using a sample of 15 container ports. The authors concluded that there is a close correlation between the size of ports and relative efficiency. The transfer of ownership from the public to

Table 2 Literature review on DEA

	Production function	Units	Inputs	Outputs
Roll and hayuth (1993)	DEA – CCR model	Hypothetical, numerical example of 20 ports	Man power, capital cargo	Cargo throughput, service level, consumer satisfaction, ship calls
Martinez budria et al. (1999)	DEA – BCC model	26 Spanish ports	Labor expenditure, depreciation charges, other expenditure	Cargo moved in docks, revenue from port facilities rent
Tongzon (2001)	DEA – CCR addictive model	4 Australian and 12 other international port	No: cranes, tugs, container berths. Terminal area, delay time, labor	Cargo throughput, ship working rate
Barros (2003a)	DEA – technical and allocative efficiency	5 Portugese sea ports	Number of employees, book value of asset	Movement of freight, market share, cargo: break bulk, container cargo, solid bulk, liquid bulk, price of capital
Barros (2003b)	DEA – Malmquist index and tobit model	10 Portugese sea ports	Number of employees, book value of asset	Movement of freight, market share, cargo: break bulk, container cargo, solid bulk, liquid bulk, price of capital
Barros and Athanassiou (2004)	DEA – CCR and BCC	2 Greek and 4 Portugese seaports	Movement of freight, cargo handled, container handled. Number of ships	Labour and capital
Cullinane et al. (2004)	DEA – CCR and BCC	World major container ports	Container throughput	Terminal length, terminal area, quay side gantry, yard gantry and straddle carriers
Cullinane et al. (2005)	DEA – BCC, CCR and FHD	57 international sea ports	Container throughput	Terminal length, terminal area, quay side gantry, yard gantry and straddle carriers
Cullinane et al. (2006)	DEA – BCC and CCR	67 Europe port with 10,000 teu	Terminal length, area, terminal equipment number	Output container throughput
Rios and Maçada (2006)	DEA –BCC and CCR	23 Mercosur seaports	TEU; Movements hour/ship	Cranes, berths, area, employees, yard equipment
Managi (2007)	Luenberger productivity indicator	11 Japanese shipping firms 1995–2005	Capital and labour expenditure	Revenue

private sector had also an impact on ports efficiency. In a follow-up study, Cullinane and Song (2003) assessed the effect of privatization on the productive efficiency of container terminals. The authors concluded that privatization and regulation can improve productive efficiency.

Some other studies include Tongzon (2001), Barros (2005), Tongzon and Heng (2005), Diaz Hernandez et al. (2008), and Barros (2009). More details on these studies as well as the above studies are available in Table 3.

4 Methodology

This paper uses the DEA-bootstrap methodology to derive the efficiency of African seaports. One of the major limitations of the standard DEA method is that it is a deterministic technique and thus does not provide any statistical properties for the derived efficiency scores. SFA is also subject to criticism, as it requires a pre-specification of the functional form in the estimation of cost or production frontier technologies. SFA also requires larger sample size than DEA. An arguably more attractive solution to the DEA problem is to use the bootstrapping methodology, in which the researcher can keep the advantages of DEA, and also obtain statistical properties for the DEA efficiency scores. In the next subsections we provide more details on the DEA and the bootstrap methodology used in this paper.

4.1 Data Envelopment Analysis

DEA involves the use of linear programming to construct a non-parametric frontier over the data, Farrell (1957). Technical efficiency measures are then derived relative to this frontier. The motivation and early versions of the DEA models have appeared in several previous studies in the literature, so they will not be reiterated here. For a detailed review refer to Coelli et al. (2005). The model used in this study follows an output-oriented assumption and can be derived for the i-th firm by solving the following linear programming:

$$\hat{\delta}_i = \max_{\hat{\delta}_i,\lambda} \left\{ \delta > 0 \middle| \hat{\delta}_i y_i \leqslant \sum_{i=1}^{n} y_i \lambda; x_i \geqslant \sum_{i=1}^{n} x_i \lambda; \sum_{i=1}^{n} \lambda = 1; \lambda \geqslant 0 \right\},$$
$$i = 1 \ldots n \text{ firms} \tag{1}$$

where y_i is vector of outputs, x_i is s vector of inputs, λ is a $I \times 1$ vector of constants. The value of $\hat{\delta}_i$ obtained is the technical efficiency score for the i-th firm. A measure of $\hat{\delta}_i = 1$ indicates that the firm is technically efficient, and inefficient if $\hat{\delta}_i > 1$. This linear programming problem must be solved n times, once for each firm in the sample. Note that the DEA model can also be estimated using either the constant

Table 3 Literature review on SFA

Author	Production function	Units	Inputs	Outputs
Liu (1995)	Translog	28 Important UK ports from 1983 to 1990	Labor, capital	Turnover
Notteboom et al. (2000)	Bayesian stochastic frontier	Sample of 36 container terminal and 2 Asian container ports in 1994	Terminal quay length, terminal area, number of gantry cranes, average number of workers per crane	Container traffic
Coto-Millán et al. (2000)	Cobb-Douglas translog	27 Spanish ports 1985–1989	Price of: capital, labor, intermediate consumption	Total cost
Estache et al. (2001)	Cobb-douglas translog	14 Ports in Mexico from 1996–1999	Labor, capital approximated	Merchandise volume handled
Cullinane et al. (2002)	Stochastic cobb-douglas	15 Container ports in Asia for the period 1989–1998	Terminal quay length, terminal area, number of pieces of cargo equipment	Annual container throughput
Cullinane et al. (2005)	Stochastic cobb-douglas	5 Container terminal from UK and Korea with a total of 62 observations	Labor and capital	Turnover (excluding property sales)
Barros (2005)	Stochastic translog frontier	10 Portuguese ports for 1990–2000	Operational cost, price: labor and capital. Number of ships	Annual throughput
Cullinane et al. (2006)	Stochastic frontier, cob douglas	74 European ports	Terminal quay length, terminal area, number of pieces of cargo equipment	Annual container throughput
Al-Eraqi et al. (2008)	DEA – CCR and DEA – BCC windows analysis	22 Middle Eastern and East African seaports	Berth length, storage area, handling equipment	Ship calls and throughput
Barros (2009)	Random frontier model	10 Portuguese ports for 1990–2002	Price of labour, price of capital, price of materials	Ships, cargo
Barros et al. (2009)	Malmquist and Luenberger methods	15 Middle East and West African Seaports from 2005 to 2007	No. of Employees, Total Cost and No. of cranes	Throughput, number of ships

returns to scale (CRS) or variable returns to scale (VRS) assumptions and the shape of the frontier will differ depending on the scale assumptions that underline the model. In this paper we rely on the VRS assumption, as the CRS is only correct as long as it is appropriate to assume that firms are operating at an optimal level of scale. Technological advances and regulatory changes might vary across firms in various size groups, so allowing for VRS would permit modelling of the entire range of technology.

4.2 *The Bootstrap*

The DEA model in (1) is relatively simple to estimate and has several strengths, mainly related to the flexibility in (1) accounting of multiple inputs and outputs, and (2) estimating the frontier without imposing any functional form on the best practice frontier. However, it is also known that DEA has several weaknesses, one of which relates to its statistical limitations. Simar and Wilson (1998) have recently raised the issue that given DEA is an empirically based estimation technique, it is sensible to outliers, error measurements and random influences in the data. In other words, DEA deems any deviation from the efficiency frontier to be the result of inefficiency.

To correct for this problem we use in this paper the bootstrap approach, developed by Simar and Wilson (1998, 1999). We do no provide here the technical details of the approach as this can be easily obtained from the original papers. In general, the aim of the bootstrap approach is to simulate the original sample B times, each time recalculating the parameter of interest, which is in our case is the DEA efficiency score. This will allow in B estimates of the parameter, and thus makes it possible to generate an empirical distribution for the parameter of interest. The empirical distribution can then be used to construct confidence intervals for the DEA efficiency scores, and also obtain other statistical properties.

5 Data and Results

We compiled our dataset on African seaports using direct contacts and other means such as fax, email and phone. Our dataset covers the 2004–2006 period and includes 23 seaports ($23 \times 3 = 69$ observations). Based on the literature review and data availability, we collected data on three outputs: (1) Number of ships call, (2) total tons embarked and (3) total number of containers embarked and disembarked, and four inputs: (4) depths of berths, (5) total area, (6) number of quay cranes and (7) number of employees. Table 4 presents descriptive statistics of the data.

Table 5 reports the results. The paper adopts the advice of Simar and Wilson (1999) and uses 2,000 bootstrap replications ($B = 2{,}000$) in obtaining the results.

Table 4 Descriptive statistics of the data

Variable	Mean	St. Dev	Median	Min	Max
Number of ships	255.1	176.1	226	49	755.1
Total tons	697,189	1,233,410	227,644	20,580	8,570,348
Containers	55,328	52,062	34,336	20,580	8,570,348
Depth of berths (m)	11.05	2.55	11	5.7	20
Total area (m^2)	57,158	61,844	32,834	1,185	200,000
Number of cranes	11.08	7.76	10	2	34
Number of employees	660	396.7	538	183	1,612

Table 5 Average efficiency scores (2004–2006)

Port	Country	$\hat{\theta}_i$	$\hat{\hat{\theta}}_i$	Bias	LB	UB
Cabinda	Angola	0.4987	0.3972	0.1015	0.3685	0.4940
Lobito	Angola	0.4140	0.3371	0.0768	0.3111	0.4106
Luanda	Angola	0.6127	0.5193	0.0933	0.4877	0.6083
Namibe	Angola	0.8589	0.6139	0.2450	0.5703	0.8506
Port Amboim	Angola	0.6980	0.4124	0.2856	0.4375	0.6902
Soyo	Angola	0.2316	0.1885	0.0430	0.1752	0.2297
Beira	Mozambique	0.6944	0.6063	0.0881	0.5594	0.6880
Chinde	Mozambique	0.3068	0.2658	0.0409	0.2491	0.3045
Maputo	Mozambique	0.7072	0.5999	0.1072	0.5676	0.7002
Mocimboa de Praia	Mozambique	0.6520	0.5423	0.1097	0.5012	0.6453
Mozambique Island	Mozambique	0.4436	0.3842	0.0594	0.3633	0.4400
Nacala	Mozambique	0.6446	0.5319	0.1127	0.4897	0.6383
Pembane	Mozambique	0.3739	0.3240	0.0498	0.3061	0.3701
Quelimane	Mozambique	0.5297	0.4568	0.0728	0.4291	0.5255
Apapa Port – Lagos	Nigeria	1.0000	0.5169	0.4830	0.6357	0.9837
Calabar Port – Calabar	Nigeria	1.0000	0.6487	0.3513	0.6542	0.9835
Container terminal – Lagos	Nigeria	0.9759	0.3899	0.5859	0.5924	0.9567
Delta Ports – Warri	Nigeria	1.0000	0.6047	0.3953	0.6426	0.9860
Federal lighter terminal – Onne	Nigeria	0.9693	0.5973	0.3719	0.6292	0.9521
Federal Ocean Termina – Onne	Nigeria	0.9824	0.5899	0.3925	0.6293	0.9632
Port Harcourt – Port Harcourt	Nigeria	0.5253	0.4275	0.0977	0.3965	0.5174
RoRo Port – Lagos	Nigeria	1.0000	0.6734	0.3265	0.6592	0.9844
Tin Can Island Port – Lagos	Nigeria	0.6471	0.5048	0.1422	0.4836	0.6375

$\hat{\theta}_i$ original DEA, $\hat{\hat{\theta}}_i$ bootstrapped DEA, *LB* lower bound of the confidence interval, *UB* Upper bound of the confidence interval

According to the authors, this should provide an adequate coverage of the confidence intervals.

The first column in Table 5 presents the DEA technical efficiency, the second column presents the DEA bootstrapped efficiency scores, the third column presents the BIAS of the original DEA, and the fourth and fifth columns present the lower and upper bound of the DEA confidence intervals.

It is evident from the first column in Table 5 that over the 2004–2006 period, only four seaports are on the frontier of best practices with a technical efficiency score equal to 1: Apapa Port – Lagos, Calabar Port, Delta Port – Warri and Roro

Port – Lagos. However, when taking into consideration the bootstrapped efficiency scores none of the analysed ports is close to being fully efficient. Additionally, the bootstrapped efficiency scores are lower than the standard efficiency scores, mainly due to the bias in the original DEA, which was discussed previously. Based on the bootstrapped efficiency results, the best performing seaport is the Roro Port of Lagos, a modern and well-developed seaport.

6 Discussions and Implications

This paper has proposed a simple framework to measure the technical efficiency of African seaports and the rationalisation of their management activities, with a bootstrapped DEA approach. It was clear from the technical efficiency is higher for Nigerian seaports than for Angolan or Mozambique. The least efficient seaport is Soyo in Angola, a small seaport in either the standard efficiency scores or in the bootstrapped efficiency scores (Englebert 2000).

What should be done to improve efficiency? At the African level, the general complaints with seaports relate to the missing linkage between seaports and rail, the lack of equipments such as cranes and other handling equipments; and the extended duration of a ship in the seaport to be unloaded. These complaints signify that African governments are encouraged to upgrade the seaport management in contemporary logistics and adequate equipments.

The three countries analysed have different democratic durations signifying that the newer democratic governments may not yet have considered some institutions like seaports as strategic assets. This situation might have particularly affected Angola, which earned its independence in 1975 after a long period of Portuguese colonisation. Soon after, the country entered a period of costly civil war that only came to an end in 2002 (Ferreira and Barros 1988). With the end of the civil war, Angola was in a condition of macro-economic turmoil, with rising inflation and a devalued national currency (the kwanza). The intervention of the International Monetary Fund (IMF) was reinforced in 2000 with the adoption of a macro-economic stabilization program that has started to achieve its aims. Similarly, Mozambique seaports have also failed to attract a strong level of capital investments, despite the high reputation of some ports such as Maputo and Beira. In contrast to Mozambique and Angola, Nigerian seaports have attracted innovation based on specialisation and capital investments. Nigerian seaports have also recently enjoyed a strong increase in the volume of imports and exports, especially post the government reform of 2000.

What are the managerial implications of the present research? The study provided each port with an efficiency score, which can be used for benchmarking purpose or to identify sources of technical inefficiency (Estache et al. 2001). The inefficient seaports should focus on using their resources more efficiently (Rios and Maçada 2006). The results should not however be generalised. The paper has one limitation related to the data set. The data set is short and might not provide a comprehensive picture. In order for the conclusions to be more generalised, a larger panel data set should be used. A variety of methodological extensions such as the introduction of the Luenberger or Malmquist productivity indexes are also encouraged.

7 Conclusions

In this paper, the technical efficiency in a representative sample of African seaports from Nigeria, Angola and Mozambique was estimated between 2004 and 2006, a period of increasing volatility in the sector. The analysis is based on a bootstrapped DEA.

Benchmarks are provided for improving the operations of inefficient performing African seaports. Several interesting and useful managerial insights and implications from the study are discussed. The general conclusion is that a public policy aiming to increase efficiency should rely on upgrading the performance of seaports by establishing an adequate contextual setting. Seaport managers are also encouraged to adopt sound and efficient managerial practices.

References

Aigner DJ, Lovell CAK, Schmidt P (1977) Formulation and estimation of stochastic frontier production function models. J Econom 6:21–37

Al-Eraqi AS, Mustafa A, Khader AT, Barros CP (2008) Efficiency of Middle Eastern and East African seaports: application of DEA Windows analysis. Eur J Sci Res 23:597–612

Al-Eraqi AS, Barros CP, Khader AT, Mustafa A (2009) Analysing the efficiency of Arabian and Western Indian Ocean seaports using a two-stage procedure. Marit Econ Logist (forthcoming)

Alvarez A, Arias C, Greene W (2004) Accounting for unobservables in production models: management and inefficiency. Working Paper 72, Centro de Estudios Andaluces

Andersen P, Petersen NC (1993) A procedure for ranking units in data envelopment analysis. Manage Sci 39:1261–1264

Banker RD, Charnes A, Cooper WW (1984) Some models for estimating technical and scale inefficiencies in data envelopment analysis. Manage Sci 30:1078–1092

Barros CP (2003a) Incentive regulation and efficiency of Portuguese port authorities. Marit Econ Logist 5:55–69

Barros CP (2003b) The measurement of efficiency of Portuguese seaport authorities with DEA. Int J Transp Econ 30:335–354

Barros CP (2005) Decomposing growth in Portuguese seaports: a frontier cost approach. Marit Econ Logist 7:315

Barros CP (2009) Analysing seaport efficiency with a random stochastic cost frontier. Marit Econ Logist (forthcoming)

Barros CP, Athanassiou M (2004) Efficiency in European seaports with DEA: evidence from Greece and Portugal. Marit Econ Logist 6:122–140

Barros CP, Peypoch N (2007) Comparing productivity change in Italian and Portuguese seaports using the Luenberger indicator approach. Marit Econ Logist 9:138–147

Barros CP, Haralambides H, Hussain M, Peypoch N (2009) A DEA analysis of seaports efficiency and productivity growth. In: Talley W, Cullinane K (eds) The international handbook of maritime economics and business. Edward Elgar, UK

Battese G, Coelli TJ (1992) Frontier production functions, technical efficiency and panel data: with applications to paddy farmers in India. J Prod Anal 3:153–169

Charnes A, Cooper WW, Rhodes E (1978) Measuring the efficiency of decision-making unites. Eur J Oper Res 2:429–444

Coelli TJ, Prasada Rao DS, O'Donell CJ, Battesse GE (2005) An introduction to efficiency and productivity analysis. Springer, New York

Cornwell C, Schmidt P, Sickles RC (1990) Production frontiers with cross-sectional and time-series variation in efficiency levels. J Econom 46:185–200

Coto-Millán P, Baños-Pino J, Rodríguez A (2000) Economic efficiency in Spanish ports: some empirical evidence. Marit policy Manage 27:170–175

Cullinane KPB, Song DW (2003) A stochastic frontier model of the productive efficiency of Korean container terminals. Appl Econ 35:251–267

Cullinane KPB, Song DW, Gray R (2002) A stochastic frontier model of the efficiency of major container terminals in Asia: assessing the influence of administrative and ownership structures. Transp Res A 36:734–762

Cullinane K, Ji Ping, Wang Teng-fei (2004) An application of DEA Windows analysis to container port production. J Rev Netw Econ 3:184–206

Cullinane K, Ji Ping, Wang Teng-fei (2005) The relationship between privatization and DEA estimates of efficiency in the container port industry. J Econ Bus 57:433–462

Cullinane K, Wang TF, Song DW, Ji P (2006) The technical efficiency of container ports: comparing data envelopment analysis and stochastic frontier analysis. Transp Res A 40:354–374

Diaz Hernandez JJ, Martinez-Budria E, Jara Diaz S (2008) The effects of ignoring inefficiency in the analysis of production: the case of cargo handling in Spanish ports. Transp Res A 42:321–329

Doyle J, Green R (1994) Efficiency and cross-efficiency in DEA. Derivations, meanings and uses. J Oper Res Soc 45:567–578

Englebert P (2000) Pre-colonial institutions, post-colonial states, and economic development in Tropical Africa. Polit Res Q 53:7–36

Estache A, Gonzales M, Trujillo L (2001) Efficiency gains from port reforms and the potential for yardstick competition: lessen from Mexico. World Dev 30:545–560

Farrell M (1957) The measurement of productive efficiency. J R Stat Soc 120:253–281

Ferreira ME, Barros CP (1988) From war to economic recovery: peace as a public good in Angola. Defence Peace Econ 9:283–297

Greene W (2004) Distinguishing between heterogeneity and efficiency: stochastic frontier analysis of the World Health Organisation's panel on national health care systems. Health Econ 13:959–980

Greene W (2005) Fixed and random effects in stochastic frontier models. J Prod Anal 23:7–32

Kumbhakar S, Lovell K (2000) Stochastic frontier analysis. Cambridge University Press, Cambridge

Liu Z (1995) The comparative performance of public and private enterprises. J Transp Econ Policy 29:263–274

Lovell CAK, Rouse APB (2003) Equivalent standard DEA models to provide super-efficiency scores. J Oper Res Soc 54:101–108

Lozano S, Villas G (2009) Computing Luenberger productivity indicators of Spanish ports with integer DEA. Marit Econ Logist (forthcoming)

Managi S (2007) Maritime shipping industry and productivity in Japan. Marit Econ Logist 9:291–301

Martinez Budria E, Diaz Armas R, Navarro Ibafiez M, Ravelo Mesa T (1999) A study of the efficiency of Spanish port authorities using data envelopment analysis. Int J Transp Econ 2:237–253

Meeusen W, van den Broeck J (1977) Efficiency estimation from a Cobb-Douglas production function with composed error. Int Econ Rev 18:435–444

Notteboom T, Coeck C, van den Broeck J (2000) Measuring and explaining the relative efficiency of container terminals by means of Bayesian stochastic frontiers models. Int J Marit Econ 2:83–106

Orea L, Kumbhakar S (2004) Efficiency measurement using stochastic frontier latent class model. Empirical Econ 29:169–183

Panayides PM, Maxoulis CN, Wang TF, Koi Yu, Ng KYA (2009) A critical analysis of DEA applications to seaport economic efficiency measurement. Transp Rev 29:183–206

Pitt M, Lee F (1981) The measurement and sources of technical inefficiency in the Indonesian weaving industry. J Dev Econ 9:43–64

Rios LR, Maçada ACG (2006) Analysing the relative efficiency of container terminals of Mercosur using DEA. Marit Econ Logist 8:331–346

Roll Y, Hayuth Y (1993) Port performance comparison applying data envelopment analysis (DEA). Marit policy Manage 20:153–161

Simar L, Wilson PW (1998) Sensitivity analysis of efficiency scores: how to bootstrap in nonparametric frontier models. Manage Sci 44:49–61

Simar L, Wilson PW (1999) Estimating and bootstrapping Malmquist indices. Eur J Oper Res 115:459–471

Simar L, Wilson PW (2007) Estimation and inference in two stage, semi-parametric models of productive efficiency. J Econom 136:31–64

Song DW, Cullinane K (2001) Asian container seaports and productivity: implications for the New Millenium. Singapore Marit Port, 149–161

Thanassoullis E (1999) Setting achievement targets for school children. Educ Econ 7:101–119

Tongzon J (1995) Systematizing international benchmarking for ports. Marit Policy Manage 2:171–177

Tongzon J (2001) Efficiency measurement of selected Australian and other international ports using data DEA. Transp Res A, Policy Pract 35:113–128

Tongzon J, Heng W (2005) Port privatization, efficiency and competitiveness: some empirical evidence from container ports (terminal). Transp Res A 39:405–424

Impact of New Technology on Port Administration

Pablo Coto-Millán, Miguel Ángel Pesquera, and Juan Castanedo

Abstract This article presents an approach to the impact of new technology on the management of port services. After an introduction on motivation, there are three sections of a descriptive nature in which the methods used to incorporate new technology in the management of the port sector are presented. In the fourth section an empirical approximation is carried out, focusing on efficiency frontiers, on the impact of investment in new technology in Spanish ports for the period 1985–1995. The research is conclusive on the existence of technological progress in the Spanish port sector from the 1990s onwards, with more technical efficiency in ports that have been pioneers in applying new information technology such as Barcelona, Valencia, Bilbao, Las Palmas and Santa Cruz de Tenerife.

1 Introduction

Port management, as a further element of the overall transport system, has been segregated from the constructive aspects of the port infrastructure, with its own identity and object of another discipline. At the same time, there is evidence that port managers should promote the role of the port as a node of modal transfer or service, like any of the other modes (such a phenomenon is still not fully adopted in daily practice). It has also been difficult to accept that traffic is not already captive. A quality plan, marketing plan or appropriate new market strategy can divert trade, preserving or even expanding the business.

In port management, once acceptable standards in construction procedures, the design of infrastructures and rolling stock have been reached, they have also made their way through the fields of strategic planning, marketing, logistics development

P. Coto-Millán (✉), M. Ángel Pesquera, and J. Castanedo
University of Cantabria, Santander, Spain
e-mail: cotop@unican.es

P. Coto-Millán et al. (eds.), *Essays on Port Economics*, Contributions to Economics,
DOI 10.1007/978-3-7908-2425-4_16,

and business management, following the evolution of the production systems and markets, because ports are not strangers to this evolution. In short, ports constitute strategic nodes to facilitate the international flow of goods, forming part of an extensive logistics network on which commercial exchanges between points and/or geographically distant areas are set up. In international goods transportation, the delivery time of the product becomes the key element. The traditional transport chains: port – maritime distance – port, have been substituted by others which are a lot more complex having been expanded in the direction of domestic door-to-door collections/deliveries, giving rise to inland distribution centres or "hubs" in the port hinterlands that supply "feeders" and, from these, delivery to the points of reception is made (cities, factories, etc.).

The development of ports has forced the modernisation of existing port management methods. In this sense, the United Nations Conference on Trade and Development (UNCTAD) prepared a report in 1990 entitled "Port Marketing and the Challenge of the Third Generation Port", where the development of ports was phased in over three generations, with this classification being accepted globally. In 1999 the UNCTAD published "Technical note: Fourth generation port", that incorporated a fourth generation to the previous classification (See Table 1).

Table 1 Ports spanning four generations

Generation	General definition and some characteristics
1st	The port with exchange functions between two modes of transport – Has no development strategies drawn up – Does not have traditional tasks of storage and its management organized – Port activity is developed on the quay – Authorities and agencies overlap – Supply dominates (it is possible to listen to the needs of the users)
2nd	The port as a transport centre for the commercial and industrial activities of its environment – With development and expansion strategies – With transformation activities (heavy industry), services to the ships – Expansion of the port area – Closeness of the port to its users; indications of a port community – Precise relationships between the city and the port
3rd	The port as an integrated transport centre and logistics platform for international trade – Development strategies of a commercial orientation – Distribution of goods, logistics activities and distribution centre – Implementation of information systems in the port (EDI) – Rational port spaces – United and active port community, coordinating activities – Close relationships between the city and the port
4th	Network port – Internationalisation strategies and diversification of activities – Organization of logistics services by dockers – EDI networks integrated into port areas – Search for port spaces distributed abroad – Cooperation between port communities

Source: Compiled by author

The objective of first generation ports was to carry out the land–sea transportation functions and vice versa of small items for a local or regional hinterland. This type of port was characterized by its lack of connection to the socio-economic environment. Second generation ports, also called "industrial ports", begin to be considered as a mode of transport and centre of commercial and industrial activity; services continue to be limited to ships and goods, although operations of added value appear. Heavy industry and processing plants with activity connected to port traffic are set up close to these ports. Third generation ports incorporate logistic functions, linked with the distribution of goods, with information processing and telecommunications contributing to generate added value. There are also companies related to maritime transport and warehousing. (UNCTAD 2000 TD/BC4/AC.7/14).

Current needs lead us to take a step further when talking about fourth and fifth generation ports, characterized by the implementation of telematic networks between port areas, by collaboration between port communities and by the internationalization and diversification of activities. A "network port" is integrated into the international transport logistics chain, with "door-to-door" services, along with other logistics operators that operate in various ports from the same facade, operating this as a centre for attracting traffic and goods. To do this requires a logistics platform (dry ports shared by various ports), linked to the ports by railways. In an environment of high consumption and/or production (preferably both), this railway system should allow a regular and competitively priced transportation of bulk volumes which ships then transport to the final distribution phase carried out by road or even to distribution points from where they are sent for shipment. The objective is to try to concentrate loads to be exported by sea from the land side and the distribution of goods unloaded on the maritime side so that transport (normally by rail) is balanced between the commercial quays of the maritime ports and the dry ports which are part of the network.

The new network or digital economy transforms classical logistics chains into logistics networks of added value, corresponding to information flow and human interactions, with the generation of knowledge as a basic resource.

This new context allows important changes to happen in the quality of port operations, given that they allow the reduction of operational times, and their real-time knowledge of door-to-door port logistics chains.

2 Management Changes in the Spanish Stated-Owned Port System

In the twenty-first century we find that the state-owned port system has developed a new culture based on liberal politics that allow free and fair competition. The functions of the port authority are, on the one hand, to promote free competition and to make sure that there is not a monopoly of the various port services, with a globally shared view of the services offered by the port and the port community; and, on the other, to adopt a decisive policy of innovation and promotion of the port's activities by pushing forward global strategies, based on the continuous improvement of procedures and services, in the permanent search for ways in which to improve the port's competitiveness. In short, this means a model of public–private contribution, which has combined general and private interests in the port, allowing it to be considered an integrated infrastructure as a fundamental part of a general intermodal transport system which is both sustainable and competitive. This makes it a modal interconnection node and logistics platform with an important role both in the transport and value chains.

Law 48/2003, of November 26, regarding the economic system and provision of services at ports of general interest, modifies the concept of port services in the 1992 Law of State Ports and the Merchant Navy, that linked its ownership and responsibilities to the port authority, in which the role of public authority and its provision was carried out by means of direct or indirect management. In the new law the port service is a more extensive concept and can be classified by:

- General port services.
- Basic port services.
- Commercial services within the port confines.
- Maritime signalling service.

In this way, there is a significant change in the management model of general interest ports, through the port authority, which is aimed at the provision and management of public domain areas, the regulation of economic activity that makes up the basic port services developed in the port, which are basically run by the private sector, leaving the port authority a beneficial subsidiary role where there is an absence or shortage of private initiative. All of this, without detriment to the fundamental role, comes from the application of a system incorporating the principles of efficiency, agility and flexibility in business management.

The port authority should also develop promotional activities and global strategies aimed at the common client to ensure the constant adaptation of the port's global contribution to the needs and strategies of the economy so that these serve and contribute to create an environment of cooperation among all the port community agents, aimed at improving the promotion and services of the port.

In short, the port authority stops owning traditional port services as these activities become more diversified and complex. Instead these become entrusted to private initiative in conditions of free competition, wherever possible, provided, in all cases, that security in the ports is guaranteed, in particular the technical-nautical services. The change in the management model tries to introduce greater flexibility and competition in port services, which is translated in a determined commitment to reinforce private involvement in service type activities in the ports, in comparison to the public role derived up till now from the port authority's reserved ownership of port services.

In this context, the old premise was that the transmission of information in the port environment was slow but sure, in many cases imposed by the necessary commercial validity of the documentation (an example was the indispensable loading manifest for the ship's office, a manifest that could only be delivered physically, by mail, or on board the ship itself); this is now making way for other forms of transmitting information, because a port needs to serve its final clients by extending its services and being integrated into the logistics chains derived from its commercial relationships. In the port the physical and documentary flows of the transport chains come together; the implementation of new digital technology substantially allows the improvement of services on offer, generating value for the clients and other economic agents as well as acting as an engine for innovation in the port community.

3 The Incorporation of New Information Technology

The port environment inside an intermodal transport chain is made up of multiple activities, making it very complex to integrate its agents and operators in its multiple economic and documentary relationships internally and with the organizations and agents that carry out functions in the intermodal transport chains. Thus, the land operators, the port authority, the consigners, dockers, stevedores, customs, consignees of the merchandise, shipping agents, customs agents, etc. come to make up a web of difficult optimization, if their relationships are not made clear in an open framework using Internet type connections in the network.

Port operations and their multiple agents use different types of documents in their procedures that should be available real time in different places, something which can be carried out employing new information technology or using the most advanced "software agent" technology on the Internet, which act as facilitating agents for port operations, although with the difficulties motivated by non-availability of all the desirable information in the "network", its use is not widespread.

The electronic exchange of data between the different industrial sectors and state, European and even world administrations, requires international standardization, essential in the port sector. To talk about EDI and standards or standardization of various sectors and institutions at local, national, and international level, is to talk about the same thing. EDIFACT (Electronic Data Interchange For Administration, Commerce and Transport) is a language that facilitates dialogue between commercial businesses on a broad level, which can be assimilated to written language; within a text the words are organized into phrases and the phrases into paragraphs.

To participate in EDI environments, some method of electronic communication is necessary; there are two methods that can be used: direct transmission or value added network (VAN). Direct transmission takes place when a company connects to its client directly through a computer. Communication using value added network is different; instead of connecting directly with its client or commercial agent, it connects to the value added network which acts as a server or intermediary; the company and its business associates, with whom it sets up trade relationships, have "mailboxes" on the network where EDI transactions can be stored. If the company wants to receive invoices from its business partner, it collects them in its network mailbox. If it needs a product, it sends a purchase order to the appropriate mailbox of its business partner.

The implementation of technological developments such as EDI, clearly affects its organization in trying to automate information processing at all levels of the business structure, from marketing to the computer service, via the commercial service, accounting and research and development.

The organizational changes are carried out facilitating the systematic use of EDI technology. By switching to smaller computers (PC), and with the development of local networks (the downsizing phenomenon), there is a big change from combining the advantages of centralization and decentralization of information. The contribution of technology on the computer hardware is crucial for new organization and EDI.

The EDI instrument should not only be considered as a mere management support tool, but also in itself it is more powerful than this call to transform many activities, by contributing to the emergence of a new socio-technical system. It would be what C. Freeman has identified as a technical mutation which transforms in depth the system of relationships between social, economic and technical. It is what in Daniel Bell's terms would make up the knowledge society or post-industrial society.

Working with EDI language goes beyond applying a mere tool. In other words, it is working with a code of conduct between multiple diverse agents and/or companies that translates a new culture appropriate to the mutations of new communication technology. It would then be possible to speak of the "EDI Culture", understanding this to be the set of habits and standards that allow work to be done without paper (paperless) avoiding any kind of break in information flows and therefore giving real time responses.

The evolution of the Internet has been very rapid, with the security problems that initially arose having now been overcome, and the XML language having been developed that has become the standard for e-commerce, databases, etc on the Internet. Translation software is not necessary since it is integrated in the browsers/explorers where computers are understood in the broadcast/reception of documents regardless of the operating system being used. Value added networks and companies that offer their services are provided by Internet service providers.

The Internet offers a ubiquitous data transportation network at low cost, so that the model established for this type of business communication which is EDI is changing. As such, Internet offers new opportunities to companies for communicating with their partners. It is possible that using EDI on the Internet makes it more accessible to small and medium businesses, but it is not probable that it is replacing existing VANs.

Currently, EDI continues to grow as the core technology which allows Business-to-Business commerce or commerce between companies. Only a decreasing percentage continues to be carried out through traditional EDI. The Internet through its Extranets is the most universal and cheap medium to carry out EDI transactions. For the moment, many large companies are designing EDI transactions via the Web, though they maintain their traditional EDI traffic based on VAN. In many cases, the large companies are starting from scratch with the implementation of linked networks that support Internet traffic and VAN.

The cost associated with the use of VAN to send EDI transactions is quite high, but VAN offers a secure transmission of EDI traffic and the associates can opt to encrypt and authenticate the traffic; a way of ensuring EDI traffic is to send it on Virtual Private Networks (VPN) that use point to point tunnels to minimize the risk of data being intercepted and improve the characteristics of data transfer; if a VPN is used, it can compress and encrypt the traffic that it transmits and receives and a tunnel between the sender and the recipient is created and a specific IP port is opened at the receiving end only for the time it takes to receive data. At present, the quantity of EDI traffic that is carried out using VPN networks is limited, but it is a technology that is gaining acceptance and will no doubt grow over the next few years.

Many manufacturers are incorporating EDI in e-commerce packages and increasingly, companies are generating new EDI traffic on the Internet network.

Without any doubt, high volume EDI traffic on VAN will continue to be essential for a long time, but e-commerce applications over the Internet based on EDI represent the area of highest growth for exchanging documents, and especially for low volume traffic that medium and small businesses require.

EDI on the Internet offers new methods for exchanging documents and an economic way for small businesses to participate and administer EDI traffic. By providing a standard with which to work, EDI would be able to help to make possible data exchange from company to company over the Internet.

4 Applications for the Port Environment

At the end of the 1980s, Spain introduced the European Community Single Administrative Document form into its customs systems for the assumptions of importing and/or entry of goods into the Peninsula and Balearic Islands, as well as for the entry of shipments to free ports, depots and warehouses under the control of customs and for the export and/or exit of goods from national territory.

In successive subsequent circulars the instructions were laid down to compliment or deal with the Single Administrative Document to Customs (SAD), among which the possibility of presenting it by means of electronic broadcast (EDI) using EDIFACT messages was included, a mode collected by the CUSDEC Community Customs Code for the shipment on behalf of operators of the import, export and transit customs declaration and CUSRES for the response from customs.

In 1995 the company PORTEL was created by the state public port organisation and Telefónica (it is 51% owned by state ports and 49% by Telefónica) and began to act as a value added network (VAN) between the port community, the port authorities and Inland Revenue in processing customs manifests and inspection services via EDI.

With the creation of an EDI Clearing and Translation Centre for the whole port community and the agreements reached with customs, port authorities and customs services to achieve the "single window" for receiving documents. PORTEL has been developing its web transportation/services platform (http://www.portel.es), where all its on-line management services and monitoring, as much documentary as physical and graphic, for the state owned port sector, includes:

- Sending loading and unloading manifests from any port (IFCSUM message).
- Sending CUSREP and CUSCAR messages to customs.
- Sending notification of hazardous materials, according to Royal Decree 1253/1997 (Transposition of HAZMAT Community Directive).
- "Cargo Tracking Facility" service, by which means a customs agent can know the degree of availability (for dispatch) of each B/L of a specific manifest.
- Import, export and transit SADs, as well as reception of transits and changes of location (CUSDEC messages).
- Requests for customs inspections (CUSDEC message).
- Shipment instructions, IFTMCS message.

The port authority of Barcelona was a pioneer in the telematic transmission of data. In 1993, a commission was created within the framework of the Quality Plan of the port, named after the guarantee of information with the purpose of speeding up and modernizing the port community of Barcelona with the extension of electronic data exchange (EDI) in all the documentary procedures of the companies and official agencies which make up this community. The Commission for the Assurance of Information itself carried to term the works of re-engineering the documentary processes of the Port of Barcelona. After a rethink of the Commission's objectives it was dissolved to form the Telematic Forum which is the spiritual heir of the Commission.

The main actions carried out by the Commission were:

- An analysis carried out by the Commission for the Assurance of Information of the starting situation that was reflected in the document "Los circuitos documentales del Puerto de Barcelona" (July 1994).
- Later a study of the priorities in restructuring the documentation circuits of the port of Barcelona was carried out (December 1994). At the same time a monitoring of the activities related to EDI was carried out in the port community and the improvement of the most troubled circuits was initiated.
- Both the Commission and the Forum have been groups that have had a great influence in defining the new documentation procedures, both for the port of Barcelona and the State.
- In this sense, the Commission for the Assurance of Information worked very actively in the new model for concise declaration of maritime traffic (1995) and in defining associated EDI messages and implementing EDI for processing associated documents on the entry of dangerous goods into the port area.
- It also carried out the re-engineering of procedures associated to requests for putting in and notification of arrivals, both from the port authorities and

maritime headquarters, simplifying the documentation procedures and introducing new presentation methods for documentation.
- They have implemented services for different agents of the port community and official agencies so that they can access the information necessary for its documentary management more easily, such as the information service for unloaded goods that puts at the disposal of the customs agents the departure data declared by the consignee on the manifest so that they can prepare the customs declaration; or the different information services that were introduced to notify that the summary declaration has already been activated and goods can now be dispatched.

Currently, the Telematic Forum is carrying out the re-engineering of all the documentation procedures associated to the entry/exit of goods to/from the port area, in order to eliminate the need for certain documents to travel with the lorries that come to collect/deliver goods from/to the port and the problems associated with inspecting goods. t the same time a complete review of all docking request procedures is being carried out with the aim of making this compatible with the new state environmental procedure which is being defined jointly with the Directorate General of the Merchant Navy and state ports. These working parties of the Quality Plan have tried to make room for all the groups of the port community and already, since their inception, there is representation from shippers, stevedore companies, customs agents, freight forwarders, the Inland Revenue (tax and customs computing) and departments of quality, strategic planning and information systems of the port authority.

With the Telematic Forum, land transport companies also have permanent representation. Furthermore, in certain cases, other groups are invited to the meetings when aspects are discussed in which they are involved, for example, maritime headquarters, the inspection services or other departments of the port authority.

The Telematic Forum meets every 3 weeks to propose new actions, discuss them and approve the work to be completed over the following 3 weeks. Since the existence of PortIC (http://www.portic.net), one of the missions of the Telematic Forum continues to be the redefinition of the documentation procedures of the port of Barcelona, but now the execution of the work specified will be completed by PortIC. In this way the work of the Forum will not just be theoretical studies that only some companies may be able to complete.

Finally Serviport Canarias S.A, (http://www.serviport.com) should be highlighted, a service company who develop their activities in the information technology and communications sector, specialising in ports. Taking advantage of their experience and knowledge in this sector, the services they offer extend to companies in other economic sectors of the Canary Islands, especially to the hotel sector, public administrations and small and medium companies. It was founded with the shareholder support of Telefónica (through PORTEL), the port authority of Las Palmas, the port authority of Tenerife, financial organizations and the major companies and port operators in the Canaries.

In parallel to the development of telematic data transmission over value added networks (VAN), in 2001 the National Tax Administration Agency opened up the possibility of formulating SAD via Internet, while maintaining the presentation of written and electronically transmitted data using EDIFACT messages over value added networks. In this way the security offered by VAN is considered equal to EDI on Internet and the electronic signature provided by the Fábrica Nacional de Moneda y Timbre (the Spanish mint) and offers a wide range of possibilities to fulfil the tax obligations derived from foreign trade operations to facilitate and speed up documentation procedures as well as reduce the costs associated with them.

All of this is framed in a logical evolution appropriate to the situation of technology associated to the Internet to facilitate and speed up communications between agencies, companies and individuals. The implementation of this new technology required collaboration between the port agencies involved, the port authorities, Customs, SOIVRE and the border inspection services, to allow the introduction of a system to authenticate the certificates issued to replace the corresponding paperwork. In this way the management of depots and other warehouses under customs control and submitted to inspections has improved.

New technology has allowed connections to be made with the port authorities, customs and customs services, forming a network of electronic communication and permitting electronic processing of data concerning the goods (via value added networks or the Internet) that speeds up the management of these processes in the port environment. The Maritime Headquarters, which is the proper agency to authorize the entry and exit of the ships that transport goods in Spanish waters, remained unconnected to this port network (web); therefore the documentation processing of ships at Maritime Headquarters (office) could only be done by paper or fax.

Traditionally, when the shipping agent presented the docking request to the port authorities, long in advance and in the form that the respective port authorities established for this purpose, the port authorities assigned a docking number that allowed the shipping agent to continue the administrative process associated to the goods transported by ship in the customs administration.

The docking request was a document that every port authority prepared for its own needs and did not correspond to a standardized national model. To make maritime transport easier and speed up the stay of ships and cargoes in port, the need to modernize and simplify the procedures and processes was presented to the port authority, Maritime Headquarters and Customs, being approved by order of the Ministry of Public Works on November 29, 2002, *the establishment of the integrated procedure for the docking of ships (PIDE) in ports of general interest.* This is where the unique docking document (DUE) standardized nationally is specified, whose purpose is to facilitate the management of the docking of ships by the port authorities and their offices by Maritime Headquarters.

The integrated procedure for the docking of ships (PIDE) creates a single window between the maritime and port administrations which strengthens the channels of collaboration between both administrations in the port environment and simplifies the procedures to be carried out, integrating the processing of documents which the shipping agents need to provide to the port authorities and maritime headquarters in only one procedure, so that they can authorise the entry and exit of ships in waters under Spanish sovereignty and jurisdiction as well as their entry, berthing/mooring in port and their subsequent dispatch. This integration is carried out by using a *docking number* which has already been used to coordinate the customs and port administrations in relation to the customs procedures for the goods.

In the new integrated procedure for the docking of ships (PIDE) a unique docking document (DUE) model is established, and its presentation, whether it be on paper or by means of electronic transmission, the place in which to do it, who is obliged to present it, time limit, etc. are regulated. The content of the unique docking document (DUE) is very comprehensive and allows all the information associated to the ship and the load it is transporting to be made available in advance.

The new regulation develops the possibility of presenting the unique docking document (DUE) and its acceptance by the port authorities and maritime headquarters by means of electronic transmission via the Internet through PORTEL, using standard messages. Moreover, it establishes channels of collaboration between the port and maritime administrations, providing the opening of single windows in the port authorities that, beginning with the new computer system for the processing of the docking of ships, will gradually allow agents of the port community to do the entry request and the shipping office from their own offices by connecting to the PORTEL portal, permitting the delivery of information on-line, to the agents who are involved in the dispatch of the ship in port and need to make decisions, in line with the criteria of efficacy, efficiency and competitiveness that should rule in the port environment.

5 Impact of the Technological Change and new Technology in the Spanish Port System

The literature on the impact of investment in new technology tends to use the focus of productivity and efficiency measures through the functions of production and costs, for example: Becchetti et al. (2003). Research on the subject tries to evaluate the impact of the specific investments in new information and communication technology in a series of intermediate variables (introduction of new products and processes, use of the production capacity....) as well as in the productivity of each company (measured through work productivity, total work productivity or level of technical inefficiency). In this way it measures the contribution of new technology to the production process. This is the focus that we consider most appropriate and constitutes a future line of research on which the authors of this article are currently working.

In this article, with the research we have available at the present time, we will try to carry out an approximation of the impact of new technology on the Spanish port sector. Obviously, this does not mean a doing a review of the literature, nor of the more ambitious investigation that we will develop in the near future. Nevertheless, we think that there are already some promising indications in the work carried out to date, which should be highlighted in this article and passed forward to the scientific community and the economic agents concerned.

Amongst the works on production and cost functions that try to study the port activity, cost structure, technical, economic and allocative efficiency and the impact of technological change in the ports, the following are noteworthy: Chang (1978); Reckers et al. (1990); Tongzon (1993) and Baños-Pino et al. (1999), who use production functions; and the work of Kim and Sachis (1986); Martínez-Budría (1996); Jara-Díaz et al. (1997, 2000); Baños-Pino et al. (1999); Martínez-Budría et al. (1998, 1999); Coto-Millán et al. (2000) and Tovar (2002) who use cost functions; finally, Talley (1994) uses performance indicators.

Kim and Sachis (1986), are pioneering authors in trying to estimate the resultant technological progress of introducing new technology in the port activity. With the purpose of estimating the technical change and technology in the port operation, these authors specified a long term total cost function of the type:

$$
\begin{aligned}
Ln\ \mathrm{C} = {} & \alpha_0 + \alpha_y Ln\ \mathrm{y} + \sum_{\mathrm{i}} \alpha_{\mathrm{i}} \left(n\mathrm{w}_{\mathrm{i}} + \frac{1}{2}\gamma_{yy}(Ln\ \mathrm{y})^2 \right) + \frac{1}{2}\sum_i \sum_j \gamma_{ij} Ln\mathrm{w}_{\mathrm{i}} Ln\mathrm{w}_{\mathrm{j}} \\
& + \sum_i \gamma_{iy} Ln\mathrm{w}_{\mathrm{i}}\ \mathrm{Ln\ y} + \sum_i \theta_{it}\ \mathrm{Ln\ w_i\ Ln\ t} + \theta_{yt}\ \mathrm{Ln\ y\ Ln\ t} + \beta_{\mathrm{t}}\ \mathrm{Ln\ t} \\
& + \frac{1}{2}\beta_{tt}(Ln\ \mathrm{t})^2
\end{aligned}
$$

where the variables are in natural logarithms (Ln); the variable C represents the total minimum cost; the variable y, shows the tonnage of the load; $\mathrm{w_i}$ indicates the price of factor i; and, t, is an index of technology for each instant of time, specifically the percentage of special merchandise; and the indices i, j = 1, …., n, represent indices of the n-factors of production.

Lastly, the parameters to estimate are:

$$
\alpha_0;\ \alpha_y;\ \alpha_i;\ \gamma_{yy};\ \gamma_{ij};\ \theta_{it};\ \theta_{yt};\ \beta_t\ \mathrm{y}\ \beta_{\mathrm{tt}}.
$$

The technology index used here by Kim and Sachis, being a percentage of containerized cargo, has initial sample values (the 1960s) that are zeros, due to the non-existence of such types of load. The application that Kim and Sachis carry out for the port of Ashdod (Israel), begins to handle containerized loads in 1970. This fact generated problems when estimating the translog function which the authors resolved with the Box-Cox transformation for technology index t, doing:

$$
t = \frac{T^{\theta} - 1}{\theta};\ con\theta = 0,01
$$

In this manner, the authors show the presence of a smooth technological innovation in port activity in the port of Ashdod.

The work of Kim and Sachis (1986) is one of the first applications of this methodology on ports. Some years later in Spain, Coto-Millán et al. (2000) presented a primary study of efficiency through translog cost functions such as the following:

$$LnCV_{SR} = \beta_0 + \sum_{r=1}^{m} \alpha_t Lny_{rpt} + \frac{1}{2}\sum_{r=1}^{m}\sum_{s=1}^{m} \alpha_{rs} Lny_{rpt} Lny_{spt} + \sum_{i=1}^{n} \beta_i Lnw_{ipt}$$
$$+ \frac{1}{2}\sum_{i=1}^{n}\sum_{j=1}^{n} \beta_{ij} Lnw_{ipt} Lnw_{jpt} + \sum_{t=1}^{m}\sum_{i=1}^{n} \rho_{ri} Lny_{rpt} Lnw_{ipt} + \sum_{t=1}^{T} D_T T + \varepsilon_{pt}$$

where: r, s = 1, ..., m; is the number of outputs and technological variablesi, j = 1, ..., n; is the number of inputs p = 1, ..., p; is the number of ports ε_0 is the error of random disturbance DT is the coefficient of the dummy variable T, which shows the specific annual effects of time, just as they can be affected by technology.

In the above equation it is usual to set the condition of homogeneity at degree one in the variable prices of the inputs on the cost function:

$$\sum_{i=1}^{n} \beta_i = 1; \sum_{\mathrm{j}=1}^{\mathrm{n}} \beta_{\mathrm{ij}} = 0; \sum_{\mathrm{r}=1}^{\mathrm{m}} \rho_{\mathrm{ri}} = 0$$

as well as the condition of symmetry:

$$\beta_{ij} = \beta_{ji}$$

The estimate carried out by Baños-Pino et al. (1999) of the cost equation presents the impossibility of rejecting the hypothesis due to a "lack" of technical progress for the panel of data from the 27 Spanish ports over the 5 years when the study was carried out (1985–1989) (in this respect see Table 2). On the other hand, significant economies of scale are detected: "A 1% increase of output only produces a cost increase of 0.26%".

Nevertheless there is another study by the previous authors, Baños-Pino et al. (1999), which, although it was carried out later, was published before the previous one, and covers a more extensive period: 1985–1995. In the estimates carried out by Baños-Pino et al. (1999) on the cost equation, for the period 1985–1995, the results obtained show that from the year 1992, a significant technological progress has existed. Also detected was the existence of economies of scale and a certain overcapitalization after the Ports Act of 1992.

A different approach is that of measuring the impact of technological innovation from the production functions. Production functions of port services usually take the form:

$$Q = B_0 L^{\alpha} K^{\beta} e^{\gamma(T/L)}$$

Table 2 Estimated cost function (sample period: 1985–1989)

Variable	Coefficient	Error	t-statistic
Constant	7.004	0.038	183.09
Log (Y)	0.260	0.031	8.297
Log (w_k)	0.350	0.048	7.293
Log (w_L)	0.430	0.038	11.263
Log (w_E)	0.218	0.029	7.526
Log (y) Log (y)	0.085	0.045	1.874
Log (w_K) Log (w_K)	0.219	0.088	2.492
Log (w_L) Log (w_L)	0.268	0.096	2.772
Log (w_E) Log (w_E)	0.092	0.027	3.344
Log (y) Log (w_K)	0.021	0.027	0.761
Log (y) Log (w_E)	−0.054	0.023	−2.257
Log (y) Log (w_L)	−0.057	0.029	−2.534
Log (w_K) Log (w_L)	−0.197	0.064	−3.076
Log (w_K) Log (w_E)	−0.021	0.033	−0.652
t	0.074	0.037	0.195
t × t	0.016	0.034	0.484
t × Log (y)	−0.024	0.018	−1.330
t × Log (w_K)	−0.094	0.023	−4.007
t × Log (w_L)	0.102	0.032	3.185
t × Log (w_E)	−0.076	0.020	−0.383

$R^2 = 0.997$; DW = 2.02

Fixed effects vs. Stochastic effects (Hausman Test): CHISQ (12) = 171.75

Source: Coto-Millán et al. (2000)

Where Q = value added or gross profitL = quantity of workK = quantity of capital, approximated by net fixed assets pertaining to the operation $e^{\gamma(T/L)}$ shows the technological progress T/L represents the tons by unit of work (α, β, γ) are the parameters to estimate.

The previous equation assumes a Cobb-Douglas production function, which is a special case of the following transcendental logarithmic specification:

$$LnQ = \beta_0 + \beta_1 LnK + \beta_2 \text{ LnL} + \frac{1}{2}\beta_3(LnL)^2 + \frac{1}{2}\beta_4(LnK)^2 + \beta_5 LnL\ LnK + \varepsilon$$

Normally, in port activity it is usual to include the work inputs, L, and capital K, as well as the input energy, E. Using this, the previous equation is transformed into the following one, plus the corresponding factorial participation equations or proportion of factors:

$$LnQ = \beta_0 + \beta_1 LnK + \beta_2 LnL + \beta_3 LnE + \frac{1}{2}\beta_4(LnL)^2 + \frac{1}{2}\beta_5(LnK)^2 + \frac{1}{2}\beta_6(LnE)^2 + \beta_7(LnL)(LnK) + \beta_8(LnL)(LnE) + \beta_9(LnK)(LnE) + \varepsilon$$

$$S_L = \frac{\partial LnQ}{\partial LnL} = \beta_1 + \beta_5 LnL + \beta_8 LnK + \beta_9 LnE$$

$$S_K = \frac{\partial LnQ}{\partial LnK} = \beta_2 + \beta_5 LnK + \beta_7 LnL + \beta_9 LnE$$

$$S_E = \frac{\partial LnQ}{\partial LnE} = \beta_3 + \beta_6 LnE + \beta_8 LnL + \beta_9 LnK$$

where S_L, S_K and S_E, represent the factorial proportions, or elasticities of output regarding the capital, in such a way that if:

$$S_L + S_K + S_E \underset{\geqslant}{\overset{\leqslant}{=}} 1$$

we will have economies of scale (>1), constant returns to scale and diseconomies of scale (<1).

In the work of Baños-Pino et al. (1999) the empirical estimate of a production function has been carried out as above for the Spanish ports, trying to check if there was any technological progress in the period (1985–1995). The results of the estimates provide some values for the technological progress parameter which are not significantly different from zero (see Table 3). This suggests that the ports would have had no exogenous technological progress during the period studied. Something similar occurs in many manufacturing sectors of Spanish industry. In this same study the average technical efficiency detected is 0.4063 and the interpretation of this indicator is that the port sector, in average values, could have expanded production by 59.36% during the period (1985–1995). This is an average level of low efficiency. Nevertheless, as can be seen in Table 4, this average level of efficiency has a very high heterogeneity and varies from 1 in the port of Barcelona to 0.04 in the port of Melilla.

Table 3 Estimated production function (sample period: 1985–1995)

Variable	Coefficient	Error std.	t-statistic
Constant	3.9622	0.1615	24.31
Log (L)	0.3143	0.1356	2.31
Log (K)	0.5773	0.04967	11.62
Log (E)	0.2499	0.0279	8.95
Log (L) Log (L)	0.5474	0.2024	2.7046
Log (K) Log (K)	0.1279	0.0794	1.6103
Log (E) Log (E)	−0.0470	0.0718	−0.6549
Log (L) Log (K)	−0.6522	0.0916	−7.1163
Log (L) Log (E)	0.1576	0.0532	2.9554
Log (K) Log (E)	0.0525	0.0804	0.6536

$R^2 = 0.969$

Hausman Test, Fixed effects vs. Random effects; CHISQ (9) = 25.41

Verisimilitude Ratio Test, Translog Functional Form vs. Cobb Douglas: CHIQ (6) = 60.01

Heteroskedasticity-robust t statistics

Source: Baños-Pino et al. (1999)

Table 4 Grade index of technical efficiency by port authority (sample period: 1985–1995)

Port	Technical efficiency index (TE)
Algeciras	0.209
Alicanten	0.568
Almería	0.593
Barcelona[a]	1
Bilbao	0.986
Cádiz	0.561
Cartagena	0.987
Castellón	0.315
Ceuta	0.999
Ferrol	0.682
Gijón-Avilés	0.997
Huelva	0.842
Coruña	0.988
Las Palmas	0.987
Málaga	0.512
Melilla	0.108
Palma de Mallorca	0.834
Pasajes	0.615
Pontevedra	0.47
S. Cruz de Tenerife	0.988
Santander	0.603
Sevilla	0.47
Tarragona	0.989
Valencia	0.998
Vigo	0.507
Villagarcía	0.354

[a]Port of reference for technical efficiency
Source: Baños-Pino et al. (1999) and compiled by author

6 Conclusions

It can be concluded that from 1992, the presence of technical change due to processes of technological innovation seems to be detected in the Spanish port system, essentially in new technologies of information and communication.

On the other hand, there seems to be conclusive evidence on the existence of economies of scale in the Spanish port sector.

In particular, with the provisional evidence that we currently have available, it seems that Valencia, Bilbao, Las Palmas and Santa Cruz de Tenerife, the pioneering ports, with the significant involvement of general and large sized merchandise in their traffic, such as Barcelona, Valencia and Bilbao, have benefited more from technological innovation than the small and medium size ports, which in turn have been later in introducing new technology.

As for technical efficiency, three groups stand out:

1. Barcelona, Bilbao, Cartagena, Ceuta, Gijón-Avilés, La Coruña, Las Palmas, Santa Cruz de Tenerife, Tarragona and Valencia, with maximum levels of

technical efficiency. That is, the ports that, technologically, combine factors of production better, so that none are used superfluously.

2. Alicante, Almería, Cádiz, El Ferrol, Huelva, Málaga, Palma de Mallorca, Pasajes, Santander and Vigo, with medium-high levels of technical efficiency.
3. Algeciras, Castellón, Melilla Pontevedra, Seville and Villagarcía, with medium-low levels of technical efficiency.

The economic interpretation that can be made in light of the previous results is the following. The pioneering ports, in introducing new technologies, such as Barcelona, Las Palmas, Santa Cruz de Tenerife, these last related to Serviport Canarias S.A, a service company in the information technology and communications sector, with a strong presence in the port sector; and, large sized ports, such as: Bilbao, Gijón-Avilés, Valencia, etc, have incorporated new technology in a technically more efficient manner. This process has become more widespread in the small and medium size ports, though they are lagging behind with respect to the pioneers. The singularity of the port of Algeciras is surprising, nevertheless, with a medium-low level of efficiency, when one would expect, given its large size and type of traffic (highly specialized in containers), a higher level of technical efficiency. A possible explanation, which can be found in Jara-Díaz et al. (2002), is that in the Spanish port system, the specialization in traffic is not advisable since the economies of diversity reveal savings by having a joint production of merchandise that could be lost by specialization of trade. Nevertheless, in the research that we are carrying out and in that outlined and presented above for the near future, with explicit incorporation within the estimated production and cost functions of a "proxy" of investment in new technology, in multi-production contexts, more appropriate for this type of research, it is possible that this anomalous result, as well as unexpected behaviour in other ports, can be corrected.

References

Baños-Pino J, Coto-Millán P, Rodríguez-Álvarez A (1999) Allocative efficiency and over-capitalization: an application. Int J Transp Econ 26(2):181–199

Becchetti L, Bedoya DAL, Paganetto L (2003) ICT investment, productivity and efficiency: evidence at firm level using a stochastic frontier approach. J Prod Anal 20:143–167

Chang S (1978) Production function and capacity utilization of the port of Mobile. Marit Policy Manage 5:297–305

Coto-Millán P, Baños-Pino J, Rodríguez-Álvarez A (2000) Economic efficiency in Spanish Ports: some empirical evidence. Marit Policy Manage 2(2):169–174

Jara-Díaz S, Cortes C, Vargas A, Martínez-Budría E (1997) Marginal costs and scale economies in Spanish ports. 25th European transport forum, proceedings seminal L, PTRC, London, pp 137–147

Jara-Díaz S, Martínez-Budría E, Cortes C, Basso L (2002) A multioutput cost function for the services of Spanish ports infrastructure. Transportation 29(4):415–437

Kim M, Sachis A (1986) The structure of production, technical change and productivity in a port. Int J Ind Econ 35(2):209–223

Martínez-Budría E (1996) Un estudio econométrico de los costes del sistema portuario español. Revista Asturiana de Economía 16:135–149

Martínez-Budría E, González-Marrero R, Díaz J (1998) Análisis económico de las Sociedades Estatales de Estiba y Desetiba. Documento de Trabajo 97/98-1. Universidad de La Laguna, Tenerife

Martínez-Budría E, Díaz-Armas R, Navarro-Ibañez M, Ravelo-Mesa T (1999) A study of the efficiency of Spanish port authorities using data envelopment analysis. Int J Transp Econ 26(2):237–253

Reckers RA, Connell DY, Ross DI (1990) The development of a production function for a container terminal in the port of Melbourne. Papers of the Australian Transport Research Forum, vol. 15, pp 205–218

Talley WK (1994) Performance indicators and port performance evaluation. Logist Transp Rev 30(4):339–352

Tongzon JL (1993) The port of Melbourne authority's pricing policy: its efficiency and distribution implications. Marit Policy Manage 20(3):197–203

Tovar B (2002) Análisis multiproductivo de los costes de manipulación de mercancías en terminales portuarias. El Puerto de La Luz y de Las Palmas. Tesis Doctoral, Departamento de Análisis Económico Aplicado, Universidad de Las Palmas de Gran Canaria, España

UNCTAD (1990) Port marketing and the challenge of the third generation port. UNCTAD, Geneva and New York

UNCTAD (1999) Technical note: fourth generation port. UNCTAD, Geneva and New York

UNCTAD (2000) TD/BC4/AC.7/14

Excess Capacity, Economic Efficiency and Technical Change in a Public-Owned Port System: An Application to the Infrastructure Services of Spanish Ports

Ramón Núñez-Sánchez and Pablo Coto-Millán

Abstract In this paper we estimate both a technical efficiency measure and the evolution of allocative efficiency in the infrastructure services of Spanish ports during the period 1986–2005. To achieve this aim, we estimate a system of equations consisting of a multioutput translog input distance function and cost shares equations. The results show that Spanish ports do not minimize their costs. Additionally, from 1992, we observe a process of under-utilization of capital relative to labour, coinciding with a decentralized process of the organizations that own and control the land and activities at Spanish ports.

1 Introduction

From 1992, infrastructure services of Spanish ports are managed by Port Authorities, state-owned organizations that own and control the land and some activities at Spanish ports.

The estimation of productivity and efficiency measures in the port industry is relatively new and scarce. However, in recent years significant progress have been made in studies related to this topic.[1] In this way, it is important to distinguish two types of ports studies, those related to infrastructure and navigation services of ports (Liu 1995; Baños-Pino et al. 1999; Coto-Millán et al. 2000; Barros 2003; Estache et al. 2004; González and Trujillo 2008), and those related to private terminals or cargo handling firms (Cullinane et al. 2002; Rodríguez-Álvarez et al. 2007; Díaz-Hernández et al. 2008a, b).

[1] A good survey related to productivity and efficiency measurement in the port industry comes from González and Trujillo (2007).

R. Núñez-Sánchez (✉), P. Coto-Millán
University of Cantabria, Santander, Spain
e-mail: ramon.nunez@unican.es

P. Coto-Millán et al. (eds.), *Essays on Port Economics*, Contributions to Economics,
DOI 10.1007/978-3-7908-2425-4_17, © Springer-Verlag Berlin Heidelberg 2010

Some of these studies use a parametric approach (Liu 1995; Baños-Pino et al. 1999; Coto-Millán et al. 2000; González and Trujillo 2008; Cullinane et al. 2002; Rodríguez-Álvarez et al. 2007 and Díaz-Hernández et al. 2008b), and some of them use a non-parametric approach (Barros2003; Estache et al. 2004; Díaz-Hernández et al. 2008a).

Liu (1995) studies the performance of the British port transport industry, evaluating the determinants of relative efficiency from the different ports. It concludes that neither the form of ownership nor size are important factors that can explain technical inefficiency.

Baños et al. (1999) use both the cost function approach and its dual approach, the input distance function, to demonstrate the existence of allocative inefficiency in the Spanish Port Authorities. The study shows that Spanish ports do not minimize their costs and the existence of an amount of the quasi-fixed input larger than the optimal if they were allocative efficient.

Coto-Millán et al. (2000) estimate the economic efficiency of Spanish Port Authorities using a stochastic cost frontier. Their results show: the relatively larger ports are more economically inefficient, the existence of large economies of scale, and the inexistence of technical progress during the last 1980s.

Barros (2003) analyses the technical and allocative efficiencies of Portuguese Port Authorities in order to investigate if the state's policy is achieving its aims of regional development. He concludes that the incentive regulation carried out by the government's regulatory body, based on direct subsidies for non-profitable ports, is not achieving the improvement of economic efficiency.

Estache et al. (2004) evaluate the sources of efficiency gains since the port reforms, based on the liberalization and decentralization of the port system. They show that the public reforms facilitated the improvement of technical change through the adoption of new technologies.

González and Trujillo (2008) analyze the extent to which port reforms that took place in the 1990s had an impact on the technical efficiency in port infrastructure provision in the major Spanish Port Authorities involved in container traffic. Their results show that technical efficiency has changed little on average, even though there is a significant movement of efficiency within ports over time as a result of reforms.

Culliname et al. (2002) apply a model to analyse the administrative and ownership structures of the terminals of the main container ports in Asia. The results of the analysis conclude that the size of a port or terminal is closely correlated with its efficiency, and that both existence of programmes in Asian ports, which aim to attract private capital, and the level of market deregulation improve economic efficiency.

Rodríguez-Álvarez et al. (2007) present a model to calculate both, the technical and the allocative efficiency in cargo handling firms in a Spanish port. They apply this model to a frontier input distance system. The results obtained show a positive relationship between firm size and technical efficiency, and the existence of allocative inefficiency, which suggests that the port labour-specific regulatory environment impedes adjustments that are needed by operators.

Finally, Díaz-Hernández et al. (2008a) evaluate the deregulation process which took place in port cargo handling sector in Spain during the 1990s. They find that technical change is the element that has caused the increase in productivity, as technical change has remained constant. Moreover, ports with a relatively large traffic volume exhibit the larger average efficiency index and rates of technical change than the average of the rest.

The main purpose of this paper is to calculate, by means of the estimation of an input distance function, both a technical and an allocative efficiency measure. After that, we have studied the evolution of the allocative efficiency in port infrastructure service provision in the Spanish ports and we have observed whether port reforms have had an effect on the allocative efficiency, or not.

The use of distance functions has, basically, three advantages with regard to the use of traditional cost functions. Firstly, it is not necessary to assume that Port Authorities minimize costs regard to input market prices. This feature is especially important in the context of Spanish Port System characterized to be a state-owned sector highly regulated. In this way, the main aim of Port Authorities is the creation of employment and the regional development of its hinterland (Puertos del Estado 2006). Secondly, input distance functions do not assume the exogeneity on input prices. Thus, the existence of price regulation and monopsonistic markets in some inputs make important this feature for the estimation of efficiency in the infrastructure services of Spanish ports. Finally, unlike the cost function, the decomposition of economic inefficiency into technical and allocative does not need restrictive assumptions.

On the other hand, the choice of an input distance function, instead of an output distance function can be justified by the conditions under which Port Authorities develop their activities. We consider that Port Authorities do not have control over the traffic of cargo that uses their facilities. However, the managers have most control over inputs such as: labour, capital, intermediate consumption and deposit area. In our study, the last input, deposit area, is considered as a quasi-fixed input.

The structure of the paper is structured as follows. Section 2 describes the Spanish port system during the period 1986–2005. Section 3 presents the theoretical model of the allocative efficiency measure. Section 4 establishes the econometric specification of the input distance function. Section 5 presents the descriptive analysis of the data and the results of the estimation and the most important results. Finally Sect. 6 concludes.

2 Description of Spanish Port System (1986–2005)

During the period 1986–2005, Spanish port system has suffered some important regulatory changes.[2] Both Law 27/1992, Law 62/1997 and Law 48/2003 of General Interest Ports have produced important changes in the structure of management of the Spanish ports. Thus, before of 1992 the port system was characterized by the existence of two different management models. Most of the ports were managed by

---

[2]Other papers which study the Spanish port reform during the 1990s are Coto-Millán (1996) and Castillo-Manzano et al. (2008).

"Juntas del Puerto" (Port Assemblies), organizations characterized by having a centralized and bureaucratic structure, dependent on Ministerio de Obras Públicas (Spanish Ministry of PublicWorks). On the other hand, some of the largest ports (Barcelona, Bilbao, Valencia and Huelva) were defined as "Puertos Autónomos" (Autonomous Ports) with a greater autonomy level. Additionally, from 1986 the Spanish Ministry of PublicWorks, through the 2546/1985 act, set a minimum rate of return based on the net investment of fixed assets for the entire port system, in order to control the current expenditures and improve the internal management of the Port Assemblies. In fact, before 1985, in cases of losses for Port Assemblies, they were covered by direct public subsidies.

Law 27/1992 created the "Ente Público Puertos del Estado" (State-owned Enterprise of National Ports) and transformed the Ports Assemblies and Autonomous Ports into new public entities less centralized, called "Autoridades Portuarias" (Port Authorities). Independent business criteria management with individual budgets and the abolition of direct public subsidies were some of the features of these new entities. One of the aims of these reforms was the encouragement of inter-port competition among the different Port Authorities. On the other hand, State-owned Enterprise of National Ports was the responsible of coordination and efficiency control for the entire system Moreover, the new legislation established the creation, at every port, of a "Sociedad Estatal de Estiba y Desestiba" (State-owned stevedore company), trying to de-regulate the composition of work teams for port services, such as port cargo handling. In this way, the Law allowed the replacement of a civil servant status of workers by private sector workers.

Other minor public regulation reforms were in 1997 (Law 62/1997) and 2003 (Law 48/2003). The first reform in 1997, established that public Regional Governments were allowed to name members of a Port Authority governing board. The second one, in 2003, encouraged private investment in port infrastructures, trying to improve the intra-port competition and the competitive position of Spanish ports.

Is important to stress that, from 1992, the Spanish port system is based on financial self-sufficiency for Port Authorities, not receiving any direct subsidy from the national government. In this way, current and investment expenditures are covered by current incomes, EU special subsidies and, occasionally, by external debt.

Within this framework, the General Interest Ports are intended to respond to the *landlord* model, whereby the Port Authority does no more than provide the port land and infrastructure and regulate the use of this public property, whereas the port services are essentially provided by private sector operators under an authorization or concession regime.

3 Theoretical Model

An input distance function, $D_I(x, y)$, can be defined through of input set, L(y), as:

$$D_I(x, y) = \max_{\rho} \{\rho : (x/\rho) \in \mathrm{L(y)}\} \tag{1}$$

where the input set L(y) represent the set of input vectors, $x \in R^K_+$, that produce the output vector, $y \in R^M_+$:

$$L(y) = \{x \in R^K_+ : x \text{ can produce } y \in R^M_+\} \tag{2}$$

An input distance function, $D_I(x, y)$, is non-decreasing, homogeneous of degree one and concave in inputs, and non-increasing in outputs. Moreover, for $x \in L(y)$, $D_I(x, y) \geqslant 1$, being $D_I(x, y) = 1$, if x is technically efficient for y.

On the other hand, one important theoretical property is that input distance function is dual of the cost function.

Following Färe and Grosskopf (1990) it is possible the estimation of an allocative efficiency measure using an input distance function. To do this, firstly we define a shadow cost function as:

$$C(y, w^s) = \min_x \{w^s x : D_I(x, y) \geqslant 1\} \tag{3}$$

where w^s is a shadow input price vector which may diverge from market input price vector, w.[3]

The first order condition with respect to the input h is:

$$w^s_h = \lambda \left(\frac{\partial D_I(x, y)}{\partial x_h} \right) \tag{4}$$

Jacobsen (1972) shows that in the optimum:

$$C(y, w^s) = \lambda \tag{5}$$

Using equations (4) and (5), we demonstrate the Shephard's Lemma:

$$\frac{\partial D_I(x, y)}{\partial x} = \frac{w^s}{C(y, w^s)} \tag{6}$$

Thus, the ratio of shadow prices for two any inputs will be:

$$\frac{\frac{\partial D_I(x,y)}{\partial x_h}}{\frac{\partial D_I(x,y)}{\partial x_j}} = \frac{w^s_h}{w^s_j} \tag{7}$$

[3]In the literature deviation between shadow input prices and market (actual) input prices may arise due to regulation, as in Atkinson and Primont (2002), as well as due to other types of noncompetitive environments, see, for example, Grosskopf et al. (1995).

If the firms minimized costs, the ratio of shadow prices for two any inputs would be equal to the ratio of market input prices. However, if the firm behaviour causes allocative inefficiency, the two price ratios differ.

To study such deviations, a relationship between the shadow prices ratios and the actual market prices is defined:

$$\frac{w_h^s}{w_j^s} = Z_{hj}\frac{w_h}{w_j} \tag{8}$$

If $Z_{hj} = 1$, shadow prices ratio is equal to actual market prices, firms minimize costs regard to actual market prices; if $Z_{hj} > 1$, means that the ratio of the shadow prices of input h to that input j is larger than the corresponding ratio of actual prices. So, h is under-utilized relative to input j; if $Z_{hj} < 1$, means that the ratio of the shadow prices of input h to that input j is lower than the corresponding ratio of actual prices. So, input h is over-utilized relative to input j.

4 Econometric Specification

For the estimation of the input distance function we have chosen a flexible functional form, the multiproduct translog input distance function. The translog form is based upon a second-order Taylor series expansion around the mean values:

$$\begin{aligned}\ln 1 &= \ln \mathrm{D}(\mathrm{y_{it}}, x_{it}, \mathrm{sup}_{it}) + u_i + \varepsilon_{it} \\ &= \alpha_0 + \sum_{r=1}^{M}\beta_r \ln y_{rit} + \frac{1}{2}\sum_{r=1}^{M}\sum_{s=1}^{M}\beta_{rs}\ln y_{rit}\ln y_{sit} + \sum_{j=1}^{N}\gamma_j \ln x_{jit} \\ &\quad + \frac{1}{2}\sum_{j=1}^{N}\sum_{h=1}^{N}\gamma_{jh}\ln x_{jit}\ln x_{hit} + \zeta_f \ln \mathrm{sup}_{it} + \zeta_{ff}\ln \mathrm{sup}_{it}^2 + \sum_{r=1}^{M}\zeta_{fr}\ln \mathrm{sup}_{it}\ln y_{rit} \\ &\quad + \sum_{j=1}^{N}\zeta_{fj}\ln \mathrm{sup}_{it}\ln x_{jit} + \sum_{r=1}^{M}\sum_{j=1}^{N}\rho_{rj}\ln y_{rit}\ln x_{jit} + u_i + \varepsilon_{it}\end{aligned} \tag{9}$$

where $\mathrm{D}(\mathrm{y_{it}}, x_{it}, \mathrm{sup}_{it})$ is the short-run input distance function, y is the output vector, x is the input vector, sup is a quasi-fixed input variable, i represents port and t time. The u_i component represents the magnitude of technical efficiency, which is allowed to vary across different Port Authorities, and ε_{it} represents statistical noise, distributed as multivariate normal with zero mean.

Applying the dual Shephard Lemma of equation (6):

$$\frac{x_{jit}w_{jit}}{C_{it}} = \gamma_j + \sum_{h=1}^{N}\gamma_{jh}\ln x_{hit} + \sum_{r=1}^{M}\rho_{rj}\ln y_{rit} + \zeta_{fj}\ln \mathrm{sup}_{it} + \mu_i + v_{it} \tag{10}$$

where μ_i represents the magnitude of allocative inefficiency with zero mean, and v_{it} represents statistical noise, distributed as multivariate normal with zero mean.

We impose homogeneity of degree one of the input distance function in variable inputs:

$$\sum_{j=1}^{N}\gamma_j = 1, \sum_{h=1}^{N}\gamma_{\mathrm{jh}} = 0, \sum_{j=1}^{N}\rho_{\mathrm{rj}} = 0\ \forall r = 1, \ldots.., M, \sum_{j=1}^{N}\zeta_{\mathrm{fj}} = 0 \tag{11}$$

We also impose the symmetry condition:

$$\beta_{rs} = \beta_{sr}, \gamma_{\mathrm{jh}} = \gamma_{\mathrm{hj}}, \zeta_{\mathrm{fj}} = \zeta_{\mathrm{jf}}, \zeta_{fr} = \zeta_{\mathrm{rf}}, \rho_{\mathrm{rj}} = \rho_{\mathrm{jr}} \tag{12}$$

4.1 Estimating Technical Efficiency

As it was said before, u_i component represents the magnitude of technical efficiency, which is allowed to vary across different Port Authorities. If we denote $\alpha_i = \alpha_0 + u_i = \alpha_p D_p$ where D_p is a dummy variable for the Port Authority p and α_p is a parameter to be estimated, the technical efficiency of a Port Authority is obtained from the estimated intercepts as $TE_i = \exp(-u_i)$ where $u_i = \alpha_i - \min(\alpha_i)$, being, $0 < TE_i \leq 1$. Thus, the most technically efficient Port Authority obtains a value 1.

4.2 Estimating Allocative Efficiency

Using a translog input distance function, we obtain coefficients z_{hj}, defined by equations (7) and (8), using the expression:

$$Z_{hj} = \frac{w_{jit}x_{jit}\left(\hat{\gamma}_h + \sum_{j=1}^{N}\hat{\gamma}_{jh}\ln x_{hit} + \sum_{r=1}^{M}\hat{\rho}_{rh}\ln y_{rit} + \hat{\zeta}_{fh}\ln \mathrm{sup}_{it}\right)}{w_{hit}x_{hit}\left(\hat{\gamma}_j + \sum_{h=1}^{N}\hat{\gamma}_{jh}\ln x_{jit} + \sum_{r=1}^{M}\hat{\rho}_{rj}\ln y_{rit} + \hat{\zeta}_{fj}\ln \mathrm{sup}_{it}\right)} \tag{13}$$

In this way, with the distance system, for the firm i at time t, we can estimate relative over- and under-utilization of any pair of inputs, x_{jit} and x_{hit}, in comparison to the cost-minimizing ratio, as we discussed in Sect. 3.

5 Empirical Results

5.1 Data Description

The sample consists of 27 Port Authorities, management entities of 50 ports considered as of general interest in Spain. We have annual data from 1986 until 2005, being the complete panel data set of 540 observations. The data was gathered from the annual reports of *Puertos del Estado* and every Port Authority. Other studies which have used this source were Baños-Pino et al. (1999), Coto-Millán et al. (2000), Jara-Díaz et al. (2002), González and Trujillo (2007).

The explanatory variables of the multiproduct translog input distance function for the infrastructure services of Spanish ports comprises two outputs, three inputs and one quasi-input.

The output variables of ports represent the movements of solid and liquid bulk (*gra*), and general cargo (*mg*). Regarding input variables, labour (*l*) is defined as the total annual number of employees. Capital (*k*) is calculated as annual depreciation expenses plus the return on the capital of the period. Finally, intermediate consumption (*ci*) is defined as annual consumption expense, services externally provided plus other expenses.[4] The stocking surface (*sup*) includes both open air and closed areas and, in our model, is considered as a quasi-fixed input. Following the *landlord* model, Port Authorities grant this area under concession to private sector operators.

Table 1 shows the mean of each variable previously defined for every port during the period 1986–2005. Thus, considering the aggregated product volume, Algeciras is the largest Port Authority, followed by Barcelona, Tarragona and Bilbao. It is to be noted that, disaggregating total product into bulk and general cargo, we observe differences into the specialization of the different Port Authorities. In this way, Tarragona, Bilbao and Algeciras are the largest ones with regard to bulk, being Algeciras, Valencia and Barcelona the largest ones for general cargo. On the other hand, Valencia is the Port Authority with a higher number of employees, followed by Bilbao, Palmas, Barcelona and Cadiz. The largest ports in terms of capital stock are Barcelona, Bilbao and Valencia. Finally, the Port Authorities with a higher intermediate consumption are: Bilbao, Valencia, Tarragona and Palmas. This first approximation shows the heterogeneity of the different Port Authorities and makes a likely existence of technical inefficiencies among them.

Average of Port Authorities input cost shares shows little variability among them. The mean of labour and capital cost share was both of 0.38 and of intermediate consumption was 0.24. The highest mean relative to labour cost share corresponds to Villa with 0.5, being Tenerife the Port Authority with the maximum mean

---

[4]As we pointed on Sect. 2, in 1985 the Spanish Ministry of Public Works set a rate of return based on the net investment of fixed assets for the port system (4%).

Table 1 Description of the variables. Mean for every port during the period 1986–2005

Port	gra (Tons)	mg (Tons)	l (Number)	k (Constant €)	ci (Constant €)	sup (m^2)	Sl (%)	Sk (%)	Sci (%)
Algeciras	19,685,527	17,599,442	279	9,465,956	5,565,810	483,878	0.342	0.414	0.244
Alicante	1,669,948	1,000,414	176	2,882,790	2,105,828	166,719	0.476	0.303	0.221
Almeria	7,535,404	477,951	145	2,803,638	1,677,317	243,115	0.440	0.350	0.210
Aviles	2,516,358	1,343,001	155	2,434,448	2,092,553	228,646	0.470	0.285	0.245
Cadiz	1,709,854	2,135,219	297	6,716,675	2,775,912	1,301,110	0.442	0.395	0.163
Barcelona	12,496,346	12,434,914	297	19,200,000	2,775,912	1,301,110	0.366	0.330	0.304
Bilbao	20,074,894	6,238,041	344	17,400,000	11,400,000	1,392,806	0.306	0.419	0.275
Cartagena	13,963,293	568,961	191	4,900,580	2,051,484	285,410	0.438	0.396	0.166
Castellon	7,953,286	596,035	97	2,530,585	1,478,808	291,271	0.383	0.389	0.228
Ceuta	2,213,863	787,909	135	2,862,241	1,871,733	50,084	0.485	0.312	0.204
Ferrol	5,397,151	332,753	146	1,579,028	849,992	141,713	0.438	0.365	0.197
Gijon	14,543,887	460,158	398	10,300,000	5,321,173	756,330	0.418	0.384	0.198
Huelva	13,613,041	557,247	219	7,357,229	4,431,093	228,657	0.380	0.387	0.233
Coruna	11,211,793	353,230	207	5,982,527	3,082,139	285,030	0.355	0.425	0.219
Palmas	4,607,825	6,984,139	299	11,100,000	6,057,937	522,727	0.337	0.429	0.234
Malaga	5,913,207	588,208	194	4,882,437	2,286,935	165,753	0.418	0.397	0.186
Melilla	540,689	503,337	83	1,596,859	897,774	27,186	0.522	0.306	0.172
Baleares	3,540,200	4,916,405	285	8,479,100	4,069,025	233,015	0.354	0.437	0.210
Pasajes	2,491,400	1,822,238	245	4,834,664	3,574,125	420,923	0.485	0.296	0.219
Pontevedra	566,465	599,202	121	1,659,645	849,579	85,183	0.431	0.376	0.193
Tenerife	9,195,904	4,633,484	219	9,815,523	5,076,732	454,339	0.297	0.463	0.240
Santander	3,639,486	938,619	242	6,343,766	3,935,852	614,321	0.378	0.384	0.238
Sevilla	2,591,500	1,066,766	249	5,712,072	3,501,355	240,349	0.402	0.371	0.227
Tarragona	25,615,066	819,839	315	8,903,090	6,135,939	653,865	0.338	0.392	0.270
Valencia	5,708,390	14,047,197	363	13,500,000	9,604,792	1,998,228	0.347	0.381	0.271
Vigo	930,114	2,177,139	223	5,135,533	3,793,209	447,989	0.421	0.333	0.246
Villagarcia	508,849	169,029	90	1,221,168	532,808	147,902	0.500	0.348	0.152
Mean	7,420,981	3,116,699	233	6,651,474	4,175,096	516,580	0.380	0.381	0.239

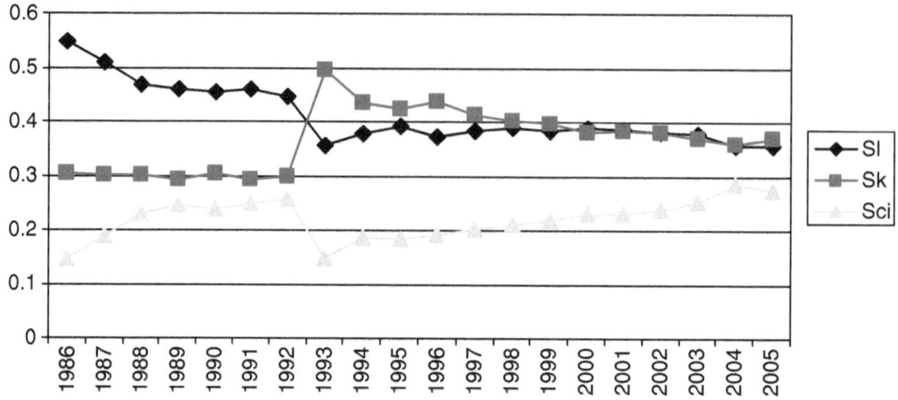

Fig. 1 Average input cost shares for the period 1986–2005

relative to capital cost share, with 0.46. Barcelona is the Port Authority with a higher mean of intermediate consumption cost share, with 0.3.

Figure 1 presents the evolution of input cost shares during the period 1986–2005. We observe the higher importance of labour regard to the other inputs in terms of total costs from 1986 to 1992, the period of the existence of Port Assemblies, characterized by a higher degree of central regulation. In the same way, there was a steady decline in the costs shares of labour, and a steady increase of intermediate consumption. Meanwhile, the cost share of capital held constant.

During the 1992 and 1993, there was an important increase of capital and a decrease of intermediate consumption and labour, so capital turned into the most important input in terms of total costs. During the next years, the importance of capital declined regard to the other inputs. As we see, new public regulations in 1997 and 2003 inside of Port Authorities does not seem to affect of input cost shares trends.

5.2 Results

We have estimated (7)–(8) by means of iterative seemingly unrelated regressions (ITSUR), which is invariant to the omitted share equation.

The variables have been divided by the mean. Therefore, the first-order coefficients can be interpreted as elasticities. As we can see in Table 2, the regularity conditions are satisfied at the sample mean.

It can be seen that all first-order coefficients are statistically significant and have the correct sign. Thus, the parameters of inputs are positive and significant, including the quasi-fixed input total deposit area (*sup*) with a positive and significant parameter. This means that the marginal productivity of total area is positive as predict the economic theory. Other studies related to port infrastructure obtain that

Table 2 Input distance function coefficients

Parameter	Coefficient	Std. error	t-statistic	p-value
L(gra)	−0.0924	0.0249	−3.700	0.000
L(mg)	−0.0394	0.0210	−1.870	0.061
L(k)	0.3748	0.0022	167.720	0.000
L(ci)	0.2309	0.0013	174.400	0.000
L(l)	0.3944	0.0034	117.270	0.000
L(sup)	0.0726	0.0264	2.750	0.006
L(gra) × L(gra)	−0.0306	0.0144	−2.120	0.034
L(mg) × L(mg)	−0.0543	0.0131	−4.140	0.000
L(k) × L(ci)	−0.0928	0.0016	−59.370	0.000
L(k) × L(l)	−0.1193	0.0032	−37.030	0.000
L(k) × L(k)	0.2120	0.0023	90.540	0.000
L(ci) × L(l)	−0.0560	0.0023	−24.460	0.000
L(ci) × L(ci)	0.1488	0.0015	101.380	0.000
L(l) × L(l)	0.1753	0.0052	33.850	0.000
L(gra) × L(mg)	−0.0186	0.0103	−1.810	0.070
L(k) × L(gra)	0.0024	0.0016	1.460	0.143
L(k) × L(mg)	−0.0036	0.0018	−2.060	0.039
L(ci) × L(gra)	0.0034	0.0010	3.530	0.000
L(ci) × L(mg)	−0.0006	0.0011	−0.610	0.545
L(l) × L(gra)	−0.0057	0.0024	−2.360	0.018
L(l) × L(mg)	0.0043	0.0027	1.610	0.108
L(sup) × L(sup)	−0.0902	0.0215	−4.190	0.000
L(gra) × L(sup)	−0.0163	0.0122	−1.340	0.181
L(sup) × L(mg)	0.0754	0.0147	5.140	0.000
L(l) × L(sup)	0.0038	0.0036	1.050	0.293
L(k) × L(sup)	−0.0040	0.0024	−1.680	0.094
L(ci) × L(sup)	0.0002	0.0014	0.160	0.871
a87	−0.1682	0.0325	−5.170	0.000
a88	−0.2569	0.0329	−7.810	0.000
a89	−0.2755	0.0332	−8.300	0.000
a90	−0.2862	0.0332	−8.630	0.000
a91	−0.3039	0.0335	−9.070	0.000
a92	−0.3029	0.0338	−8.960	0.000
a93	−0.5855	0.0344	−17.000	0.000
a94	−0.5154	0.0344	−14.970	0.000
a95	−0.4882	0.0347	−14.050	0.000
a96	−0.5372	0.0348	−15.430	0.000
a97	−0.5080	0.0358	−14.180	0.000
a98	−0.5069	0.0366	−13.870	0.000
a99	−0.5080	0.0379	−13.420	0.000
a00	−0.5285	0.0383	−13.800	0.000
a01	−0.5218	0.0390	−13.380	0.000
a02	−0.5399	0.0393	−13.720	0.000
a03	−0.5298	0.0400	−13.250	0.000
a04	−0.5641	0.0413	−13.650	0.000
a05	−0.5888	0.0438	−13.440	0.000
Equation	RMSE	R^2	Squared-Chi	P-value
Input distance function	0.1896381		119241.32	0
Capital share equation	0.0414737	0.7689	17576.25	0
Intermediate consumption share equation	0.024379	0.8581	16517.97	0
Labour share equation	0.05461	0.6418	191.32	0

marginal productivity of total area is negative but not statistically significant (Rodríguez-Álvarez et al. 2007). These authors justify that result to the fact that port deposit area has an indivisibility nature, so it is not possible to enlarge a deposit area in a continuous way. In this process, ports authorities are having more total area than previously needed. In this way, Baños-Pino et al. (1999) estimate the deviation between the optimal quasi-fixed input[5] if ports were allocative efficient and that, actually, employed in the short-run. The results obtained show that Spanish ports during the 1985–1997 period were allocative inefficient, with an amount of the quasi-fixed input larger than the optimal.

In Table 3 we present the estimated technical efficiency index (TE_i) following Sect. 4.1. As we can see, all of individual indexes are statistically significant. The most efficient Port Authority is Valencia, followed by Bilbao, Tenerife, and Las Palmas. The less efficient Port Authorities are Melilla, Ceuta and Villagarcia. It is worth mentioning that three of these Ports Authorities were considered as Ports Autonomous before of Law 27/1992. Moreover, this ranking shows that the most efficient Port Authorities are those which are larger in capacity and total cargo loaded.

In Fig. 2 we can see the relationship between technical efficiency and size, measured by total cargo, for all of Port Authorities. These results are similar to those obtained by González and Trujillo (2008).

Regarding to allocative inefficiency indexes (Z_{hj}) we follow Sect. 4.2. in order to demonstrate whether Port Authorities minimize costs or not. Table 4 presents the mean values of port authority-specific measures of allocative efficiency for the 1986–2005 period. Thus, 12 Port Authorities exhibit mean values greater than 1, showing the existence of under-utilization of capital relative to labour (Algeciras, Barcelona, Bilbao, Castellon, Gijon, Huelva, Coruna, Palmas, Baleares, Tenerife, Tarragona and Valencia). These Port Authorities present similar characteristics. In this way, Palmas, Baleares and Tenerife are located in islands. The rest of Port Authorities are characterized to be the largest ones (see Fig. 3). From the second and third columns of Table 4 we conclude that intermediate consumption is over-utilized relative to capital and labour. Furthermore, as we can observe in Figs. 4 and 5, for the largest Port Authorities these allocative inefficiencies are smaller.

Finally, in Fig. 6 we present the evolution of allocative efficiency in the infrastructure services of Spanish ports during the period 1986–2005. As we see, the trend of the three series changes in 1992, coinciding with the new public regulation reform which transformed the centralized ports councils into new more decentralized public entities, as it was said in Sect. 2. In this way, before of 1993, capital and intermediate consumption were over-utilized relative to labour Port Authorities, and intermediate consumption was over-utilized relative to capital. The trend of these three series (Z_{ij}s) in this period tended to reach an allocative

[5]Baños-Pino et al. (1999) considered as quasi-input number of lineal meters of a port land surface of more than 4 meters, instead of total deposit area. However, with our data, both variables are highly correlated.

Table 3 Port Authorities technical inefficiencies

	Port Authority-specific parameter	Technical efficiency score	t-statistic
Algeciras	0.595	0.644	11.54
Alicante	0.618	0.630	12.14
Almeria	0.620	0.629	11.82
Aviles	0.579	0.655	13.25
Cadiz	0.345	0.828	15.48
Barcelona	0.306	0.861	21.44
Bilbao	0.164	0.992	16.25
Cartagena	0.617	0.631	11.71
Castellon	0.481	0.723	12.26
Ceuta	0.933	0.460	9.77
Ferrol	0.695	0.583	10.35
Gijon	0.579	0.655	12.31
Huelva	0.408	0.777	11.39
Coruna	0.441	0.752	11.19
Palmas	0.257	0.903	17.66
Malaga	0.543	0.679	11.42
Melilla	1.227	0.342	7.36
Baleares	0.426	0.763	14.44
Pasajes	0.489	0.716	15.18
Pontevedra	0.730	0.563	10.01
Tenerife	0.208	0.949	15.89
Santander	0.357	0.817	14.59
Sevilla	0.495	0.713	13.00
Tarragona	0.473	0.728	11.83
Valencia	0.156	1.000	2.85
Vigo	0.353	0.821	14.20
Villagarcia	0.897	0.477	8.76

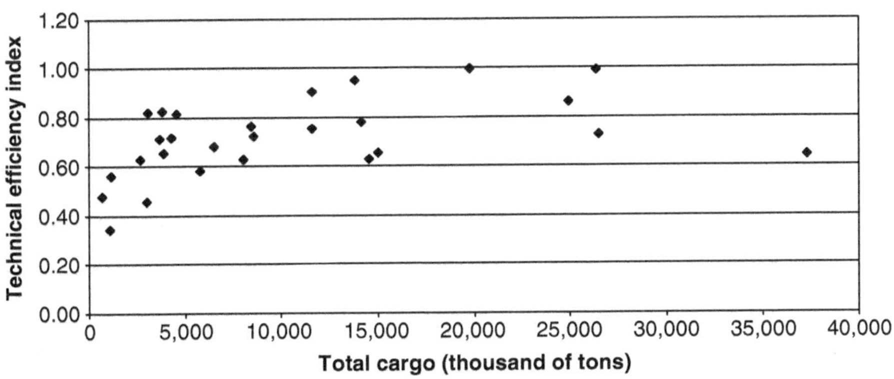

Fig. 2 Relation between technical inefficiency scores and size of Port Authorities

efficiency situation (mean value of Z_{ij}s equal to 1). However, after 1993 capital became under-utilized relative to labour, and the over-utilization of intermediate consumption relative to capital and labour increased. The reforms initiated in 1997

Table 4 Average Port Authority allocative inefficiencies Z_{hj}

	Zkl	Zcik	Zcil
Algeciras	1.224	0.398	0.449
Alicante	0.458	0.600	0.239
Almeria	0.572	0.426	0.235
Aviles	0.549	0.635	0.288
Cadiz	0.846	0.246	0.177
Barcelona	1.105	0.806	0.848
Bilbao	2.148	0.513	1.026
Cartagena	0.892	0.242	0.180
Castellon	1.002	0.384	0.388
Ceuta	0.498	0.510	0.235
Ferrol	0.675	0.264	0.202
Gijon	1.150	0.298	0.350
Huelva	1.269	0.423	0.493
Coruna	1.202	0.315	0.390
Palmas	1.343	0.453	0.463
Malaga	0.779	0.289	0.218
Melilla	0.515	0.332	0.158
Baleares	1.050	0.358	0.410
Pasajes	0.550	0.603	0.262
Pontevedra	0.531	0.335	0.164
Tenerife	1.779	0.372	0.642
Santander	0.922	0.494	0.404
Sevilla	0.876	0.408	0.387
Tarragona	1.288	0.483	0.612
Valencia	1.329	0.524	0.684
Vigo	0.642	0.687	0.376
Villagarcia	0.396	0.330	0.106

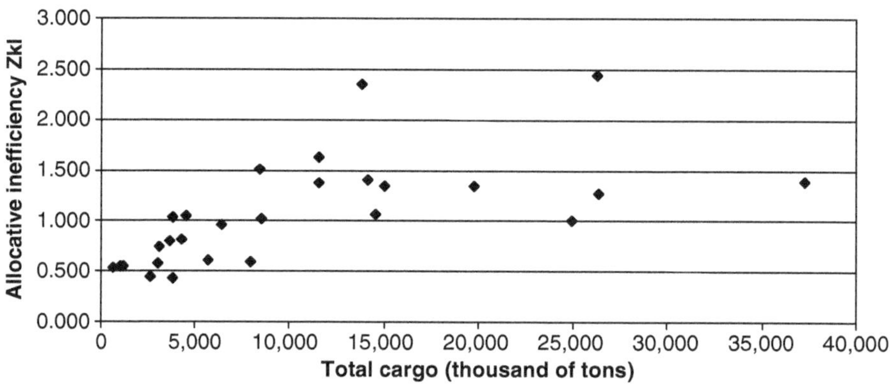

Fig. 3 Relation between allocative inefficiency scores Z_{kl} and size of Port Authorities

and 2003 did not suppose a significant change on the trend of Z_{ij}s, continuing the overcapitalization process relative to labour. The only change that we can observe is the reduction of allocative inefficiency of the intermediate consumption with regard to the rest of the inputs (Table 5).

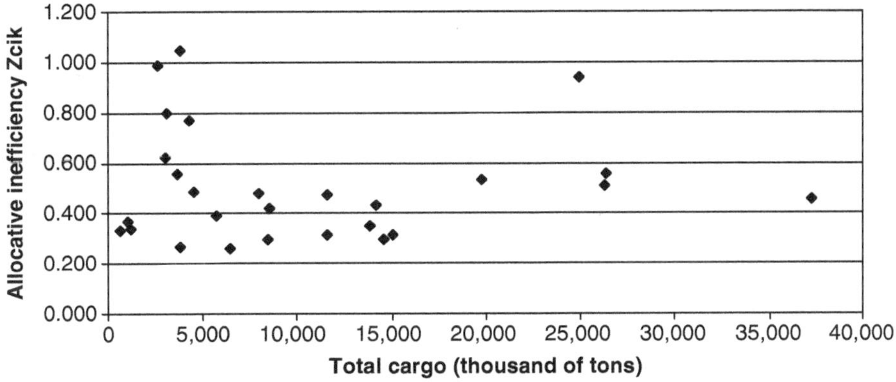

Fig. 4 Relation between allocative inefficiency Z_{cik} scores and size of Port Authorities

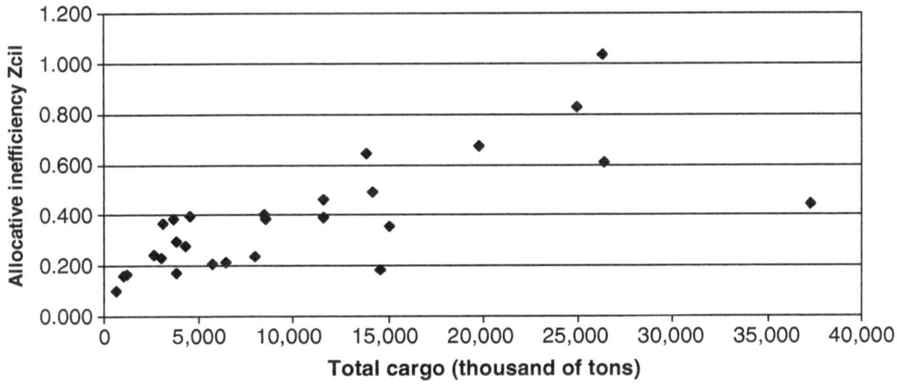

Fig. 5 Relation between allocative inefficiency scores Z_{cil} and size of Port Authorities

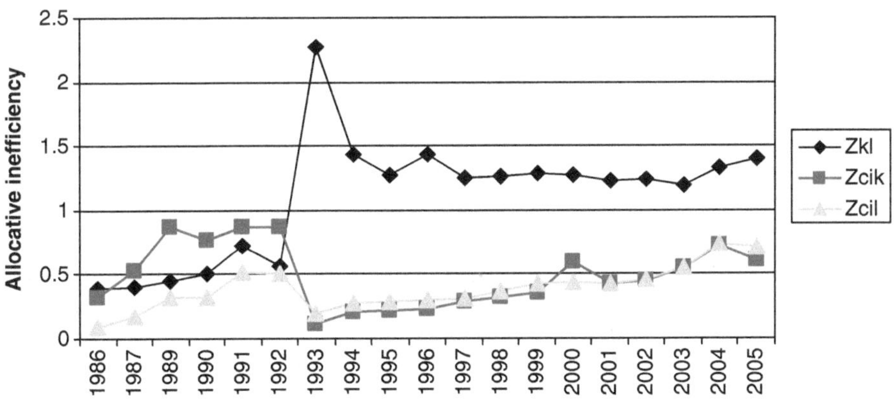

Fig. 6 Average allocative inefficiencies for each pair of inputs for the period 1986–2005

Table 5 Port Authority technical change for the period 1986–2005

	Technical change	t-statistic
1986–1987	−0.168	−5.170
1987–1988	−0.089	−2.730
1988–1989	−0.019	−0.570
1989–1990	−0.011	−0.330
1990–1991	−0.018	−0.550
1991–1992	0.001	0.030
1992–1993	−0.283	−8.430
1993–1994	0.070	2.150
1994–1995	0.027	0.840
1995–1996	−0.049	−1.510
1996–1997	0.029	0.900
1997–1998	0.001	0.030
1998–1999	−0.001	−0.030
1999–2000	−0.021	−0.630
2000–2001	0.007	0.210
2001–2002	−0.018	−0.560
2002–2003	0.010	0.310
2003–2004	−0.034	−1.060
2004–2005	−0.025	−0.760

6 Conclusions

In this paper we have calculated technical and allocative efficiency measures in the infrastructure services of Spanish ports during 1986–2005. Such years did coincide with a period of public reforms that affected the structure of the sector. We have been able to estimate a system of equations consisting of a multioutput translog input distance function and cost shares equations. We have used data of 27 Port Authorities from 1986 to 2005, being the final panel data set of 540 observations.

In a descriptive analysis we showed that average of Port Authorities input cost shares showed little variability among them. However, during the 1992 and 1993, there was an important increase of capital cost share and a decrease of intermediate consumption and labour cost shares, so capital turned into the most important input in terms of total costs. During the next years, the importance of capital declined regard to the other inputs.

We obtained as a result that marginal productivity of the quasi-fixed input, proxied by total deposit area, is positive.

The calculation of a technical efficiency measure shows that the most efficient Port Authorities are those which are larger in size.

Regarding to the allocative efficiency measure we demonstrate that Port Authorities do not minimize their costs. Furthermore, the largest Port Authorities and those located in islands present an under-utilization process of capital relative to labour, and the smallest Port Authorities an over-use of capital relative to labour.

Finally the trend of allocative inefficiencies for the mean of Port Authorities changes after the main public reform of 1992. In this way after 1992 capital becomes under-utilized relative to labour and intermediate consumption is over-utilized relative to capital.

References

Atkinson SE, Primont D (2002) Stochastic estimation of firm technology inefficiency, and productivity growth using shadow cost and distance functions. J Econom 108:203–225

Baños-Pino J, Coto-Millán P, Rodríguez-Álvarez A (1999) Allocative efficiency and over-capitalization: an application. Int J Transp Econ 26(2):181–199

Barros C (2003) Incentive regulation and efficiency of Portuguese port authorities. Marit Econ Logist 5(1):55–60

Castillo-Manzano JI, López-Valpuesta L, Pérez JJ (2008) Economic analysis of the Spanish port sector reform during the 1990s. Transp Res A 42:1056–1063

Coto-Millán P (1996) Maritime transport policy in Spain (1974–1995). Transp Policy 3(1):37–41

Coto-Millán P, Baños-Pino J, Rodríguez-Álvarez A (2000) Economic efficiency in Spanish ports: some empirical evidence. Marit Policy Manage 27(2):169–174

Cullinane KPB, Song DW, Gray R (2002) A stochastic frontier model of the efficiency of major container terminals in Asia: assessing the influence of administrative and ownership structures. Transp Res A 36:743–762

Díaz-Hernández JJ, Martínez-Budría E, Jara-Díaz S (2008a) Productivity in cargo handling in Spanish ports during a period of regulatory reforms. Netw Spat Econ 8:287–295

Díaz-Hernández JJ, Martínez-Budría E, Jara-Díaz S (2008b) The effects of ignoring inefficiency in the analysis of production: the case of cargo handling in Spanish ports. Transp Res A 42:321–329

Estache A, Tovar de la Fé B, Trujillo L (2004) Sources of efficiency gains in port reform: a DEA decomposition of a Malmquist TFP index for Mexico. Util Policy 12:221–230

Färe R, Grosskopf S (1990) A distance function approach to price efficiency. J Public Econ 43:123–126

González MM, Trujillo L (2007) Efficiency measurement in the port industry: a survey of the empirical evidence. Discussion Paper Series No. 07/08. City University London

González MM, Trujillo L (2008) Reforms and infrastructure efficiency in Spain's container ports. Transp Res A 42(1):243–257

Grosskopf S, Hayes K, Hirschberg J (1995) Fiscal stress and the production of public safety: a distance function approach. J Public Econ 57:277–296

Jacobsen S (1972) On Shephard's duality theorem. J Econ Theory 4:458–464

Jara-Díaz SR, Martínez-Budría E, Cortés CE, Basso L (2002) A multioutput cost function for the services of Spanish port's infrastructure. Transportation 29:419–437

Liu Z (1995) The comparative performance of public and private enterprise. the case of British ports. J Transp Econ Policy 29(3):263–274

Puertos del Estado (2006) Informe de Gestión del Sistema Portuario de Titularidad Estatal 2005. Ministerio de Fomento

Rodríguez-Álvarez A, Tovar B, Trujillo L (2007) Firm and time varying technical and allocative efficiency: an application to port cargo handling firms. Int J Prod Econ 109:149–161

Analysis of Technical Efficiency and Rate of Return on Investment in Ports

Vicente Inglada and Pablo Coto-Millán

Abstract This article carries out a comparative technical efficiency analysis of investment in 27 Spanish ports using panel data over 20 years (1986–2005) and Data Envelopment Analysis (DEA) methodology. Results from the Tobit model suggest that the size of port has a negative impact on the total technical and scale efficiency of the ports. The results also show that Spanish ports have experienced an improvement in the total production factor which is due exclusively to the technical efficiency change since the average annual change corresponding to the technological change is negative. Finally, another result is that ports with a greater specialisation in general cargo (containerized and non-containerized general cargo) make significantly greater progress in technological change during the period analysed.

1 Introduction

The basic objective of this work is to analyse the efficiency and productivity of investment in Spanish ports and its development over time, as well as the returns to scale in each port corresponding to this investment.

Spanish ports are characterised by the existence of a high differentiation in size. Furthermore, in the last few decades significant amounts of investment have been set aside for their expansion and improvement. For this reason there is great interest in analysing, not only the efficiency in the distribution of these investments, through estimating the degree of efficiency of the productive stock of each port, by breaking

V. Inglada (✉)
University of Complutense, Madrid, Spain
e-mail: vinglada@ccee.ucm.es

P. Coto-Millán
University of Cantabria, Santander, Spain

P. Coto-Millán et al. (eds.), *Essays on Port Economics*, Contributions to Economics,
DOI 10.1007/978-3-7908-2425-4_18,

down the efficiency into its two components, associated, respectively, with the scale and the purely technical, but also in studying the relation between efficiency and the characteristics of each port.

In this research the magnitude of the total technical efficiency is obtained for Spanish ports in the period 1986–2005, as well as that of their two components: pure technical efficiency and scale efficiency. All of this allows us to determine the type of returns to scale in each port and serves as an invaluable aid in analysing the relationships between the levels of efficiency obtained, with the dimensions and other specific characteristics of each port, carrying out the estimation using a Tobit model.

Also, the changes in productivity produced in investment, technological change, pure technical efficiency change and scale efficiency change during the period 1986–2005 for the 27 ports are analyzed, using a non-parametric methodology of analysis, that allows the estimation of Malmquist indices of total factor productivity (TFP) change and its subsequent breaking down into indicators of technical efficiency change and technological change.

Although there are numerous works which analyse the technical efficiency of ports, in Spain as well as in other countries,[1] there are very few studies which, like this article, consider a time span as long as 20 years, use such a high number of outputs, limit themselves exclusively to the input of investment[2] and lastly, which study in detail the relationships between the characteristics of each port and the results obtained concerning their levels of efficiency and the returns of scale of the productive port stock.

The methodological tool used is called the Data Envelopment Analysis (DEA), a technique of non-parametric linear programming appropriate for estimating efficiency and returns of scale by means of constructing the most efficient frontier. This methodology has, among other things, an advantage in its flexibility, by not imposing any functional forms on the technological frontier, which is built from the most efficient practices and their linear combinations.

With these objectives, the structure of this research is as follows. In Sect. 2 the methodology used for calculating the change in productivity and the indices of technical efficiency and of scale is described as well as a brief description and analysis on sources of dates used in this study. Results on efficiency and temporal variation of the productivity of port investment, respectively, and their relationships with the specific characteristics of each port, are presented and discussed in Sects. 3, 4 and 5. In Sect. 6 the main conclusions are presented.

[1] In González-Serrano and Trujillo (2005) there is an exhaustive review of the work and research carried out on port efficiency.

[2] In this line, the article by Baños-Pino et al. (1999) on the dimension of port stock should be highlighted.

2 Methodology and Data Used

2.1 DEA Methodology

A company that employs K inputs is considered: $\overline{X} =$ (X1, X2 .... Xk), which are available at fixed prices, to produce an output Y, that can be sold at a fixed price p. The efficiency measurements of the units investigated are obtained by comparing the values observed for each productive unit with the optimum defined by the estimated border. In literature, this function $f(\overline{X})$ is called the frontier function since it characterises the optimizing behaviour of an efficient producer and, therefore, marks the limits of the possible values of its respective dependent variable (output).

A production process can be based on the hypothesis of constant returns to scale (CRS), although in the majority of situations it is more appropriate to assume a technology with variable returns to scale. This occurs, for example, because the size of the units can affect the average productivity. When there are large disparities in the size of the production units, a fact that we will see verified as happening in Spanish ports, it is advisable to compare each unit with others that are similar in production scale since the existing difference in the magnitude of inefficiency between units could be due to scale.

If we assume that the company produces the quantity Y_0 with a quantity of production units $\overline{X}^o$, a measurement of the technical efficiency of this allocation would be $\frac{Y^o}{f(\overline{X}^o)} \in [0, 1]$.

The technical efficiency index supports two specifications, depending on what is taken as a reference for the level of output or inputs. In the first model which is output orientated, the proportion in which the output level can be increased with the quantities of inputs used is measured while in the second model, which is input-orientated, the proportion in which the inputs can be reduced to continue producing the same quantity of output is measured. In a technology of constant returns to scale, the two versions of the index would coincide.

Ultimately, it is feasible to minimize the use of inputs, given the outputs, or to maximize the outputs, given the inputs. For example, in the case of utilities, infrastructures or public utilities, the output is related to the existing demand which is, in great measure, an exogenous variable to the company, while saving costs is one of the prime objectives of the company. Because of this, these cases provide the conditions suitable to develop an input-orientated projection model which is the one that will be used in this research.

Considering a broader context, technical efficiency has two multiplicative components $TE_{CRS} = TE_{VRS} \times SE$; with TE_{VRS} being the pure technical efficiency and SE the scale efficiency. This procedure to break down the technical efficiency index into two parts is based on the comparison of the measurements obtained by imposing the assumption of constant returns to scale and those resulting from allowing the existence of variable returns. In the case that the two measurements coincide, the company would be operating at optimum scale.

Charnes et al. (1978) provides the methodological technique called Data Envelopment Analysis (DEA) that allows the relaxation of assumptions such as that of constant returns and the contemplation of more general cases like that of multi-product technology. By means of this technique a frontier production of reference is built using mathematical programming methods from efficient productive units and linear combinations of the same. The measurements of efficiency will be the distances that separate each company from the frontier. This DEA methodology is also used to determine the type of returns to scale. It consists basically in identifying the non-parametric linear polygonal frontier that would represent the optimum process in the transformation of a set of inputs into the final output. A description of the DEA methodology is carried out in Cooper et al. (2000) and there are numerous studies in literature on DEA methodology which identify economies of scale.[3]

This type of non-parametric frontier has been used profusely in empirical literature, fundamentally in the banking sector, the railways and airlines.[4]

Below the exercise of linear programming which underlies the input-orientated DEA models is presented with variable returns to scale which will be the one used in this research.

Given N productive units (j = 1,. . ., N) that produce S outputs (r = 1,. . ., S) using M inputs (i = 1,. . ., M), the magnitudes of the efficiency of each unit j_0 under variable returns to scale (VRS) and an input-orientated DEA model can be estimated by solving the following linear programming problem:

$$Min\,\vartheta_{\mathrm{j_0}}^{\mathrm{VRS}}$$

$$Subjectto:$$

$$\sum_{\mathrm{j=1}}^{\mathrm{N}} \lambda_{\mathrm{j}} y_{rj} - y_{rj0} \geqslant 0$$

$$\theta_{j_0}^{VRS} x_{ij_0} - \sum_{\mathrm{j=1}}^{\mathrm{N}} \lambda_{\mathrm{j}} y_{rj} \geqslant 0$$

$$\sum_{\mathrm{j=1}}^{\mathrm{N}} \lambda_{\mathrm{j}} = 1$$

$$\lambda_j \geqslant 0 \;\mathrm{j} = 1, \ldots\ldots, \mathrm{N},\; \theta_{j_0}^{VRS} free$$

The non-negative weights λ measure the contribution of the efficient units selected to define a point of reference for the inefficient unit j_0. The convexity

[3]For example, Färe et al. (1994,1998); Charnes et al. (1996); Coelli (1998); Nishimizu and Page (1982); and Bauer (1990).

[4]For example, Inglada et al. (2004); Coto-Millán et al. (1999); Rey et al. (2009); Domenech (1992); Pastor (1995); Forsund and Hjalmarsson (1979) and Simar (1992).

restriction $\sum_{j=1}^{N} \lambda_j = 1$ *requires the evaluation of the efficiency under* variable returns to scale (VRS).

The optimum solution $\hat{\theta}_{j_0}^{VRS}$ is non-negative and less than 1, being equal to the unit when the total efficiency is obtained. The magnitude $1 - \hat{\theta}_{j_0}^{VRS}$ represents the potential reduction in the inputs, given the output.

2.2 Malmquist Indices

One way of studying the evolution of the factor productivity over time is by means of so-called Malmquist indices. These indices of total factor productivity (TFP) change have the advantage of being able to break down TFP change as a combination of the technological change and the technical efficiency change (relative to CRS technology). Technical efficiency change measures the ability to make the best use of available technology, while technological change refers to an improvement or deterioration in the state of technology. Also, Malmquist indices do not require data on prices in contrast with other indices of productivity.

Malmquist indices are calculated by means of the geometric average of the ratios of factor productivity in the periods t and t + 1. Thus, the Malmquist index between the periods t and $t + 1$ is:

$$M_I^{t,t+1}(X_t, Y_t; X_{t+1}, Y_{t+1}) = \left[\frac{D_I^t(X_{t+1}, Y_{t+1})}{D_I^t(X_t, Y_t)} \times \frac{D_I^{t+1}(X_{t+1}, Y_{t+1})}{D_I^{t+1}(X_t, Y_t)}\right]^{\frac{1}{2}}$$

where the X_t and Y_t are the inputs and outputs in period t, respectively.

$D_I^t(X_{t+1}, Y_{t+1})$ is the distance between the observation in period $t + 1$ and the frontier technology in period t, and is defined by: $D_I^t(X_{t+1}, Y_{t+1}) = \max\{\lambda : (X_t/\lambda) \in L_t(Y_t)\}$ where λ measures the maximum reduction of the inputs and $L_t(Y_t)$ is the set of inputs.

This index would represent the total factor productivity change of the production point (x_{t+1}, y_{t+1}) relative to the production point (x_t, y_t). A value greater than 1 indicates positive growth of productivity between the period t and t + 1. This index is in fact the geometric mean of two Malmquist TFP indices. One index uses the technology of period t and the other that of period t + 1. To solve this equation it is therefore necessary to estimate the four functions of corresponding distance, which brings with it four problems of similar linear programming to those existing in the calculation of the technical efficiency measurements.

The Malmquist index of total factor productivity change can also be expressed in the following way, which allows us to break it down into two components.

$$M_I^{t,t+1}(X_t, Y_t; X_{t+1}, Y_{t+1}) = \frac{D_I^t(X_{t+1}, Y_{t+1})}{D_I^t(X_t, Y_t)} \left[\frac{D_I^{t+1}(X_{t+1}, Y_{t+1})}{D_I^{t+1}(X_{t+1}, Y_{t+1})} \frac{D_I^t(X_t, Y_t)}{D_I^{t+1}(X_t, Y_t)}\right]^{\frac{1}{2}}$$

The first component is the technical efficiency change (relative to CRS technology): $\frac{D_I^t(X_{t+1},Y_{t+1})}{D_I^t(X_t,Y_t)}$

The second component is the technological change: $\left[\frac{D_I^{t+1}(X_{t+1},Y_{t+1})}{D_I^{t+1}(X_{t+1},Y_{t+1})}\frac{D_I^t(X_t,Y_t)}{D_I^{t+1}(X_t,Y_t)}\right]^{\frac{1}{2}}$

The expression above can be expanded by breaking down the total technical efficiency change (CRS) into its two components of scale efficiency change and pure technical efficiency change (VRS). For this it is necessary to estimate two new linear programming exercises by considering the distance functions relating to technology of variable returns to scale instead of constant returns to scale.

2.3 Analysis of the Data Used

The data used in this research is made up of a panel of 27 ports observed during the period 1986 to 2005. The following outputs have been considered: liquid bulk; solid bulk; containerized general cargo; non-containerized general cargo and number of passengers.[5] The port stock measured by assets has been considered as input.

Tables 1 and 2 show that there is a large concentration of investment and different types of traffic in certain ports. Thus, for example, the stock of the ports of Bilbao, Barcelona and Valencia represents almost 30% of the total port stock. In relation to the traffic, it would be worth highlighting the high specialisation that exists as shown by the fact that for each type of traffic there is a different leading port.

These results regarding the great disparity of dimensions and specialization in Spanish ports suggests the advisability of analysing in detail the implications of the dimension scale variable on the technical efficiency.

3 Efficiency Analysis

3.1 Returns to Scale and Scale Efficiency

The DEA methodological focus has been used which is described in Sect. 2 of this research to determine the type of returns to scale and the magnitudes of the scale efficiency in each port.

Table 3 shows that the ports with constant returns to scale are those with higher scores in their scale efficiency. Likewise, Tables 3 and 4 show the ranking and levels of scale efficiency obtained in the ports analyzed, as an average for the period

[5] Series are taken from "Puertos del Estado" (Spain's Public Works Ministry; http://www.fomento.es/).

Table 1 Magnitudes of the various types of outputs used expressed in % with respect to the total ports and classified by size (assets) (average for 1986–2005)

Ports	Assets (2001 constant Euros)	Liquid bulk (ton)	Solid bulk (ton)	Containerized general cargo (ton)	Non-containerized general cargo (ton)	Passengers (number)
Bilbao	11.05	12.18	6.54	6.42	8.85	0.57
Barcelona	10.84	6.80	5.34	18.26	9.69	6.43
Valencia	8.04	1.25	5.39	20.58	11.05	1.83
Las Palmas	6.90	2.93	1.30	8.99	7.33	5.27
Tenerife	5.53	6.55	1.49	4.54	6.91	20.60
Huelva	5.12	7.20	6.14	0.00	1.63	3.24
Gijon	5.06	1.09	16.98	0.15	1.12	0.00
Tarragona	4.81	14.48	10.08	0.54	1.60	0.04
Sevilla	4.65	0.19	3.03	0.93	1.76	0.01
Algeciras	4.14	14.36	2.69	29.80	7.94	24.77
Santander	4.03	0.41	4.05	0.06	2.66	0.82
Baleares	3.68	1.76	1.77	2.86	10.17	11.62
Cadiz	3.59	0.38	1.60	1.52	4.02	0.61
Coruña	2.68	6.81	3.68	0.01	1.02	0.13
Pasajes	2.49	0.34	2.67	0.12	5.14	0.00
Cartagena	2.40	9.35	3.18	0.55	0.86	0.03
Vigo	2.24	0.35	0.64	2.03	3.40	0.15
Malaga	2.08	3.85	1.53	0.09	1.35	1.72
Alicante	1.86	0.36	1.58	1.30	1.03	1.05
Aviles	1.50	0.49	2.48	0.03	3.87	0.00
Castellon	1.44	5.60	1.38	0.45	1.15	0.00
Almeria	1.35	0.60	8.78	0.01	1.35	3.33
Ceuta	1.27	1.76	0.07	0.10	2.16	15.16
Ferrol	1.03	0.40	6.32	0.00	0.97	0.01
Melilla	0.97	0.35	0.14	0.23	1.33	2.48
Pontevedra	0.76	0.01	0.72	0.42	1.16	0.12
Villagarcia	0.49	0.15	0.43	0.00	0.49	0.02

Note: *CRS* Constant returns to scale, *VRS* Variable returns to scale

1986–2005, assuming an input-orientated model. It is worth noting that the greatest magnitudes (1) are reached for the ports of Algeciras, Almería, Castellón, Ceuta and Pasajes. These five ports, where the score of scale efficiency is maximum, are the only ones that show constant returns to scale. Likewise, the majority of the nine ports with decreasing returns to scale are among the group of larger ports (most stock).

To find out the impact of size and other variables such as port characteristics, the Tobit analysis is carried out on efficiency scores. Therefore, the DEA efficiency scores were submitted to the Tobit model, whereby the regressors are size and other port characteristics, in order to test that, for example, the size might have an adverse impact on technical efficiency. That model is chosen because the dependent variable (efficiency scores) is restricted by having a value between zero and one.

Table 2 Magnitudes of the various types of cargos, expressed in % with respect to the total of cargo for each port and classified by containerized general cargo. (average for 1986–2005)

Ports	Liquid bulk (ton)	Solid bulk (ton)	Containerized general cargo (ton)	Non-containerized general cargo (ton)
Valencia	7.75	21.13	51.97	19.16
Algeciras	47.26	5.59	39.86	7.29
Las Palmas	31.00	8.68	38.67	21.65
Barcelona	33.51	16.61	36.57	13.32
Vigo	13.89	16.03	32.56	37.51
Alicante	16.68	45.83	24.26	13.23
Cadiz	12.12	32.34	19.73	35.81
Pontevedra	0.93	47.28	17.93	33.86
Baleares	25.62	16.24	16.88	41.25
Tenerife	58.15	8.33	16.40	17.12
Sevilla	6.52	64.33	12.68	16.47
Bilbao	56.94	19.32	12.20	11.55
Melilla	38.85	9.68	10.48	40.99
Castellon	80.26	12.51	2.62	4.61
Cartagena	79.10	16.97	1.89	2.03
Ceuta	71.93	1.77	1.65	24.66
Pasajes	9.73	48.03	1.42	40.82
Tarragona	67.32	29.58	1.02	2.08
Malaga	73.57	18.50	0.71	7.22
Santander	10.89	68.61	0.60	19.90
Gijon	8.98	87.95	0.51	2.56
Aviles	15.47	49.73	0.42	34.38
Almeria	9.13	85.02	0.05	5.80
Coruna	72.29	24.65	0.03	3.02
Ferrol	8.62	85.58	0.02	5.79
Huelva	62.44	33.62	0.00	3.93
Villagarcia	26.39	48.86	0.00	24.75

For estimation purposes, the technique of maximum likelihood is used. The results of the model are presented in Table 6.

It can be seen in Table 6 that the size of the port has a negative influence on scale efficiency while, in contrast, the specialization of a port in passenger and liquid bulk traffic has a positive effect.

In order to corroborate the relevance of the size variable in explaining the magnitude of the scale efficiency, Fig. 1, where both variables and the linear trend are represented, shows that there is a negative relationship between the levels of scale efficiency obtained by the ports analyzed and the magnitudes of their stock. In this sense, the less efficient ports correspond in general to those of greater size.

Another relevant magnitude for the study of the scale efficiency is its variation in each port during the period analyzed, obtained from the Malmquist indices as described in Sect. 2. As can be seen in Table 7 the average magnitude of the scale efficiency has increased during this period with an annual growth rate of 0.4%. Likewise, from Table 8, which shows the main results from the estimated

Table 3 Classification of ports in each type of investment efficiency in order of their total technical efficiency (constant returns) (panel 1986–2005)

Ports	Total technical efficiency (constant)	Pure technical efficiency (variable)	Scale efficiency	Returns to scale
Algeciras	1	1	1	crs
Almeria	2	2	2	crs
Castellon	3	5	3	crs
Ceuta	4	6	4	crs
Pasajes	5	8	5	crs
Tarragona	6	10	13	drs
Aviles	7	13	7	irs
Cartagena	8	14	8	drs
Valencia	9	11	21	drs
Baleares	10	18	11	irs
Pontevedra	11	9	23	irs
Santander	12	19	14	irs
Malaga	13	21	9	irs
Las Palmas	14	22	10	irs
Alicante	15	16	19	irs
Melilla	16	17	20	irs
Vigo	17	20	16	irs
Ferrol	18	15	22	irs
Barcelona	19	3	24	drs
Tenerife	20	24	6	drs
Coruña	21	23	15	drs
Bilbao	22	4	25	drs
Villagarcia	23	12	26	irs
Huelva	24	26	12	drs
Cadiz	25	25	17	irs
Gijon	26	7	27	drs
Sevilla	27	27	18	irs

Tobit model to obtain the effects of size and specialization of the port, it can be deduced that size has an adverse effect on scale efficiency change while the specialization of the port in liquid bulk and passenger traffic has a positive effect on this variable.

4 Technical Efficiency

4.1 Pure Technical Efficiency

Below, the analysis of the so-called pure technical efficiency variable for each port is presented, which corresponds to the technical efficiency once the effects of scale have been eliminated.

Tables 3 and 4 show the ranking and scores of pure technical efficiency obtained for the ports analyzed, as an average for the period 1986–2005, assuming technology

Table 4 Scores for the various types of investment efficiency in ports classified by their total technical efficiency (constant returns) (panel 1986–2005)

Ports	Total technical efficiency (constant)	Pure technical efficiency (variable)	Scale efficiency	Returns to scale
Algeciras	1	1	1	crs
Almeria	1	1	1	crs
Castellon	1	1	1	crs
Ceuta	1	1	1	crs
Pasajes	1	1	1	crs
Tarragona	0.942	1	0.942	drs
Aviles	0.904	0.927	0.975	irs
Cartagena	0.875	0.898	0.974	drs
Valencia	0.7	1	0.7	drs
Baleares	0.63	0.656	0.96	irs
Pontevedra	0.552	1	0.552	irs
Santander	0.544	0.588	0.925	irs
Malaga	0.534	0.549	0.974	irs
Las Palmas	0.515	0.53	0.972	irs
Alicante	0.507	0.686	0.739	irs
Melilla	0.479	0.666	0.719	irs
Vigo	0.479	0.556	0.862	irs
Ferrol	0.467	0.824	0.567	irs
Barcelona	0.465	1	0.465	drs
Tenerife	0.437	0.44	0.993	drs
Coruña	0.435	0.475	0.917	drs
Bilbao	0.426	1	0.426	drs
Villagarcia	0.383	1	0.383	irs
Huelva	0.303	0.322	0.942	drs
Cadiz	0.275	0.325	0.845	irs
Gijon	0.246	1	0.246	drs
Sevilla	0.142	0.172	0.827	irs
Media	0.601	0.764	0.811	

and a production frontier with variable returns to scale, as well as input-orientated. It should be noted that 3 of the 27 ports analyzed reach the maximum level of pure technical efficiency (1) and that the average score is very high (0.811), which shows the high level of pure technical efficiency achieved by Spanish ports associated to optimum port management, without taking into account the effects of scale.

Likewise, Table 5 shows that during the period 1986–2005 an increase in the average pure technical efficiency is generated with an annual growth rate of 0.3%. Also, Tables 7 and 8 show that there is a convergence in relation to the levels of pure technical efficiency in the ports, since in the Tobit model of pure technical efficiency change the coefficient of this variable is negative.

4.2 Total Technical Efficiency

DEA methodology has been applied, described in Sect. 2, from a model with variable returns to scale and input-orientated to determine the magnitudes in each

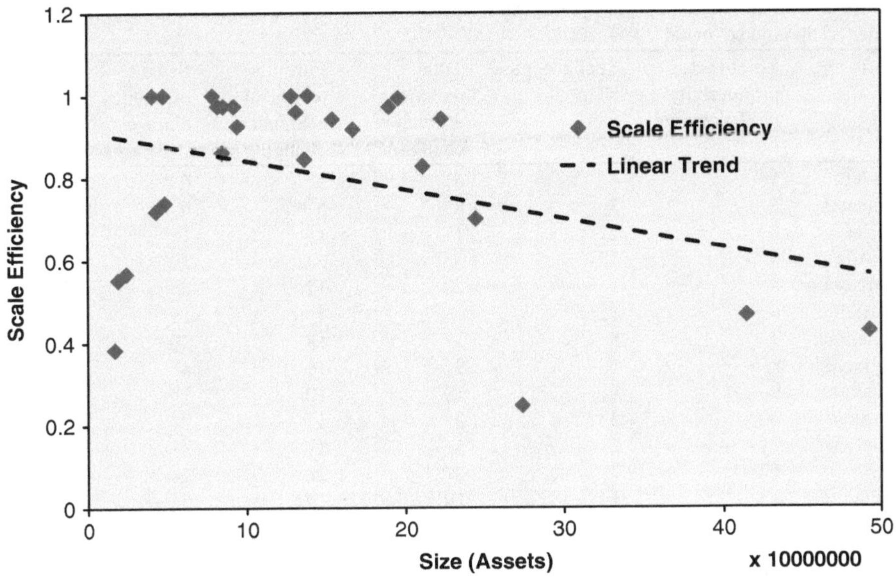

Fig. 1 Relationship between scale efficiency and size of port

port of the total technical efficiency, which is a combination of the two variables described and analyzed previously: pure technical efficiency and scale efficiency.

Tables 3 and 4 show the ranking and levels of total technical efficiency obtained in the ports analyzed, as an average for the period 1986–2005, assuming an input-orientated model. It is worth noting that the highest magnitudes (1) are reached for the same ports as for scale efficiency, which is to say Algeciras, Almería, Castellón, Ceuta and Pasajes.

The results of the Tobit model performed on total efficiency scores are presented in Table 6. It can be seen that the size of the port has a negative influence on total technical efficiency while the specialization of the port in passengers and liquid bulk traffic has a positive effect.

In order to corroborate the relevance of the size variable in explaining the magnitude of the total technical efficiency, both variables and the linear trend are represented in Fig. 2, which shows that there is a negative relationship between the levels of total technical efficiency obtained by the ports analyzed and the magnitudes of their stock. In this sense, the less efficient ports correspond in general to those of greater size.

Another relevant magnitude for the study of the total technical efficiency is its variation in each port during the period analyzed, obtained from the Malmquist indices as described in Sect. 2. As can be seen in Table 7 the average magnitude of the total technical efficiency has increased during this period with an annual growth rate of 0.7%. Likewise, from Table 8, which shows the main results from the estimated Tobit model to obtain the effects of size and specialization of the port,

Table 5 Classification of ports by total factor productivity (TFP) change and its components, in order of this index (panel 1986–2005)

Ports	Total factor productivity (TFP) change	Technological change	Total technical efficiency change	Pure technical efficiency change	Scale efficiency change	Returns to scale
Cadiz	1	4	6	5	6	irs
Baleares	2	3	8	6	9	irs
Gijon	3	27	1	14	1	drs
Sevilla	4	18	3	2	10	irs
Barcelona	5	1	20	12	25	drs
Tenerife	6	15	7	4	19	drs
Valencia	7	5	17	18	20	drs
Pontevedra	8	10	9	16	4	irs
Algeciras	9	8	13	10	13	–
Coruña	10	23	2	1	7	drs
Huelva	11	21	5	3	8	drs
Vigo	12	7	21	21	24	irs
Villa	13	14	10	19	5	irs
Las Palmas	14	6	23	24	17	irs
Ceuta	15	11	15	13	16	–
Pasajes	16	12	16	15	18	–
Cartagena	17	17	11	8	11	drs
Melilla	18	2	27	20	27	irs
Bilbao	19	16	12	27	2	drs
Ferrol	20	26	4	7	3	irs
Aviles	21	13	19	9	23	irs
Alicante	22	9	25	26	14	irs
Tarragona	23	22	18	17	21	drs
Castellon	24	20	22	22	22	–
Santander	25	19	24	25	12	irs
Almeria	26	25	14	11	15	–
Malaga	27	24	26	23	26	irs

Table 6 Impact of port characteristics on scale and technical efficiency (panel 1986–2005)

	Scale efficiency	Pure technical efficiency	Total technical efficiency
Size (assets)	−0.028529**	n.s.	−0.036117***
Liquid bulk	0.002725***	n.s.	0.001625**
Solid bulk	n.s.	n.s.	n.s.
Containerized general cargo	n.s.	n.s.	n.s.
Non-containerized general cargo	n.s.	n.s.	n.s.
Passengers	0.009929	n.s.	0.006953**

,*: Significance at 10%, 5%, 1%

n.s. Non-significance

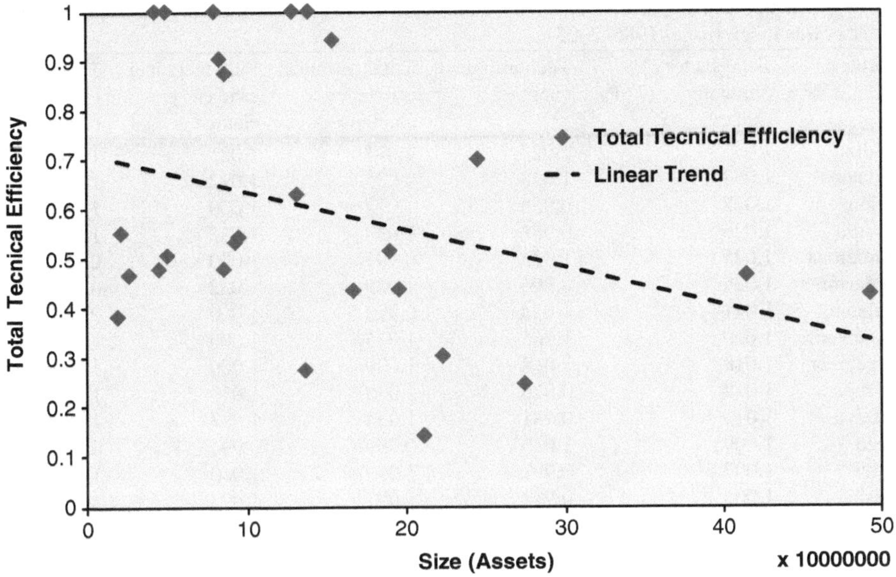

Fig. 2 Relationship between total tecnical efficiency and size of port

it can be deduced that size of port stock has an adverse effect on scale efficiency change while the specialization of the port in liquid bulk and passenger traffic has a positive effect on this variable.

5 Change in Productivity: Malmquist Indices

5.1 Malmquist Indices

To get an overall view of the productive technical efficiency of investment in Spanish ports it is necessary to complete the information about the levels of productive efficiency obtained, with the change or improvement in productivity compared to investment over the period of the study.

In this section the performance of port investment in Spain is analyzed, from a technological point of view, over the period analyzed. To investigate the evolution in time of the productivity of the capital stock of the main Spanish ports, the Malmquist indices of productivity have been obtained, using data from 1986 to 2005.[6] These indices have the advantage of being able to break down the total factor productivity (TFP) change into two components: technical efficiency change

[6]Tthe DEAP 2.1 program developed by Coelli (1996) was used.

Table 7 Magnitudes in ports of total factor productivity (TFP) change and its components, in order of this index (panel 1986–2005)

Ports	Total factor productivity (TFP) change	Technological change	Total technical efficiency change	Pure technical efficiency change	Scale efficiency change
Cádiz	1.055	1.023	1.031	1.025	1.007
Baleares	1.053	1.028	1.025	1.022	1.002
Gijón	1.032	0.958	1.077	1.000	1.077
Sevilla	1.026	0.988	1.038	1.035	1.002
Barcelona	1.025	1.035	0.991	1.000	0.991
Tenerife	1.025	0.996	1.029	1.032	0.998
Valencia	1.021	1.023	0.998	1.000	0.998
Pontevedra	1.020	1.005	1.015	1.000	1.015
Algeciras	1.018	1.018	1.000	1.000	1.000
Coruña	1.018	0.976	1.043	1.039	1.003
Huelva	1.014	0.981	1.034	1.032	1.002
Vigo	1.009	1.019	0.989	0.995	0.994
Villa	1.007	0.997	1.010	1.000	1.010
Media	1.005	0.998	1.007	1.003	1.004
Las Palmas	1.003	1.021	0.983	0.983	1.000
Ceuta	1.002	1.002	1.000	1.000	1.000
Pasajes	1.001	1.001	1.000	1.000	1.000
Cartagena	1.000	0.993	1.007	1.006	1.001
Melilla	1.000	1.029	0.971	0.999	0.972
Bilbao	0.999	0.994	1.006	0.972	1.035
Ferrol	0.999	0.964	1.036	1.009	1.027
Avilés	0.994	0.998	0.996	1.001	0.995
Alicante	0.985	1.009	0.976	0.976	1.000
Tarragona	0.978	0.981	0.997	1.000	0.997
Castellón	0.969	0.982	0.987	0.991	0.996
Santander	0.968	0.984	0.983	0.982	1.001
Almería	0.964	0.964	1.000	1.000	1.000
Málaga	0.948	0.973	0.975	0.991	0.984

relative to CRS technology and technological change (frontier shifts). Total technical efficiency change measures the change in the ability to make the best use of available technology, while the technological change refers to an improvement or deterioration in the state of technology (frontier shifts). Moreover, Malmquist indices have the advantage of not requiring data on prices, in contrast with other indices of productivity. To calculate the Malmquist indices the so-called input-orientated DEA methodology has been used which has been described in Sect. 2 of this research.

Table 5 and 7 report the ranking and results of the Malmquist efficiency change obtained from the ports analyzed, as an average for the period 1986–2005. The TDF has increased during this period with an annual growth rate of 0.5%. Likewise, from Table 8, which shows the main results of the estimated Tobit model to obtain the effects of size and specialization of the port on TDF change, the most striking result is that the specialization of a port in containerized general cargo has a positive effect on the total factor productivity change.

Table 8 Impact of port characteristics on TFP, technological and efficiency change (panel 1986–2005)

	Total factor productivity (TFP) change	Technological change	Total technical efficiency change	Pure technical efficiency change	Scale efficiency change
Size (assets)	n.s.	n.s.	−0.036117**	n.s.	−0.028529**
Liquid bulk	n.s.	n.s.	n.s.	n.s.	0.002725***
Solid bulk	n.s.	−0.000266***	0.000317*	n.s.	n.s.
Containerized general cargo	0.000494**	0.000712***	n.s.	n.s.	n.s.
Non-containerized general cargo	n.s.	0.000774***	n.s.	n.s.	n.s.
Passengers	n.s.	n.s.	n.s.	n.s.	0.009929***
Total technical efficiency	n.s.	n.s.	−0.035997**	n.s.	n.s.
Pure technical efficiency	−0.029912**	n.s.	n.s.	−0.035178**	n.s.
Scale efficiency	−0.040561**	n.s.	n.s.	n.s.	n.s.

*, **, ***: significance at 10%, 5%, 1%
n.s. Non-significance

5.2 *Breakdown of Technological and Technical Efficiency Change*

Table 7 shows the average magnitudes of the total factor productivity (TFP) change of the stock of each port in the period analyzed, as well as its breakdown into its two components: Technological change and technical efficiency change. Likewise, the technical efficiency change (relative to CRS technology) is broken down between its two components: pure technical efficiency change (relative to a VRS technology) and scale efficiency change.

The breakdown of the total factor productivity (TFP) change in the component of technological change, which represents the displacement of the frontier for the output level of each company, and the component of technical efficiency change shows us that the increase in TFP (0.5% annually) is due exclusively to the technical efficiency change that increases 0.7% annually since the other component (technological change) decreases by an annual average rate of change of 0.2%.

Likewise, the technical efficiency change (0.7% annually) is practically due in the same proportion to its two components: Pure technical efficiency change (0.3%) and scale efficiency change (0.4%).

Finally, Table 8 shows that the ports with a greater specialization in general cargo (containerized and non-containerized) are those which have a greater magnitude of technological change during the period analyzed. By contrast, the specialization of ports in solid bulk has a negative effect on technological change. The results shown in Table 7 confirm this assertion since the seven ports with the greatest improvement in technological change (Barcelona, Melilla, Baleares, Cádiz, Valencia, Las Palmas and Vigo) are characterised because more than have of their cargos are general cargos.

6 Conclusions

In this work the behaviour of returns to scale, productivity and the productive efficiency of investment in 27 Spanish ports is analysed, as well as its relationship with the characteristics of each Spanish port during 1986–2005. As a result of the analysis of the results obtained, the most notable conclusions are the following:

6.1 Investment, Capacity and Traffic

There is a great concentration of investment and of different types of traffic in certain ports. Thus, for example, the stock of the ports of Bilbao, Barcelona and Valencia represents almost 30% of the total stock.

In relation to the traffic, it would be worth highlighting the high existing specialisation as shown by the fact that for each type of traffic there is a different leading port.

6.2 Returns to Scale and Scale Efficiency

Five ports show constant returns to scale and also have the maximum magnitudes of scale efficiency and total technical efficiency.

The size of the port has a negative effect on scale efficiency while the highest magnitudes are achieved in the ports that specialise in passenger and liquid bulk traffic.

There is a negative relationship between the levels of scale efficiency obtained for the ports analyzed and the magnitudes of its stocks. In this sense, the less efficient ports correspond in general to those of greater size.

The average magnitude of scale efficiency has increased during this period with an annual growth rate of 0.4% and this growth is greater in ports specialising in liquid bulk and passenger traffic while it is smaller in the larger ports.

6.3 Pure Technical and Total Technical Efficiency

Three of the 27 ports analyzed reach the maximum level of pure technical efficiency (1) and the average score is very high (0.811), which shows the high level of pure technical efficiency achieved by Spanish ports, without taking into account the effects of scale and which implies optimum management of the port.

During the period 1986–2005 an increase in average pure technical efficiency was produced with an annual average rate of growth of 0.3%. Furthermore, there was convergence in relation to the initial levels of pure technical efficiency.

It can be seen that the size of the port has a negative influence on the total technical efficiency while the specialization of a port in passengers and liquid bulk traffic has a positive effect.

In relation to the total technical efficiency it should be noted that the size of the port has a negative influence on its magnitude while in contrast, the specialization of a port in passenger and liquid bulk traffic has a positive effect.

6.4 *Total Factor Productivity and Technological Change*

In the period 1986–2005 Spanish ports have experienced an improvement in factor productivity (TFP) that is 0.5% on average annually. This growth is greater in ports specialising in containerized general cargo. The growth of TFP is due exclusively to the technical efficiency that increases 0.7% annually since the other component (technological change) decreases with an annual average rate of change of −0.2%.

Likewise, the technical efficiency change (0.7%) is practically due in the same proportion to its two components: Pure technical efficiency change (0.3%) and scale efficiency change (0.4%).

Finally, the ports with a greater specialization in general cargos (containerized and non-containerized) have a greater magnitude of technological change during the period analyzed. By contrast, the specialization of ports in solid bulk traffic has a negative effect on technological change.

References

Baños-Pino J, Coto-Millán P, Rodríguez-Álvarez A (1999) Allocative efficiency and overcapitalization: an application. Int J Transp Econ XXVI(2):181–199

Bauer PW (1990) Discomposing TFP growth in the presence of cost inefficiency, non constant returns to scale and technological progress. J Prod Anal 1:287–289

Charnes A, Cooper WW, Rhodes E (1978) Measuring the efficiency of decision marking units. Eur J Oper Res 6(2):429–444

Charnes A, Cooper W, Lewin A, Seiford L (eds) (1996) Data envelopment analysis: theory, methodology and applications, 2nd edn. Kluwer, Boston

Coelli TJ (1996) A guide to DEAP version 2.1: a data envelopment analysis (computer) program. CEPA Working Paper 96/8. Department of Econometrics, University of New England, Armidale, NSW, Australia

Coelli T (1998) A multi-stage methodology for the solution of orientated DEA models. Oper Res Lett 23:143–149

Cooper WW, Seiford LM, Tone K (2000) Data envelopment analysis: a comprehensive text with models, applications, references and DEA-Solver software. Kluwer, Boston
Coto-Millán P, Inglada V, Rodriguez-Alvarez A (1999) Economic and technical efficiency in the world air industry. Int J Transp Econ 26:119–235
Domenech R (1992) Medición no paramétrica de la eficiencia en el sector bancario español. Revista Española de Economía 9:171–196
Färe R, Grosskopf S, Norris M, Zhang Z (1994) Productivity growth, technical progress, and efficiency change in industrialized countries. Am Econ Rev 84(1):66–83
Färe R, Grosskopf S, Roos P (1998) Malmquist productivity indexes: a survey of theory and practice. In: Färe R, Grosskopf S, Russell R (eds) Index numbers: essays in honor of Sten Malmquist. Kluwer, Boston
Forsund FR, Hjalmarsson L (1979) Generalized farrell measurement of efficiency: an application to milk processing in Swedish dairy plants. Econ J 89:294–315
González Serrano MM, Trujillo L (2005) La medición de la eficiencia en el sector portuario: Revisión de la evidencia empírica. Documento de trabajo 2005–2006. Universidad de Las Palmas de Gran Canaria.
Inglada V, Rey B, Rodríguez Álvarez A, Coto-Millán P (2004) Liberalization and efficiency in international air transport. Transp Res A 40:95–105
Nishimizu M, Page JM (1982) Total factor productivity growth, technological progress and technical efficiency change: dimensions of productivity changes in Yugoslavia, 1965–1978. Econ J 92(368):920–936
Pastor JM (1995) Eficiencia. Cambio productivo y cambio técnico en los bancos y Cajas de Ahorro españolas: un análisis de frontera no paramétrica. Revista Española de Economía 12(1):35–73
Rey B, Inglada V, Quiros C, Rodriguez-Alvarez A, Coto-Millan P (2009) From European to Asian leadership in the economic efficiency of the world air industry. Appl Econ Lett 16(2):203–209
Simar L (1992) Estimating efficiencies from frontier models with panel data: a comparison of parametric, non-parametric and semi-parametric methods with bootstrapping. J Prod Anal 3:171–203

Part V
Cost Benefit Analysis and Externalities

A Cost–Benefit Analysis of a New Container Terminal

Pablo Coto-Millán, Ramón Núñez-Sánchez, Miguel Ángel Pesquera, Vicente Inglada, and Juan Castanedo

Abstract This paper present a cost–benefit analysis approach applied to measure benefits involving the expansion of container capacity in a seaport context. The paper starts by setting out the theoretical background for identifying and measuring the project benefits. There have been identified and calculated three different benefits: benefits from existing traffic, benefits from avoided diversion costs and benefits from generated traffic. A practical application of this methodology has been included. The results for this application show that this project is viable, either if this infrastructure has congestion problems in the present or may have them in the short term.

1 Introduction

The Spanish public-owned seaport system comprises commercial seaports managed by port authorities (called *Autoridades Portuarias*), and a national regulatory agency (called *Puertos del Estado*) that coordinates the investment projects between seaports.

In order to divert the traffic from one port seaport to another and generate new traffic, substantial investments which enable the competitive position of the seaport to improve, are required.

However, in the Spanish public-owned seaport system there are a significant number of non self-financing seaports. These seaports cannot finance themselves their investment projects to improve their competitive position. In these cases, the national regulatory agency partially finance these projects. This regulatory scheme

P. Coto-Millán (✉), R. Núñez-Sánchez, M. Ángel Pesquera, and J. Castanedo
University of Cantabria, Santander, Spain
e-mail: cotop@unican.es

V. Inglada
University of Complutense, Madrid, Spain

P. Coto-Millán et al. (eds.), *Essays on Port Economics*, Contributions to Economics,
DOI 10.1007/978-3-7908-2425-4_19,

generates potential situations of *moral hazard*, changing the market capacity to allocate resources efficiently. For this reason, the national regulatory agency requires control mechanisms to minimize negative effects of *moral hazard* situations.

On the other hand, the existence of *Puertos del Estado* is necessary as national regulatory agency to make a thorough analysis of the large investments which should be carried out at each Spanish port. Otherwise, this process could end in an increasing expense on investment in each national port, and inefficient situations would take place from allocative point of view. One of the functions of this agency should be coordinate investment projects from different ports and create a detailed agreement among them to obtain an optimal capacity expansion sequence (Sharkey 1982).

With this purpose, it is required that empirical and theoretical approaches of welfare measures are given. Cost–Benefit Analysis (CBA) – based on the micro-economic theory and welfare economy – and Economic Impact Analysis are two of the most widely known tools.

Some empirical attempts in the literature for Economic Impact Analysis are Coto-Millan and Villaverde (1997); Coto-Millan and Villaverde (1998); Verbeke and Debbischop (1996); Gripaios and Gripaios (1995) and Desalvo (1994).

However, there are few empirical attempts in the literature for Cost–Benefit Analysis applied to ports. In Spain, Rus de et al. (1995) provide empirical evidence of a CBA that evaluates the construction of a new port.

This paper studies, by means of Cost–Benefit Analysis,[1] the investment project for the development of a new container terminal, which will allow to a seaport to increase its capacity to attract containerized goods traffic.

An illustration of the total social surplus approach as applied to investment project of seaports is the main contribution of this paper. The application of this approach to congestion problems in the seaports industry is novel in the literature.

The rest of the paper is arranged as follows: in Sect. 2, the methodology used for estimation of benefits and costs is discussed. Application is described in Sect. 3. Section 4 shows the sensitivity analysis using Monte-Carlo simulation. The main conclusions are presented in Sect. 5.

2 Methodology

The main benefits and costs considered in the CBA of the project for the creation of a new container terminal, are the following:

[1] In the Cost–Benefit Analysis, we will follow the methodology used by Dodgson and Gonzalez-Savignat (1998) in such a way that the national costs and benefits are taken into account. Accordingly, the costs and benefits generated by the investment project which can be located in geographic areas other than the area where the investment project is developed, are taken into account. Then, a global evaluation of the project is carried out in this way. More details about methodological aspects, Harberger (1974); Layard and Glaister (1994).

1. Time savings for containerized goods owners.
2. Variation in costs of other transport operators.
3. Investment costs.
4. Economic costs of land.
5. Differential maintenance and exploitation costs.
6. Changes in quality of service.
7. Environmental impact.

This paper is concerned with the estimation of the costs and benefits listed above, except the last two. By simplicity, the last two aspects have not been introduced in the analysis. Changes in quality of service and environmental impact would be studied through specific techniques as contingent valuation.

We set out below the procedure used for the economic estimation of the different benefits and costs.

2.1 Benefits

The main benefit of port enlargement is an increase in the supply of containerized goods transport with no restriction of capacity.

There are different methods for estimating the economic benefits of an extension of transport infrastructure: the social surplus approach and the resource use approach. The social surplus approach consists of the direct calculation of changes in consumer and producer surpluses. This requires identifying changes in prices, costs and revenues with and without project. This approach requires strong assumptions about the demand and supply curves.[2] The alternative approach, resource use approach, consists of looking at the changes in willingness to pay and the use of real resources,[3] ignoring transfers between consumers and producers. This paper focus on measuring real resource costs changes ignoring traffic revenues from existing traffic.

The NEPV of an investment in transport infrastructure can be expressed as (1), assuming that investment costs are realized in year 0 and changes in benefits and costs of the implemented project occur in year 1 onwards:

$$NEPV = -I + \delta_t \sum_{t=1}^{T} (\Delta CS_t + \Delta PS_t), \tag{1}$$

[2]When the marginal utility of the income is not constant, changes in consumer surplus are imperfect monetary measures of changes in utility.

[3]Real resources can be both outputs (quantities of production, commodities, time) and inputs (labour, capital).

where I is the investment costs, T is the project life, ΔCS_t is the change in consumer surplus in year t, ΔPS_t is the change in producer surplus in year t, δ_t is the discount factor: $(1+i)^{-t}$, and i is the social rate of discount.

The change in consumer surplus can be estimated as follows:

$$\Delta CS_t = (g_{ts} - g_{tc})q_{ts} + \frac{1}{2}(g_{ts} - g_{tc})(q_{tc} - q_{ts}) = \frac{1}{2}(g_{ts} - g_{tc})(q_{ts} + q_{tc}),$$
$$g_{ts} = p_{ts} + \tau_{ts} \qquad (2)$$
$$g_{tc} = p_{tc} + \tau_{tc}$$

where g_{ts} is the generalized cost in year t without the investment, g_{tc} is the generalized cost in year t with the investment, q_{ts} is the number of TEUS in year t without the investment, q_{tc} is the number of TEUS in year t with the investment, p_{ts} is the price per trip inclusive of port charges without the investment, p_{tc} is the price per trip inclusive of port charges with investment, and τ_{ts}, τ_{tc} is the value of total travel time without investment and with investment, respectively.

The change in producer surplus is equal to:

$$\Delta PS_t = p_{tc}q_{tc} - p_{ts}q_{ts} + C_{ts}(q_{ts}) - C_{tc}(q_{tc}), \qquad (3)$$

where $C_{ts}(q_{ts})$ and $C_{tc}(q_{tc})$ denote total variable costs without the project and with the project respectively.

In our CBA, we assume that the marginal port costs for shipping companies do not vary as the product increases.

This analysis is partially based on the analysis of Jorge and de Rus (2004). Figure 1 shows the present situation of an investment in a container terminal, which eventually leads to higher capacity. Generalized costs for port services are measured in the vertical axis and the number of containers per year in the horizontal

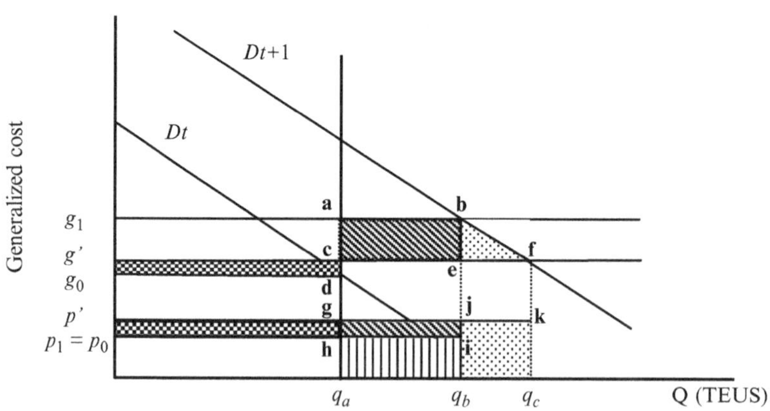

Fig. 1 Economic benefits with congestion

axis. Given a demand of port services in period t, equal to D_t. We assume a linear demand curve to facilitate the illustration of the approach.[4]

Let's assume that the current generalized cost relative to handling of containers is go and the number of containers is q_a[5] in period t. This number of containers coincide with the maximum annual capacity of the port. If in the next period, demand were lower than q_a, the generalized cost would be go. If in the next period, demand were higher than q_a, the generalized cost would be higher than go. In this case, the seaport has congestion problems.

If the investment project is implemented, seaport capacity is higher and congestion problems eventually disappear. In Fig. 1, we assume new capacity is higher than q_c with the project.

Demand of port services has an imperfect substitute available at a generalized cost of g_1, higher than g_0.

We suppose a demand increase of up to D_{t+1} for a future period $t + 1$, which will overcome maximum annual capacity without project. In this situation, if the investment is not implemented, q_a will be the amount of containers using this port (existing traffic) at a constant generalized cost equal to go. Also there will be some deviated traffic to alternative port $(q_b - q_a)$ and some deterred traffic $(q_c - q_b)$.

The construction of a new terminal allows a generalized cost of g' as demand increase of up to D_{t+1}, because congestion problems disappear. This generalized cost is higher than g_0 but lower than g_1. If g_0 is equal to $g_0 = p_0 + \tau_0$, g' is equal to $g' = p' + \tau_0$. We suppose that prices with investment have risen due to the rise of port charges for covering the investment and to attain a normal rate of return on capital. We also suppose that p_0 (initial price without investment) is equal to p_1 (price of the imperfect substitute).

Given the demand D_{t+1} for a future period $t + 1$, the quantity of containers using the seaport with investment would be q_c. $(q_b - q_a)$ would be the portion of containers retained after the construction of the project and $(q_c - q_b)$ the portion of traffic generated.

The derived benefits from the reduction in generalized cost can be expressed as:

$$(g_1 - g')(q_b - q_a) + \frac{1}{2}(g_1 - g')(q_c - q_b) - (g' - g_0)q_a + p_0'q_c - p_0q_a \quad (4)$$

In Fig. 1 it can be identified three categories of benefits: (1) benefits from existing traffic, (2) benefits from avoided diversion costs and (3) benefits from generated traffic.

Benefits from existing traffic are equal to $-(g' - g_0)q_a + (p' - p_0)q_a$ that is represented by the areas $-cdg'g_0 + ghp'p_0$. Given that $(g_0 - p_0) = (g' - p')$, benefits are equal to zero, $cdg'g_0 = ghp'p_0$.

[4]This analysis could be repeated assuming logarithmic demand or semi-logarithmic demand.

[5]For simplicity in the notation, we don't consider the sub-index for temporal period.

Benefits from avoided diversion costs are equal to $(g_1 - g')(q_b - q_a) + p'(q_b - q_a)$. This benefit is interpreted as the saving of real resources from deviated traffic plus saving in the transport costs incurred by containerized goods which have been retained and diverted to other ports or means of transport, due to the saturation suffered by facilities at the seaport in the situation without project. This benefit is represented by the areas $abce + ghji$ which corresponds with savings of real resources and the area $hiq_{ta}q_{tb}$ which corresponds with saving in the transport costs.

Benefits from generated traffic are equal to $\frac{1}{2}(g_{t1} - g'_t)(q_{tc} - q_{tb}) + p'_t(q_{tc} - q_{tb})$, this benefit can equally be interpreted as deterred traffic avoided thanks to the investment. In Fig. 1 it corresponds with the area $bef + jkq_bq_c$.

2.2 Costs

The costs incurred by the seaport in the construction of the container terminal are the costs of construction and those relating to maintenance and operation. For their inclusion into the CBA, some of these costs calculated in terms of market prices, must be expressed in terms of opportunity costs[6].

In general terms, the opportunity cost or shadow price of the inputs employed in the construction of the project is equal to the market price minus indirect taxes.

3 Application

The application of this methodology is the study of an investment project for the development of a new container terminal in the port of Santander (Spain). It will allow the seaport to increase its capacity to attract containerized goods traffic.

3.1 Estimation of Containerized Goods Traffic

As remarked by Coto-Millan and Baños-Pino (1996), port service demand is a consequence of transport services, which also depends on demand for goods. If, of all the possible alternatives to satisfy demand for goods, the cheapest one is shipping transport, there will be port service demand. Port services could be qualified as production factors of the companies.

In order to develop a CBA, it is necessary to compare the benefits and costs that would exist in the event that the project was carried out, with the benefits and costs

[6]The opportunity costs of the input express the loss in benefit produced in the best use possible of that input.

without carrying out the project. It is therefore necessary to forecast the traffic for the whole of the analyzed period in those cases when the port may increase its capacity for the reception of containerized goods, in relation to those cases where the port did not make the investment so that its capacity would remain constant.

From the forecasts of its Managing Plan, the Port Authority has forecasted the traffic of containerized goods for the 2002–2026 period with the project and without the project.

It must be pointed out that, as the traffic of goods and vessels increases, there is a gradual increase in the occupational rate at docks and of the equipment. This gives rise to a higher increase in the vessels' waiting time, and, to a certain extent when these waiting times are not tolerable, this traffic is diverted to other competing ports or transport modes. In such cases we must refer to congestion of port facilities.

The Port Authority considers that the maximum annual capacity of the existing facilities for containerized goods is 10,174 TEUS/year. The maximum annual capacity of the new container terminal would be 60,000 TEUS/year. Therefore, in the event of the project being developed, the maximum capacity of the port for this type of goods would be 70,174 TEUS/h (Autoridad Portuaria de Santander 2003).

Once the data on the capacity and traffic forecasts have been obtained, we calculate the increase in traffic produced by the improvement of the facilities, and distinguish two types of traffic, usually analyzed in CBA studies linked with projects of investment in transport infrastructure. On the one hand, there is the traffic that would use the port facilities in the event of lack of congestion in the situation with investment project. This traffic would be transferred to other ports due to the congestion existing in the situation without project, and is denominated *diverted traffic*.

On the other hand, we must take into account the traffic that, in the situation without project, wouldn't use other Spanish ports or transport chains, even though there is no congestion in the port of Santander. This traffic is denominated *generated traffic*.[7]

In this paper we assume uncertainty associated to these variables so we won't use deterministic values for the containerized traffic with project. We try to specify probability distributions for them. We assume normal distributions for every annual traffic with mean equal to the deterministic value previously estimated. We also assume that as we move out in time, variance is higher, reflecting the greater uncertainty which exists in our knowledge of the distant future.

3.2 Benefits

By means of the straightforward theoretical model explained in Sect. 2.1., we will calculate the net benefit for each temporal period. Firstly, it is necessary to estimate the values of the different parameters. Given the lack of data relating to the port of

[7]The estimation of containerized goods traffic can be viewed on tables A1 and A2.

Santander, we have taken into account the cost per container of the Port of Bilbao's services, updated to 2002 prices, in order to estimate the initial price. The price per container (included port charges) for 2002 is 144 €, which corresponds with $p_0 = p_1$. If Port Authority implements the project the new price per container would be 153 €, which corresponds with p'.

We assume that imperfect substitute is Port of Bilbao that is situated 100 km. away from Santander. The journey time is about 1 h 45′ by truck. We consider the values of time for containerized goods employed by Central Port Authority (Puertos del Estado 2003)[8]. The unit cost transport by truck is about 0.05€/Ton.km (Puertos del Estado 2003). We assume this value as uncertainty[9] so unit cost of transport is uniformly distributed between 0.04 and 0.05.

3.3 Construction Costs

The Project budget at market prices and its annual distribution is the following:

Since the evaluation covers 20 years and the useful life of some of the investment assets is less than this, we must proceed to estimate the replacement costs (Tables 1 and 2).

On the other hand, the net residual value of the assets which are part of the investment project during the last evaluation period, must be considered.

Regarding the residual value in terms of opportunity costs, the next formula is proposed when there are no consistent criteria for determining market prices for the residual value of the assets at the time when they cease to be used (Table 3).

$$VR_{CO} = VI_{CO} \times \left(\frac{1}{20}\right)^{\frac{PE}{VU}}$$

Table 1 Project budget

Object	Issue	Amount invested (€)
New docks in South Quay at Raos	Dredging, digging and rock filling	12,057,922.88
	Dock and oversize structures	3,583,945.48
	Paving	48,529.18
	Miscellaneous	188,227.71
	Dredging and supply	369,000.00
	Security and health	98,400.00
	Gantry crane	3,000.000
	Total	19,346,025.25

Source: Autoridad Portuaria de Santander (2003)

[8]Value of time (in euros of 2002) per hour of containerized goods is about 0.36 €.

[9]This cost of transport depends on the cost of fuel that is very volatile.

Table 2 Investment program at market prices

Years	Budget (€)
2002	0.00
2003	100,000.00
2004	8,143,000.00
2005	8,103,025.25
2006	3,000,000.00
Total	19,346,025.25

Source: Autoridad Portuaria de Santander (2003)

Table 3 Replacement costs and residual values at market prices

Element	Useful life	First year of service	Rep. costs and residual costs	Residual value in 2026 (%)	Initial value (euros)	Residual value (euros)
Gravity-type docks	40	2006		22,36	6,368,684	1,424,081
Landfills	100	2006		54,93	391,425	215,002
Dredging	50	2006		30,17	8,887,757	2,681,514
Bumpers and mooring devices	15	2006	178,816	36,84	188,227	69,343
Paving	15	2006	46,102	36,84	48,529	17,878
Facilities	20	2006		5,00	369,000	18,450
Security and health	0	2006		0,00	98,400	0
Gantry crane	20	2006		4,43	9,000,000	133,125
Total			224,919		16,352,02,	4,426,270

Source: Autoridad Portuaria de Santander (2003)

With PE $\leq$ VUwhere,

- *VRco*: Residual Value at Opportunity Costs;
- *VIco*: Initial Value at Opportunity Costs;
- *PE*: Evaluation Period during which the asset provides the service; and
- *VU*: Useful life of the asset.

Once the investment figures relating to the initial set up (at market prices and excluding indirect taxes, VAT) are available, we have deducted those labor costs corresponding to social security payments and also the energy costs corresponding to special taxes, in order to obtain the investment value of the initial set up in terms of opportunity costs. These taxes must be removed from the market price because they are not aimed at correcting the negative externalities generated by the project.

The estimation of the economic costs of land has been separated from the evaluation of the rest of investment costs due to its particularities.

Since the land to be incorporated within the investment has different origins (land reclaimed from the sea and land previously owned by the Port Authority), we must treat each of them differently. The cost of the land reclaimed from the sea, which accounts for a total of 11,826 m^2, is nil; however, it does have a residual value at the end of the analyzed period of the Project in 2026. In particular, it has been estimated that each square meter of surface is 97.71 euros/m^2 (Autoridad Portuaria de Santander 2003).

On the other hand, the Port Authority owns a surface area of 39,600 m^2, which must be included in the economic evaluation at the hypothetical purchase value (market prices) of the current year, leading to an estimation of residual value of the same amount but with the opposite sign during the last year of the project's evaluation period. The values for those areas have been 88.83 euros/m^2 for 22,000 m^2 while the rest have been valued at 79.95 euros/m^2, amounts obtained according to the distance away from the edge of the quay.

3.4 Maintenance and Operation Costs

The project will require a staff increase of eight people in order to control Port Authority terminal management as well as the contracting of a new crane operator (Table 4).

The annual cost of each person contracted by the Port Authority is 30,862.97 €, while the crane operator receives an annual wage of 30,000 €.

The preservation costs for the works are estimated as annual percentages with respect to the initial cost (Table 5). The expected preservation and maintenance costs are:

Table 4 Staff costs

	Port Authority	Stevedore	Total
Posts	8	1	9
Average wage	30,861.97	30,000	60,861.97
Staff expenses	−246,895.77	−30,000	−276,895.77

Source: Autoridad Portuaria de Santander (2003)

Table 5 Annual maintenance costs

	First year of service	Initial value (euros)	Annual costs (in % of the initial value)	Annual costs
Gravity-type docks	2006	6,368,684.87	1.00	63,686.85
Landfills	2006	391,425.64	0.50	1,957.13
Dredging	2006	8,887,757.84	1.00	88,877.58
Bumpers and mooring devices	2006	188,227.71	0.50	941.14
Paving	2006	48,529.18	2.00	970.58
Facilities	2006	369,000.00	9.00	33,210.00
Security and health	2006	98,400.00	0	0.00
Gantry crane	2006	3,000,000	4.5	135,000
Total		19,352,025.24		324,643.28

Source: Autoridad Portuaria de Santander (2003)

Table 6 Benefits of increasing capacity of the port of Santander (€ of 2002)

	Social benefit of carrying out port extension
Benefits	31,862,873
Investment costs	−15,856,250
Operation costs	−4,940,711
Net present value	11,065,912

Source: Autoridad Portuaria de Santander (2003)

Table 7 Potential traffic demand forecast (TEUS)

Years	Normal traffic	Traffic diverted from other ports
2002	8,378	–
2003	8,968	–
2004	9,600	–
2005	10,276	545
2006	11,000	8,000
2007	11,887	10,637
2008	12,845	14,142
2009	13,881	18,803
2010	15,000	25,000
2011	15,438	26,203
2012	15,888	27,464
2013	16,352	28,786
2014	16,829	30,171
2015	17,321	31,623
2016	17,826	33,145
2017	18,346	34,740
2018	18,882	36,411
2019	19,433	38,163
2020	20,000	40,000
2021	20,556	42,080
2022	21,147	44,267
2023	21,746	46,569
2024	22,361	47,817
2025	22,993	47,181
2026	23,644	46,530

Source: Autoridad Portuaria de Santander (2003)

Once the investment figures relating to the costs of conservation and maintenance at market prices and excluding VAT, are available, we have deducted those labor costs corresponding to social security payments and also the energy costs corresponding to special taxes, in order to obtain the value of conservation and maintenance costs in terms of opportunity costs (Tables 6–8).

The comparison between the benefits and the economic costs for the project period (from 2002 to 2026), duly updated, with a 5% social discount rate, shows a positive social benefit (NEPV) of 11,065,912 € at 2002 values, with an Internal Rate of Return (IRR) of 9.28% (Table 9). As IRR is higher than the social discount rate we conclude that this project is socially profitable.

Table 8 Traffic forecast with and without carrying out the project (TEUS)

Years	Traffic without carrying out the project		Traffic with project				
	Normal traffic	Capacity	Normal traffic	Traffic diverted from other ports	Traffic generated	Total traffic	Capacity
2002	8,378	10,174	8,378	0	0	8,378	10,174
2003	8,968	10,174	8,968	0	0	8,968	10,174
2004	9,600	10,174	9,600	0	0	9,600	10,174
2005	10,174	10,174	10,276	545	29	10,850	10,174
2006	10,174	10,174	11,000	7,600	400	19,000	70,174
2007	10,174	10,174	11,887	10,105	532	22,524	70,174
2008	10,174	10,174	12,845	13,435	707	26,987	70,174
2009	10,174	10,174	13,881	17,863	940	32,684	70,174
2010	10,174	10,174	15,000	23,750	1,250	40,000	70,174
2011	10,174	10,174	15,438	24,893	1,310	41,641	70,174
2012	10,174	10,174	15,888	26,091	1,373	43,352	70,174
2013	10,174	10,174	16,352	27,347	1,439	45,138	70,174
2014	10,174	10,174	16,829	28,662	1,509	47,000	70,174
2015	10,174	10,174	17,321	30,042	1,581	48,944	70,174
2016	10,174	10,174	17,826	31,488	1,657	50,971	70,174
2017	10,174	10,174	18,346	33,003	1,737	53,086	70,174
2018	10,174	10,174	18,882	34,590	1,821	55,293	70,174
2019	10,174	10,174	19,433	36,255	1,908	57,596	70,174
2020	10,174	10,174	20,000	38,000	2,000	60,000	70,174
2021	10,174	10,174	20,556	39,976	2,104	62,636	70,174
2022	10,174	10,174	21,147	42,054	2,213	65,414	70,174
2023	10,174	10,174	21,746	44,241	2,328	68,315	70,174
2024	10,174	10,174	22,361	45,422	2,391	70,174	70,174
2025	10,174	10,174	22,993	44,822	2,359	70,174	70,174
2026	10,174	10,174	23,644	44,204	2,327	70,174	70,174

Source: Autoridad Portuaria de Santander (2003)

However we said in previous sections that there were any uncertainty variables in this model. In the next section we implement a risk analysis applying Monte Carlo simulations considering uncertainty in our model.

4 Sensitivity Analysis

CBA often requires us to predict the future. Whether or not is desirable to begin a project depends on what we expect will happen after we have begun. But, we rarely are able to make precise predictions about the future. We always face some uncertainty about the magnitude of the impacts we predict and the values we assign to them. Our basic analysis usually submerges this uncertainty by using our most plausible estimates of these unknown quantities. These estimates comprises what is called base case. The purpose of sensitivity analysis is to acknowledge the underlying uncertainty. In particular, it should convey how sensitive predicted net

Table 9 Benefits and costs of developing the Southern terminal at the Port of Santander

Years	Benefits	Investment costs	Operation costs	Updating factor	Net present value
2002	0	−3,361,380	0	1,000	−3,361,380
2003	0	−78,178	0	0,952	−74,455
2004	0	−6,366,043	0	0,907	−5,774,188
2005	57,156	−6,334,791	0	0,864	−5,422,857
2006	796,600	−2,662,500	−446,098	0,823	−1,902,087
2007	1,059,179	0	−446,098	0,784	480,365
2008	1,408,190	0	−446,098	0,746	717,928
2009	1,872,309	0	−446,098	0,711	1,013,581
2010	2,489,375	0	−446,098	0,677	1,382,970
2011	2,609,164	0	−446,098	0,645	1,394,331
2012	2,734,728	0	−446,098	0,614	1,405,020
2013	2,866,366	0	−446,098	0,585	1,415,081
2014	3,004,277	0	−446,098	0,557	1,424,490
2015	3,148,860	0	−446,098	0,530	1,433,332
2016	3,300,413	0	−446,098	0,505	1,441,623
2017	3,459,236	0	−446,098	0,481	1,449,371
2018	3,625,625	0	−446,098	0,458	1,456,578
2019	3,800,081	0	−446,098	0,436	1,463,332
2020	3,983,000	0	−446,098	0,416	1,469,656
2021	4,190,116	−170,897	−446,098	0,396	1,414,005
2022	4,407,887	0	−446,098	0,377	1,493,156
2023	4,637,108	0	−446,098	0,359	1,504,331
2024	4,760,979	0	−446,098	0,342	1,475,042
2025	4,698,048	0	−446,098	0,326	1,384,313
2026	4,633,225	3,496,266	−446,098	0,310	2,382,373
Total	67,541,922	−15,477,523	−9,368,059	15	11,065,912
Net economic present value					
Internal rate of return (IRR)					9.28%

benefits are to changes in assumptions. If the sign of net benefits does not change when we consider the range of reasonable assumptions, then our analysis is robust and we can have greater confidence in our results.

Traditional sensitivity analyses have two major limitations. First, they may not take account of all the available information about assumed values of parameters. Second, these techniques do not directly provide information about the variance, or spread, of the statistical distribution of realized net benefits (Boardman et al. 2001). Monte Carlo analysis provides a way of overcoming these problems.

The basic steps for doing Monte Carlo analysis are as follows. First, specify probability distributions for all the important uncertain quantitative assumptions. In previous sections we have specified normal distributions for future traffic and a uniform distribution for unit cost of transport. Second, we execute a trial by taking a random draw from the distribution for each parameter to arrive at a set of specific values for computing realized net benefits. Third, we repeat the trial described in the second step many times to produce a large number of realizations of net benefits. The average of the trials provides an estimate of the expected value of net benefits. An approximation of the probability distribution of net benefits can be obtained by breaking the range of realized net benefits into a number of equal increments and

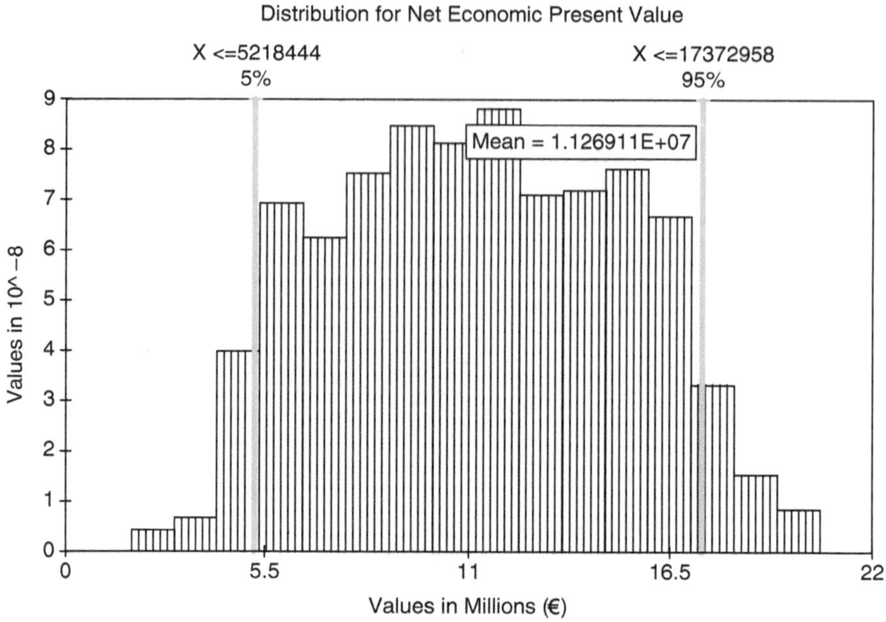

Fig. 2 Distribution of net economic present value

counting the frequency with which trials fall into each one. The resulting histogram of these counts provides a picture of the distribution. The more trials that go into the histogram, the more likely it is that the resulting picture gives a good representation of the distribution of net benefits.

Figure 2 presents a histogram of 1,000 replications of random draws considering uncertain variables traffic demand and unit cost of transport.

The average of net benefits over the 1,000 trials is 11,269,110 €. This value is close to our base-case calculation of 11,065,912 €. The histogram provides a visual display of the entire distribution of net benefits so that its spread and symmetry can be easily discerned. The trials themselves can be used to calculate directly the sample variance, standard error, and other summary statistics describing net benefits.

5 Summary

In this paper, we have developed by means of Cost–Benefit Analysis, the investment project for the development of a new container terminal, which will allow to a port to increase its capacity to attract containerized goods.

An illustration of the total social surplus approach as applied to investment project of seaports is the main contribution of this paper.

In order to estimate the demand it has been assumed that the function is linear and the movements of the function are produced as a result of increases in the amounts of goods which use containers as means of transport.

We have identified and calculated three different categories of benefits for the investment project: benefits from existing traffic, benefits from avoided diversion costs and benefits from generated traffic.

The costs incurred by the seaport in the construction of container terminal were the costs of construction and those relating to maintenance and operation, expressed in terms of opportunity costs.

The application of this methodology has been the study of an investment project for the development of a new container terminal in the port of Santander (Spain).

The result of the application reveals that the extension project should be carried out since the social benefits derived from retained traffic at the port exceed investment costs. The sensitivity analysis reveals that considering uncertainty the result obtained in the base case is robust.

References

Autoridad Portuaria de Santander (2003) Evaluación económico financiera del proyecto Nuevos muelles en el espigón central de Raos del Puerto de Santander. Santander: Unpublished Internal Document

Boardman AE, Greenberg DH, Vining AR, Weimer DL (2001) Cost-benefit análisis. concepts and practice. Prentice Hall, New Jersey

Coto-Millan P, Baños-Pino J (1996) Derived demands for general cargo shipping in Spain 1975–1993: an economic approach. Appl Econ Lett 3(3):175–178

Coto-Millan P, Baños-Pino J, Villaverde J (2005) Determinants of maritime import and export functions. Transp Res E 44(4):357–372

Coto-Millan P, Baños-Pino J, Rodriguez-Alvarez A (2000) Economic efficiency in Spain: some empirical evidence. Marit Policy Manage 27(2):1169–1174

Coto-Millan P, Villaverde J (1997) Economic impact analysis of Santander port on its hinterland. Int J Transp Econ XXIV(2):259–277

Coto-Millan P, Villaverde J (1998) Port economic impact: methodologies and application to the Port of Santander. Int J Transp Econ XXV(2):159–179

Desalvo J (1994) Measuring the direct impacts of a port. Transp J 33(4):33–42

Dodgson J, Gonzalez-Savignat M (1998) Efficiency in public resource allocation: the social appraisal of projects. Int J Transp Econ XXV(2):221–242

Gripaios P, Gripaios R (1995) The impact of a port on its local economy: the case of Plymouth. Marit Policy Manage 22(1):13–23

Harberger AC (1974) Cost-benefit analysis. Cambridge University Press, Cambridge

Jorge J-D, de Rus G (2004) Cost-benefit analysis of investment in airport infrastructure: a practical approach. J Air Transp Manage 10:311–326

Layard R, Glaister S (1994) Cost-benefit analysis. Cambridge University Press, Cambridge

Puertos del Estado (2003) Método de Evaluación de Inversiones Portuarias. Madrid: Unpublished document

Rus de G, González M, Roman C, Romero M, Tovar B, Trujillo L (1995) Análisis Coste-Beneficio del Puerto de Arinaga. Working Paper. University of Las Palmas de Gran Canaria

Sharkey WW (1982) The theory of natural monopoly. Cambridge University Press, Cambridge

Verbeke A, Debbischop K (1996) A note on the use of port economic impact studies for the evaluation of large scale port projects. Int J Transp Econ XXIII(3):247–266

Evaluation of Port Externalities: The Ecological Footprint of Port Authorities

Pablo Coto-Millán, Ingrid Mateo-Mantecón, Juan Luis Doménech Quesada, Adolfo Carballo Panela, and Miguel Ángel Pesquera

Abstract This chapter presents an analysis of the main results of the calculation of the ecological footprint and carbon footprint (or CO_2 emissions into the atmosphere) produced by the economic activity of a Port Authority of the North Coast of Spain, using a compound financial accounts method (MC3).

Furthermore, the results will be compared with those obtained for Gijón Port Authority in 2006. For the Spanish port authorities, environmental protection and sustainable development are a commitment and part of their strategic actions. Some eco-efficiency indicators for the companies under analysis are also calculated, compared and analysed.

1 Introduction

Traditionally, one of the problems we face when analysing environmental assets from an economic point of view is the absence of a market for these assets. Because of this, both producers and consumers fail to take into account environmental costs, and consequently the wellbeing achieved in the economy is not as good as it could be; or in other words, a Pareto Optimum is not reached.

When a given company, due to its business activity, produces pollution (CO_2 emission into the atmosphere, for instance) it is generating an undesired effect for other agents in society. At first, and until just a few years ago, this emission that

P. Coto-Millán (✉), I. Mateo-Mantecón, and M. Ángel Pesquera
University of Cantabria, Santander, Spain
e-mail: cotop@unican.es

J.L.D. Quesada
Port Autority of Gijón, Gijón, Spain

A.C. Panela
University of Santiago of Compostela, Galicia, Spain

P. Coto-Millán et al. (eds.), *Essays on Port Economics*, Contributions to Economics,
DOI 10.1007/978-3-7908-2425-4_20,

affected society as a whole was not compensated for economically, and therefore constituted an externality, given that, "*externality exists when the functions of the production and/or usefulness of the economic agents depend on a decision made by other agents without the intervention of economic compensation*" (Xavier Labadeira et al 2007).

Nowadays there is a market for CO_2 emissions and, in Europe specifically, the European Union stipulates to each Member State what their stake in this greenhouse gas (GHG) emissions market will be. Each Member State distributes these emission quotas between a group of companies that are assumed to be the ones that produce these emissions and are allocated an individual emission allowance. For the time being, in Spain, the companies that are assumed to be the cause of GHG emissions are regulated by Law 1/2005, of March 9, governing the GHG emission allowance system, and by Directive 2003/87/EC. In Spain, these companies are those that are exclusively involved in the following activities: combustion, generation, combined-cycle generation, coal generation, fuel generation, extra-peninsular generation, steel industry, tile and brick industry, wood pulp and paper industry, lime industry, cement industry, oil refining industry, glass industry, ceramic tile and floor tile industry, fried industry.

Only the companies that fall within the categories listed exhaustively in the above paragraph are subject to the GHG registry and consequently participate in the emission allowance market for these gases. However, in the future it is likely that other companies, including those involved in air transportation, will take part in this market.

But in reality, practically all companies and individual agents produce these emissions due to production and/or consumption. This is why, endeavouring to corroborate this assertion and as an example, this chapter will present a calculation of the carbon emissions, or ecological footprint, produced by the port authorities.

As a starting point, it is important to note that the European Union's transportation policy seeks to create transportation systems that meet the needs of society from an economic, social and environmental point of view. Various studies reveal that the transportation sector generates 7% of European GDP and 5% of EU employment, but this economic value- and employment- generating development must be coupled with an increase in technological development in order to achieve transportation that is less harmful to the environment. In particular, it must incorporate the international environmental agreements, which include the Kyoto Protocol. If we bear in mind that the transportation sector accounts for 30% of the EU's total energy consumption, achieving the objectives contracted with regard to CO_2 emissions becomes a major challenge. Thus, the environmental costs that the EU currently meets are very high, amounting to 1.1% of the GDP.[1]

[1]Figures taken from the Communication from the Commission to the Council and the European Parliament, "Keep Europe moving – Sustainable mobility for our continent, published in 2001. Mid-term review of the Transport White Paper, published in 2001 by the European Commission". COM (2006) 314 final.

To this end, at least since 2001, with the White Paper on European transportation policy, the aim has been to restore the balance between the various modes of transport as a strategy for achieving sustainable development, and redress the importance of the role that ports must play to support sustainability in the movement of both people and goods.

The ecological footprint is a single-index sustainability indicator that measures the full impact of a population expressed in hectares of ecosystems or of "nature". Commonly used for regions or countries, it has been demonstrated that it can also be used for companies or any kind of organisation, and on this occasion the "compound financial accounts method" (MC3) will be used (Doménech 2006b).

The corporate ecological footprint allows us to establish specific environmental sustainability objectives; it allows the incorporation of indicators, lifecycle and eco-labelling in a single tool and provides a new method for political decision-making to fight climate change more fairly (Doménech 2004).

The main objective pursued in this study is to apply the ecological footprint calculation methodology to a real company, such as a distribution or transportation company, and in our case the footprint method will be applied to a company operating in the port sector. Once the ecological footprint results have been obtained for a Port Authority of the North Coast of Spain, the results will be compared to those of another authority of the same coast, namely Gijón.

The study will be structured into four parts:

Part One: Port externalities and environmental sustainability.
Part Two: Application of the ecological footprint calculus to a port authority of the north coast of Spain.
Part Three: Comparison of the results obtained for the North Coast Port Authority and Gijón Port Authority.
Part four: Conclusions.

2 Port Externalities and Environmental Sustainability

Through the environmental services created at all ports, in particular during the last decade, ports have contributed more than any other sector of activity to the recovery of the quality of the water, air, land and natural spaces of major coastal urban areas, which were invariably afflicted for years by industrial development and major port and large-scale construction projects and infrastructures. Some of these infrastructures, such as lighthouses, have contributed to keeping intact Spain's best coastal areas. Thanks to these cross-sector synergies, combining economic development and growth with environmental conservation is becoming increasingly viable.

One of the challenges that European ports will have to confront to ensure their future competitiveness will be sustainability. At this time of environmental, social and economic uncertainty due to the globalisation of the emerging markets, climate change and social instability, there is no kind of administration that is currently viable, in terms of competitiveness, if it is not envisaged from a sustainable development point of view.

In the annual conference of the European Sea Ports Organisation (ESPO), held in Algeciras in May 2007, it was stressed that Europe requires a competitive port industry that responds quickly and effectively to the demands of the market. A solid legal framework for fair competition must be established. The port system must provide sustainable mobility within a context of intense international competition. The space and the intermodality with the hinterland must be planned.

Since sustainable development is a primary need, it is worth clarifying from the outset what we believe the priority of ports should be in the field. To this end, we can distinguish between environmental management, eco-efficiency and sustainability, concepts that are sometimes still confused. While environmental management refers solely to controlling the various environmental factors (air, water, waste, land, consumption etc), eco-efficiency correlates the environmental factor with the economic factor (producing more with less resources). As far as sustainable development is concerned, another component comes into play: the social factor, making environmental management an element that is fully integrated into the economy and society.

Many ports, like many companies, are still at the first stage (that of environmental management) and there is an urgent need to address the second phase (eco-efficiency) and to move quickly on to full sustainable development.

We have said that economic growth will not be possible outside of sustainable development. Sustainability will therefore be the core factor in any kind of economic success and ports must strive not only to achieve this but also, as administrators of spaces and activities, to ensure they know how to transmit it to the rest of their associated agents.

There is a very small margin by which a port can differentiate itself from the rest. They must all adapt, with large-scale investments, to new changes, such as the growing size of ships or the tendency towards to the containerisation of all kinds of goods.

This study will argue that some of the main procedures incorporated into strategies that can successfully open up the small margins of competitiveness that European ports can squeeze out of themselves will undoubtedly come from the strategies adopted for sustainable development and the fight against climate change. One of the most important roles of ports in future integrated management will be precisely to act as a *driver of sustainability* through strategic alliances with shipping companies, industry and intermediary agents. The first consideration, then, is to achieve full sustainability.

Ports must go from environmental management to eco-efficiency without delay and without generating more impact due to the increase in production. In other words, the environmental impact must be detained despite the high demand for goods and traffic expected in European ports, which makes major investments in construction projects and infrastructures a necessity.

One way to achieve this is to control our CO_2 equivalent emissions, which are the conversion into carbon of every type of impact caused by waste generation or resource consumption (the *carbon footprint*).[2]

Gijón Port Authority, for instance, has an annual carbon debt in the order of 30,000 tons of CO_2 per year, which means that the second step towards eco-efficiency can be approached with assurance in order to move on immediately to the last step which is the execution of designated sustainability projects. These include the installation of solar panels for heating water in the logistics area, as well as presence detectors in the multi-service building. Air conditioning has been regulated, lighting has been replaced and lighting on wharves has been controlled. Old transformers have been replaced, mains water leakage detection services have been hired and investments have been made to prevent losses from the mains water network. A certified environmental management system has been introduced (ISO 14001), which, among other things, reduces waste, and two anemometric towers have been set up to assess the viability of harnessing wind energy at the port.

Another example of this commitment to sustainability is provided by the Port Authority of New York and New Jersey, where a target has been set to make the port "carbon neutral" (i.e. with zero net carbon emissions) by the year 2010. To do this, the proposal is to invest in wind farms, methane capture and strategies to reduce the risks posed by climate change, and to reduce environmental demands on clients and suppliers, all with the aim of reducing a carbon footprint which stands, according to their calculations, at around 298,000 tons of CO_2. According to the authority's executive director, Anthony E. Shorris, the initiative has meant the introduction of an aggressive annual investment plan that will continue until the target is reached. To ensure that its carbon reductions are appropriate, the Port Authority will use rigorous standards and audits to ensure coherence with the international verification protocols.

Bearing all of this in mind, it comes as no surprise that the ports have put together a World Ports Climate Declaration, signed at the port of Rotterdam in July 2008, the aim of which is for all of the world's ports to introduce action plans to fight climate change and achieve a high level of air quality. To summarise, the ports that sign the Climate Declaration commit to the following:

- To acknowledge that transportation between ports, port operations and industrial port activities contribute to greenhouse gas (GHG) emissions. Ports play an important role in the supply chain, enabling them to have a positive influence on the sustainability of the chain. Ports have many opportunities for GHG reduction.

---

[2]Including the use of land, which must be managed with increasing efficiency, in particular within the context of integrated coastal management. The Belgian researcher Jacques Charlier talks about three types of sustainable port development: (a) the "green alternative" or growth on new inland spaces, which are increasingly scarce; (b) the "blue alternative" or growth out to sea, which is very expensive and has a considerable impact on the environment; and (c) the "brown alternative", which consists of growing on spaces already in use. With the 'brown alternative', the recommended alternative, the port redefines the use of its spaces and develops itself.

- Initiatives to reduce CO_2 on ships: (a) development of clean shipping (design, fuel etc); (b) develop and standardise (renewable) electrical supply at ports; (c) consider speed reductions; (d) develop transparent incentives based on environmental systems on ships; (e) accelerate the provisions to reduce CO_2 and to develop Marpol Annex VI.
- Initiatives to reduce CO_2 emissions in port operations: (a) promote CO_2 reduction measures at terminals and in cargo handling; (b) promote measures for energy efficiency in all elements and port members; (c) develop sustainable nautical services (tugboats etc); (d) install (renewable) electrical connection systems for moored ships.
- Initiatives to reduce emissions of hinterland transport: (a) use efficient and innovative logistics to reduce the need for transport in the hinterland; (b) promote clean and energy-efficient modes of transport; (c) improve the environmental performance of all transport modes.
- Promote the use of renewable energy: (a) promote the generation of renewable energy (wind, solar, geothermal) in the public and private domain; (b) use renewable energy where possible in the port authority's operations and in port operations in general; (c) promote the transport and processing of certified biomass for the production of renewable energy.
- Carbon footprint: (a) begin quantification and management of carbon footprints by creating carbon inventories for their own activities, for port operations as a whole and for the relevant parts of the supply chain; (b) create structures and reporting mechanisms to internalise CO_2 self-assessment and control; (c) determine and reduce the footprint of the port area (per unit of activity) and distinguish between cargo handling and industrial port activities; (d) develop their own CO_2 emission reduction targets.
- Implementation: (a) create institutional mechanisms and responsibilities at ports to drive continual reduction of emissions and innovation; (b) monitor and evaluate the implementation of the aforementioned initiatives; (c) advocate these initiatives actively in their region and in their networks or alliances; (d) organise and facilitate technology transfer, education and exchange of good practices and cost benefit examples.

This declaration is absolutely clear about the tendencies that we have mentioned, which all ports must aim for.

3 Port Authorities and Ecological Footprint

The Spanish state-owned port system is made up of 44 "ports of common interest", managed by 28 port authorities whose coordination and efficiency control falls to the *Puertos del Estado* ("State Ports") public body, dependant on the *Ministerio de Fomento* ("Ministry of Development") and responsible for executing the government's port-related policy.

The Spanish legal framework that the port authorities' operations are regulated by is defined by Law 27/1992, of November 24, on State Ports and the Merchant Navy, modified by Law 62/1997, of December 26, as well as Law 48/2003, of November 26, on the Economic System and Service Provision of Ports of Common Interest, which provides the Spanish port system with the instruments needed to improve its competitive position in the open and globalised market, establishing a system of autonomous management for the port authorities, which must undertake their activity with business-oriented criteria.

Within this framework, the aim is for the management of the ports of common interest to adhere to the "landlord" model, in which the port authority is limited to acting as a supplier of port infrastructure and land, and to regulating the use of the public domain, while the services are essentially provided by private operators under a system of authorisation or concession. Equally, the function of the ports is not just as mere embarkation and disembarkation points for goods and passengers, but also as centres where activities are undertaken that add value to the goods and are completely integrated into the logistical and intermodal chains.

In this context, it is especially important to consider the fact that, for ports, like any other company, their financial performance is a necessary variable for becoming an important centre of business, but is not enough to guarantee their sustainability. To ensure this, environmental and social performance must be taken into consideration.

In response to this need, the port authorities establish standardised comprehensive environmental management systems as a tool for instituting a policy of environmental protection and sustainability.

One of the tools used to measure the sustainability of a business is the ecological footprint. Gijón Port Authority has been a pioneer in the use of this indicator within the Spanish port system and in this case the ecological footprint of another port authority of the north coast has been calculated in order to compare the results obtained.

The ecological footprint is used to indicate both the impact of human activities on the ecosystem and the corrective measures that can minimise the impact taking place. Furthermore, it is a single-index indicator which, although at first applied extensively to population centres and territories, has only in recent years been adapted for calculation of the ecological footprint or carbon footprint of firms (Doménech 2004, 2006a, b, 2007).

Specifically, to calculate the ecological or carbon footprint of firms, the financial accounts method (MC3) is used. Corporate ecological footprint is defined as the environmental impact (in hectares) of any organisation, caused by: (a) the purchase of any kind of product and service clearly reflected in their financial accounts; (b) the sale of products deriving from the primary production of food and other forestry or biotic resources, or in other words when vegetables, fruit and meat enter the market chain for the first time; (c) occupation of space; and (d) generation of waste clearly reflected in their environmental report. Every type of impact considered in the corporate ecological footprint is perfectly controllable and auditable and, therefore, objective and transparent (Doménech 2006a, 2008). Moreover, this impact measured in hectares can be transformed to obtain a result in tonnes of CO_2 emitted (the carbon footprint), which allows us to be consistent with the measures that must be taken to mitigate climate change as much as possible. Because, in order to be honest, all private individuals and legal entities must be aware of the extent to which their activity affects the environment, and seek measures to reduce this impact.

As far as the calculation method (MC3) used is concerned, it is included on a simple spreadsheet, based on the majority of the spreadsheets in existence, in particular those of Wackernagel (1998). Some data on energy intensity was taken from other studies on the ecological footprint of Barcelona or Berlin's footprint. The familiar footprint of particular interest and many of the conversion indices used in the corporate ecological footprint are taken from there. In particular, this study used the spreadsheet that appears in the methodological guide for corporate ecological footprint calculation, which details in full the financial accounts method (MC3); it can be found at http://www.huellaecologica.com/ (Doménech 2006b, 2007).

3.1 Principal Ecological Footprint Results for a North Coast Port Authority

To calculate the ecological footprint of the North Coast Port Authority under analysis, the trial balance, details of the tangible fixed assets and general ledger details for certain accounts were requested from the Finance Department. Other data such as electricity, fuel, water and paper consumption were requested from those responsible for these services. The ecological footprint was calculated for the year 2006.

Having calculated the ecological footprint of the North Coast Port Authority, Table 1 shows a summary of the results obtained in the various resource consumption categories:

As can be inferred from these results, this port authority produces an ecological footprint of 5,125.8 hectares, which can be expressed in tons of CO_2 emissions into the atmosphere to give a result of 20,518.3 ton CO_2 (we have taken into account that the forests offer an absorption factor of 1.42 tC/ha/year and that for a proportion of 12:44 the CO_2 absorption is 5,2066 t/ha/year). In this case, to go from the hectares of ecological footprint to the carbon footprint (or tons of CO_2 emitted to the atmosphere), only the carbon absorption factor of the forests is taken into account, and not that of crops, the sea and pastures. Furthermore, the net ecological footprint has been calculated, which is the difference between the environmental debt and credit, or in other words the ecological footprint minus the counterfootprint. The net ecological footprint is 4,040.6 hectares.

Table 1 Main results

Ecological footprint of the North Coast Port Authority by resource type, year 2006

	Footprint		Counter footprint	
Resource consumption	ha	t CO_2	ha	t CO_2
Electricity	694.6 (13.55%)	3,175.2		
Fuel	119.6 (2.33%)	546.8		
Materials	234.1 (4.56%)	1,070.3		
Construction materials	2,978.4 (58.1%)	13,618.4		
Services	194.6 (3.79%)	889.8		
Waste	21.2 (0.41%)	97.1		
Land	55.6 (1.08%)	0	1,085.2	0
Farming and fishing	665.3 (12.98%)	378		
Forestry and water	162.4 (3.17%)	742.6		
Total	5,125.8	20,518.3	1,085.2	0

Source: compiled by author from the results obtained from the ecological footprint calculation

In the light of these results, it is clear that construction materials cause the biggest footprint, accounting for around 58% of the ecological footprint of the North Coast Port Authority under analysis. Two other categories that contribute a high percentage of footprint are electricity consumption with 13.55% and farming and fishing resources with a proportion of 13% of the total footprint.

In this case, the footprint caused by the firm under analysis due to its use of land is of particular significance. While the ecological footprint is equivalent to the hectares of land "consumed" or environmental "debt", the footprint negative is equivalent to the hectares of land that we have in "credit". The footprint that we cannot eliminate by reducing the debt (by saving energy, buying efficient materials, recycling etc) must be eliminated by increasing debt, or in other words by investing in natural resources.

In this case, two special features of ports hold true: firstly the fact that the majority of their territory is built on water, which is less productive than land. Secondly, ports have waters under their jurisdiction, needed for anchorage and entry of ships, which constitute an important natural asset that must be looked after and to which significant investments are allocated every year for controlling spillages, quality control etc.

The hectares that appear as pasture or garden areas on dry land are essentially land found around lighthouses. The rest of the land is the result of filling and a breakdown of its various uses is shown in Table 2.

As can seen from an analysis of the results appearing in Table 2, there are 55.6 hectares of footprint per land type, or in other words 1.08% of the total ecological footprint. The counterfootprint is of 1,085.2 hectares and is mainly a result of the port authority's jurisdictional waters, which constitute an important source of natural capital.

In addition to taking into account the results that refer to the ecological footprint and carbon footprint, some eco-efficiency indicators will also be analysed, having been obtained by dividing the results for the financial year by the results obtained

Table 2 North Coast Port Authority land types

Land type	ha	Total footprint [ha × ef]	Counterfootprint [ha × ef × yf]
On dry land			
Crop cultivation areas	0	0	0
Pasture or garden areas	16.66	0	9.01
Wooded areas	0	0	0
Built-up, asphalted, eroded etc.	0	0	0
Fish farming	0	0	0
On water			
Filled land used for crops	0	0	0
Filled land used for pasture or gardens	32.17	8.80	17.41
Filled land used for forest or woods	0	0	0
Filled land for construction, tracks etc.	215.37	46.78	0
Aquatic uses (without aquaculture)	3,868.9	0	1,058.77
Aquaculture in the sea	0	0	0
Subtotal 2	4,133.1	55.58	1,085.19

Note: *ha* Hectares, *ef* Equivalence factor, *yf* Yield factor
Source: compiled by author

for the environmental impact (Lehni 1999). It is worth underlining that eco-efficiency implies a commitment to clean processes which save on natural resources and reduce environmental impact.

With the aim of calculating certain eco-efficiency indicators, the analysis will take into account the income of the North Coast Port Authority that has been studied, measured by the net turnover, as well as the tons of goods handled at the facilities managed by the port authority in question. There are four ratios for measuring eco-efficiency that will be analysed. The first two are the income/net footprint (€/ha) and the income/net carbon footprint (€/tC O_2) indicators, which are often very useful for comparing the efficiency of companies: the greater the indicator the more efficient the company, given that a higher income is obtained per hectare and per ton of emitted CO_2. The third indicator is based on cargo/net ecological footprint (t/ha) and finally the cargo/net carbon footprint (t/tCO_2) has been calculated. In this case the same relationship as above occurs, since the more tons of cargo that are handled per hectare or ton of emitted CO_2, the more efficient, on principle, the port will be.

In addition, the net footprint has been calculated in m^2 per ton of cargo handled by the port facilities. This index shows the space occupied by each ton at the port and on principle the lower this ratio the more efficient.

The final indicator calculated is the port authority's sustainability index, which is essentially the relationship between the hectares that we have and those we use with our business activity, or in other words the counterfootprint divided by the net footprint. In this case, an index close to zero would indicate that no natural capital is held and absolutely nothing is done to achieve sustainability, while an index equal to one would mean that it is fully sustainable.

Table 3 shows the main results for the indicators that have been calculated:

Table 3 Main results, eco-efficiency and sustainability indicators

Ecological footprint of a North Coast Port Authority (2006)	
Gross ecological footprint (ha)	5,125.8
Counterfootprint (ha)	1,085.2
Net ecological footprint (ha)	4,040.6
Net emissions or net carbon footprint (t CO_2)	20,518.3
Income (net turnover) (€)	29,012,905.3
Cargo (t)	5,854,561
Income/net footprint (€/ha)	7,180.34
Income/net carbon footprint (t/t CO_2)	1,414
Cargo/net footprint (€/ha)	1,448.9
Cargo/net carbon footprint (t/t CO_2)	285.3
Net footprint per ton of cargo (m^2/t)	6.9
Sustainability index (Counterfootprint/footprint)	0.264

Source: compiled by author

4 Comparison Between the Ecological Footprint Indicators of a North Coast Port Authority and Gijón Port Authority

Having calculated the ecological footprint of a port authority of the north coast of Spain, the results will be compared with those of The Gijón Port Authority, which has been a pioneer in calculating ecological footprint indicators.

Table 4 shows the percentages by category of the ecological footprint of the port authority under analysis in the year 2006 and of the Gijón Port Authority for the years 2004, 2005 and 2006:

Although we are comparing different port authorities and different years, the above table illustrates the existence of a similar structure in the distribution of the ecological footprint by category. In these cases, construction materials account for the highest proportion of the total footprint at around 50% of the total, followed by the footprints from electricity and from farming and fishing resources, accounting for 13 and 11%, respectively.

The table clearly shows how in all cases the ecological footprint of the construction materials is what contributes most to the total footprint. Consequently, and with the aim of reducing the footprint of the construction materials, measures are usually adopted to guide the company towards full sustainability. These measures include using sustainable or eco-construction techniques, proposing new bidding specifications to include the use of at least 50% recycled or reused materials, and imposing environmental requirements on contractors (ISO 14001, energy and materials monitoring, waste recycling, use of low energy cement...).

One of the reasons for which the structure of the results for the various categories of ecological footprint is similar for both port authorities could be that both have introduced an Environmental Management System in accordance with Regulation UNE-EN-ISO 14001. Furthermore, as can be seen in Table 5, both port authorities have a very similar built-up surface area and their counterfootprint or natural capital (as indicated previously this natural capital is basically made up of waters under the

Table 4 Comparison of footprint percentage by category

Ecological footprint

Resource consumption	NCPA 2006 (%)	GPA 2006 (%)	GPA 2005 (%)	GPA 2004 (%)
Electricity	13.55	11.56	10.87	14.9
Fuel	2.33	2.5	1.96	2
Materials	4.56	11.27	10.88	12
Construction materials	58.1	56.74	52.83	48.2
Services	3.79	3.55	4.02	2.3
Waste	0.41	0.029	2.79	3.4
Land	1.08	0.97	1.41	1.1
Farming and fishing	12.98	10.56	11.32	9.8
Forestry and water	3.17	2.82	3.89	6.3
Total (ha)	5,125.8	7,366.2	7,865.5	6,483.0

Source: compiled by author

Table 5 Land type of the port authorities compared in 2006

Land type	ha NCPA	ha GPA	Total footprint NCPA	Total footprint GPA	Footprint negative NCPA	Footprint negative GPA
On dry land	16.66	15.8	0	13.3	9.01	15.8
On water	4,116.44	4,538.1	55.58	58.1	1,076.18	1,168.6
Total	4,133.1	4,553.9	55.58	71.4	1,085.19	1,184.4

Where: *NCPA* North Coast Port Authority and *GPA* Gijón Port Authority
Source: compiled by author

jurisdiction of the port authorities) is also very similar, which facilitates the comparison of results.

Table 6 below shows the main ecological footprint results, the eco-efficiency indicators defined above, and a sustainability indicator for both Port Authorities. Some conclusions can be drawn from a graphical analysis of the results obtained in Table 6 both for the main ecological indicators (Graph 1) and the eco-efficiency indicators (Graph 2).

As can be seen in Graph 1, the absolute values for the ecological footprint and the carbon footprint are higher for Gijón Port Authority in the year 2006. However, in the case of the sustainability index, for the NCPA it is 0.264 and for GPA it is 0.26. Both port authorities are a long way from an index equal to 1, which would indicate that they are fully sustainable.

The figures relating to the eco-efficiency indicators in Graph 2 show that the North Coast Port Authority has better results than Gijón Port Authority for the indicators relating to income to net footprint and carbon footprint. However, a comparison of the cargo to net footprint and carbon footprint reveals that Gijón Port Authority is more efficient in this regard. In the light of these results, it can be inferred that the North Coast Port Authority under analysis handles high-value goods at its facilities, which generate more income but which are also more intensive in terms of the space used than that of Gijón Port Authority. In other words, the cargo handled by the North Coast Port Authority, although generating

Table 6 The main results from the comparison of indicators

Comparison of indicators of two port authorities (2006)		
	NCPA	GPA
Gross ecological footprint (ha)	5,125.8	7,366.2
Footprint negative (ha)	1,085.2	1,184.4
Net ecological footprint (ha)	4,040.6	6,181.9
Net emissions or net carbon footprint (t CO_2)	20,518.3	30,193.8
Income (net turnover) (€)	29,012,905.3	38,752,272
Cargo (t)	5,854,561	20,488,293
Income/net footprint (€/ha)	7,180.34	6,269
Income/net carbon footprint (t/t CO_2)	1,414	1,283
Cargo/net footprint (€/ha)	1,448.9	3,314
Cargo/net carbon footprint (t/t CO_2)	285.3	679
Net footprint per ton of cargo (m^2/t)	6.9	3.02
Sustainability index (Counterfootprint/footprint)	0.264	0.26

Source: compiled by author

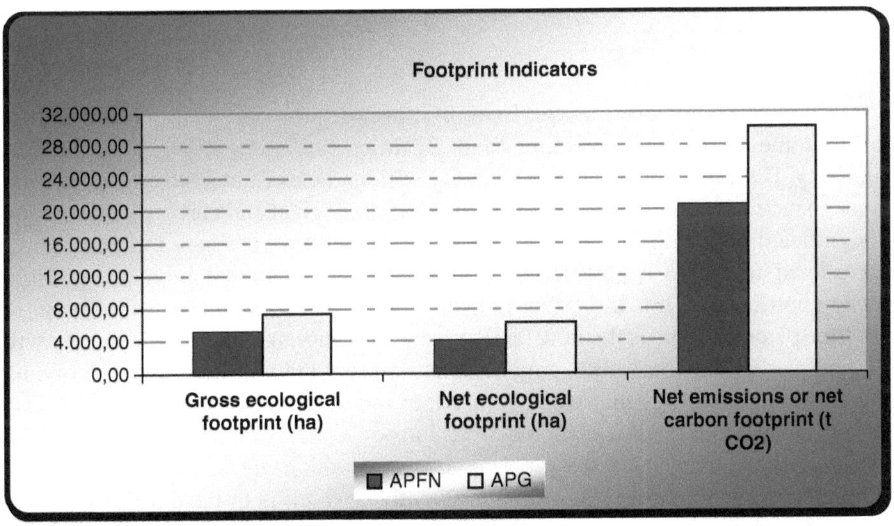

Graph 1 Graphical comparison of the ecological footprint of the two port authorities analysed in 2006
Source: compiled by author

more income, also needs more port space for handling, so that when it comes to calculating its eco-efficiency it produces worse results. In this case, the less space needed per ton handled at the port facilities, the greater the efficiency and productivity of the firm.

It is important to note that European transport policy, geared towards modal rebalancing, will produce an increase in maritime transportation, and this increase

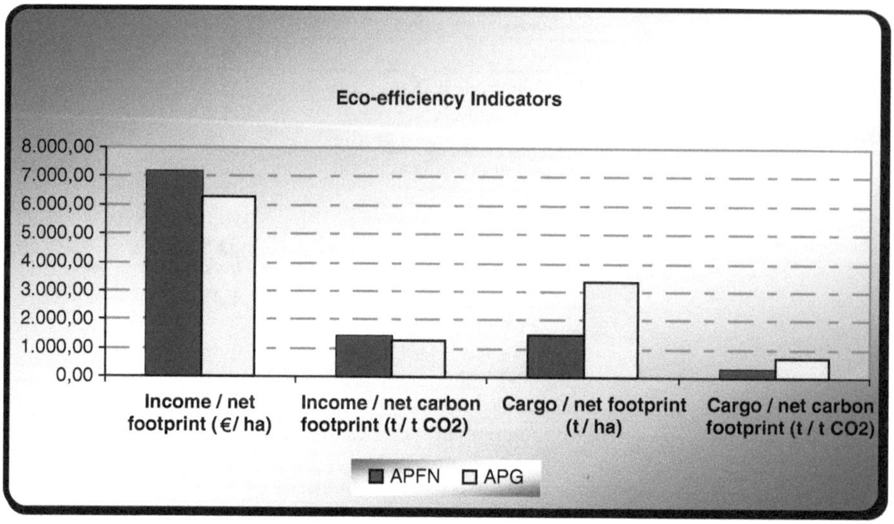

Graph 2 Graphical comparison of eco-efficiency indicators of the two port authorities under analysis
Source: compiled by author

in traffic must be absorbed by the EU ports. In order to cope with this increase in traffic, some Ports are expanding their facilities, as is the case of Gijón Port Authority. But this increase in port capacity will not be exempt from environmental impact, since it translates into an increase in the ecological and carbon footprint due to significant consumption of materials used for building the new infrastructures, constructed to increase port capacity and prevent this modal rebalancing from causing bottlenecks. It is important to bear in mind that these increases in footprint will take place as soon as the new facilities are operational, which is when they will become part of the company's tangible fixed assets, and will therefore be counted within the methodology that we have used, in the construction projects matrix.

This is why there is a clear need to introduce eco-efficiency measures that minimise the impact caused by port expansion, which are so necessary among European ports to achieve the ultimate objective, which is to have a Europe-wide transport system that is as energy efficient and as environmentally sustainable as possible.

4.1 Conclusions on the Ecological Footprint of the Two Port Authorities Under Analysis

Having calculated the ecological footprint of a North Coast Port Authority and compared the main results and indices with those of Gijón Port Authority, it is clear that for both firms the aim will be to increase cargo handling at their facilities while containing or reducing the environmental impact they cause, at the same time as increasing the income to footprint eco-efficiency indicators. The objective for the future is therefore to attempt to reduce their ecological footprint and carbon footprint while handling more tons of cargo at their facilities.

Thus, the steps that the port authorities should take to become environmentally sustainable can be summarised in four stages. The first step is to adopt the calculation of a single-index indicator, such as the ecological footprint proposed in this article. The second step would be to focus on the footprint categories that can be reduced, which in the case of the port authorities under analysis would mean attempting to minimise the footprint from construction materials, electricity and farming resources, given that these categories are the ones that produce the greatest ecological footprint and CO_2 emissions (carbon footprint). In order to obtain these results, a rigorous eco-efficiency study should be carried out (which Gijón Port Authority has already done) and lastly the measures deriving from the eco-efficiency study should be implemented; these actions should then be monitored to ensure their effectiveness.

For the purposes of this article the ecological footprint of a North Coast Port Authority has been calculated for the year 2006, and the results obtained have been compared with those of Gijón Port Authority, yielding significant results. Firstly, we developed an important tool for establishing the impact of their business activity, and secondly, on the basis of this indicator, we can verify which areas require corrective measures in order to alleviate this impact. We have also established that the structure of the ecological footprint of the two port authorities under analysis is similar when the various categories are compared. A comparison was drawn between the results for certain eco-efficiency indicators, yielding interesting results. It was shown that there can be situations in which the port authorities that obtain higher returns by handling high-value-added cargo should improve some aspects to minimise the space that these goods occupy at the port.

Evidently, nowadays it is not possible to conceive of a modern company that does not have the strategic objective of making profit while respecting the environment. To this end, many companies implement eco-efficiency measures which amount to a commitment to clean processes that save natural resources and reduce the impact on the environment. This is the case of the port authorities under analysis, where a methodology is being introduced with the aim of achieving full sustainability. This methodology comprises three stages: firstly to calculate a single-index indicator such as the ecological footprint, secondly to study the current eco-efficiency situation and lastly to implement specific actions to favour sustainability and continued monitoring of the ecological footprint.

5 Conclusions

The main conclusions that we can draw from this chapter are, firstly, that ports must continue to evolve towards complete integration with the other agents and sectors with an interest in the coast, through strategic alliances, and that the integrating factor for these alliances must be sustainability.

Close observation reveals that the history of ports is really a history of integration, resulting, above all, from their capacity to draw together various sectors, agents and interests. And if this history has gone hand in hand with the creation of wealth and the improvement of quality of life, continuing to make progress in this capacity for integration seems a logical course of action.

However, cooperation between companies, services, functions, sectors and activities, or in other words total or large-scale integration, is the last and true revolution in maritime transport and the port sector. The next evolutionary step for the port system will be to progress to what we could call the *Maritime Community*, a system that includes the port community and the other members or sectors with interests in the marine environment, such as the shipbuilding sector, the fishing sector, the tourist sector, the energy sector, global sustainable development and integrated research and development.

The basic difference between the *port community* and the *maritime community* does not lie in the number or type of members, which could just as well be the same in each case, but in the breadth of their objectives. The ecosystem of the former is the port itself and its immediate surroundings (the port-industrial sector), while that of the latter is the maritime sector as a whole and comprehensive maritime-coastal development, with a whole host of derived synergies.

There is already talk of the future fifth generation ports, with an approach featuring a cooperative sort of logistics, e-logistics networks, software agents, semantic websites, collaborative learning and other concepts in which the leading role is played by collaboration, a broad space for knowledge and new symbioses (Pesquera 2002). In fact, one piece of advice given to ports by the director-general of energy and transport of the European Commission, Jean Trestour, in early 2008, was that they should compete together, establishing alliances with each other with the aim of instituting joint strategies to address Europe's new logistical map. The steps seem clear: supra-sectorial alliances (between maritime communities), with the aim of reducing the uncertainty of conditions. The fifth generation ports will, without doubt, be the ports of the *maritime communities*.

The second conclusion is that the main differentiating factor of the new fifth generation ports must be the sustainability of all the related agents and companies in the geographical area under consideration. We could start to talk about a hinterland of sustainability, in which the port extends its sustainability to its entire catchment area.

Ports tend to grow at an increasing rate and have an increasing effect on the economy of their surroundings; indeed, port investment within the Spanish port system as a whole surpassed the billion euro milestone in 2006 (1.114 billion euros, 27% more than in 2005), in 2007 it climbed to 1.296 billion (16.3% more) and in 2008 it increased to 1.723 billion euros (33% more). All economic growth should be coupled with sustainable growth, especially when there is such a high rate of investment.

We have seen that an ecologically efficiency port must aspire to achieving a *zero carbon* target, reducing energy consumption, investing in its own renewable energy, buying green materials, contracting eco-friendly construction projects and even investing in natural capital (carbon sinks, protected spaces, biodiversity protection etc). Ports must not rule out the idea of spearheading the creation of carbon sinks, with the aim of ensuring that the associated industry (in particular that which is most badly affected by CO_2 emissions trading) can invest in this natural capital and reduce their emissions. It would be desirable for these projects even to be included within the Kyoto Protocol's Clean Development Mechanism and for them to be executed in the less developed countries with which the Spanish ports or port system are most interested in collaborating (Latin American, African countries etc). This kind of project and this type of cooperation could include the creation of protected marine areas, biodiversity conservation in degraded areas, alternative energy harnessed from port dykes, support for sustainable transportation or promoting the reef effect on breakwaters of other ports with which there is a desire to establish commercial relations.

Logically, new and interesting possibilities arise for ports that are capable of foreseeing in good time the considerable opportunities deriving from a firm and decisive commitment to meeting the challenges posed by climate change and to favouring global social responsibility.

The European Commission is aware that 90% of Europe's foreign trade and around 40% of its domestic trade take place through its ports, which is why it will propose a new port policy, taking into account the multiple functions of ports and the broader context of European logistics. In the conclusion of its Communication on a European Ports Policy of October 2007,[3] the EC establishes the framework for the development of this policy: "*As part of European maritime transport this communication provides a framework and a number of related actions to be carried out, including an extended dialogue… It will help concentrate the efforts so that Europe's ports can face the challenges of tomorrow, attract new investment and fully contribute to co-modal development. The Commission calls upon all public and private stakeholders to support this approach and looks forward to a continuation of dialogue to ensure the most harmonious development of EU ports*".

The framework for establishing a solid network of alliance is, therefore, firmly established. All that remains to do is to coordinate and develop it through appropriate joint strategic medium- and long-term plans and objectives. The ports of common interest have made a considerable effort in recent years to work more closely with the various groups of interest with which they interact and they can be expected to continue evolving in this direction. It is therefore essential to continue working jointly to achieve sustainable economic development.

---

[3]Communication from the Commission. Communication on a European Ports Policy, COM (2007) 616 final, Brussels, 18/10/2007.

References

Commission of the European Communities (2006) Communication of the commission to the European council and parliament. Keep Europe moving – sustainable mobility for our continent. Intermediate review of European Commission's White Paper on transport of 2001. COM, Brussels, p 314

Doménech JL (2004) Huella ecológica portuaria y desarrollo sostenible. Puertos 114:26–31

Doménech JL (2006a) Ecoeficiencia y sostenibilidad en Puertos: aplicaciones en el puerto de Gijón. Comunicación técnica. CONAMA 8

Doménech JL (2006b) Guía metodológica para el cálculo de la huella ecológica corporativa. In: Terceros Encuentros sobre Desarrollo sostenible y población; eumed.net. Universidad de Málaga, p 46

Doménech JL (2007) Huella ecológica y desarrollo sostenible, 1st edn. España: AENOR, Madrid, pp 400

Labadeira X, León CJ, Vázquez MX (2007) Economía ambiental. Prentice may, Pearson, p 356

Lehni M (1999) El medio ambiente como factor clave de competitividad. Ecoeficiencia. In: Eco-eficiencia, los negocios en el próximo milenio. Fundación Entorno, Madrid, pp 23–28

Pesquera MA (2002) EU research and port development. Towards the fifth generation of ports. In: Surface transport technologies for sustainable development, EU conference. Valencia, 4–6 June 2002

Wackernagel M (1998) The ecological footprint of Santiago de Chile. Local Environ 3:7–25

FSC
www.fsc.org
MIX
Papier | Fördert
gute Waldnutzung
FSC® C083411

Kolibri 360 GmbH
Ferdinand-Jühlke-Straße 7
99095 Erfurt, Deutschland
produktsicherheit@kolibri360.de